ALGEBRA EXAMPLES

CONIC 3 CIRCLES

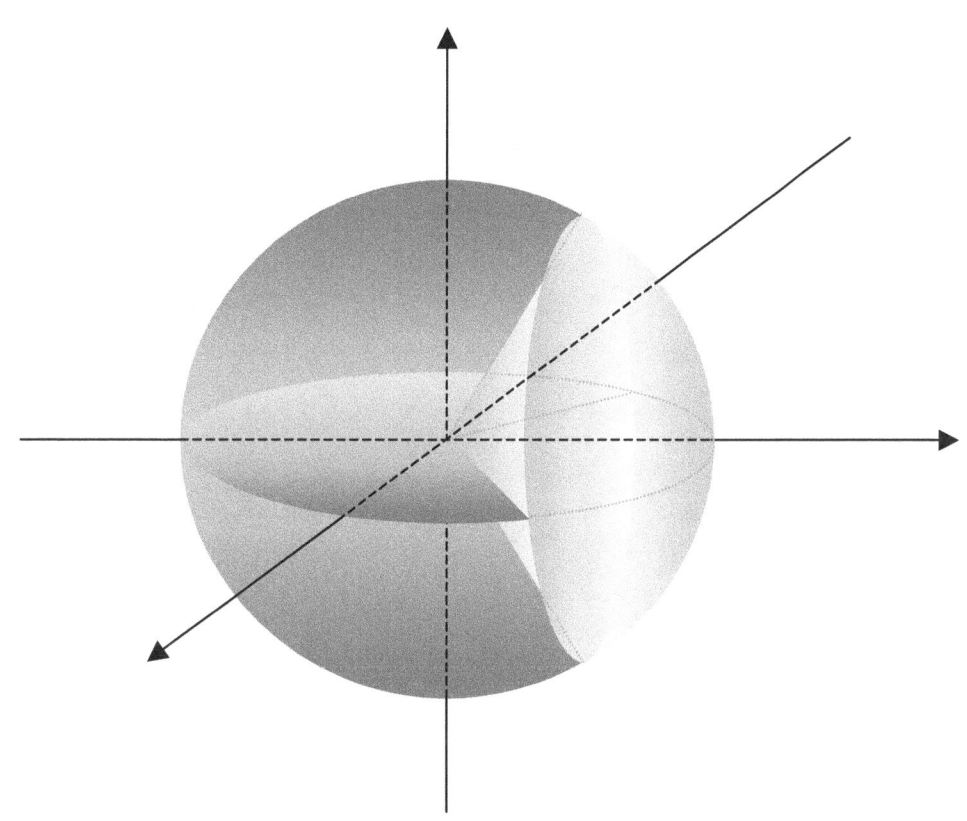

Seong R. Kim

Dear students:

Students need the best teacher, so you need examples, because examples are the best teacher. All the examples here are fully worked, and explain **how** the basic and essential tools in math are made, together with **what** they are, **how** they work, and **how** to work with them. Such tools include numbers, formulas, identities, equations, laws, etc.

Examples here begin with easy ones, of course. Covering every meter and yard properly, we can cover thousands of miles and kilometers. And it is particularly the case in math.

Of those examples therefore, some might even look too easy for you. It's not that easy though, to come up with those examples. Anyways, the bigger and the taller the tree, the deeper and the stronger the root.

Doing math, we work with ideas and run ideas, because every thing in math is an idea. A number is an idea, for instance, and the same is true for a line or circle, too. And putting ideas together, we build another, which becomes the base or an element of another, and each is connected. And that's the way your math grows. So you get to build a circuit, and sometimes, need to fill the gap or repair the circuit so that you get the sense of it.

So your calculation runs properly, and you get the problem solved.

The examples have been made and arranged so that they get tougher (or sometimes easier for some reason) as you proceed with them. In particular, similar examples with some variations are strategically repeated so that you can get the ideas or the tools tricky or complicated, and can get them mastered.

This book is however, nothing but a bunch of examples until you get it powered.   How then, to get it powered, and make it run and work for you?

Just read it, and then, do each example in writing. And it is important to note that you do it in **your** writing. Just watching someone doing it, you just only feel that you can do it. If you do it, you can do it, but if you don't, we can hardly. It's a cliché, of course, but is always true that knowing is one thing and doing is another.

I've been helping students grow, take care of, and run their own math. The area covers algebra and geometry for high school or college students, and is especially for equations (for unknowns or curves), functions, and their graphs, which are the basic elements in calculus, which's been the core of my interest from my early age in high school.

Of my students, some are quite poor in math, and thus, are afraid of or hate math, some require special education because of exceptional intelligence, some are smart enough, some are naïve and diligent, some are clever but lazy, and most behave in general. All the students are badly after though, one thing in common: a strong and secure math skill. It is of course, the prime objective of my work, and I'm always happy to and eager to help them achieve it. The problem was however, that many of them wanted it to be purchased. And the question is, can we buy it?

We can buy the means, of course. And a solid math skill is feasible, too. We know however, we can't buy love, and the same is true for the math skill, too. It's not what we can buy or sell, and not what we can give or take. It is however, what we can grow, and need to grow. Your math grows as much as you grow and take care of it. So does mine.

What math then, do students most often do or use in high schools or colleges?

It is algebra and geometry. What algebra though?

Elementary algebra, of course
Doing the algebra, we work with numbers (many in kinds), constants, variables, ratios, rates, expressions, equations, inequalities, functions, identities, formulas, laws, etc., together with signs and symbols. And if we want to do algebra properly, we want to know their natures and how they mingle with each other.

So studying math ideas or tools, you want to know **what** they are, **how** they work, and **how** to work with them or **what** to do with them.   What then, about the geometry?

Basically, the geometry has much to do with shapes, positions, and angles. The shapes begin with triangles and circles, and move on to rectangles, squares, parallelograms or rhombuses, trapezoids, tetragons, other polygons, polyhedrons, etc.

Doing the geometry, too, though, we need to do the algebra stated above. So it is analytic geometry, often called coordinate geometry, too. And doing it, we can specify positions using coordinates. So in the geometry, basically, we work with graphs. Putting a math idea in a graph, we can not only effectively think about it but actually see it, too, and therefore, can efficiently work with it. What idea then, is it?

The idea begins with a point, line, parabola, circle, ellipse, and hyperbola, called a conic section or basic curve, and then, moves on to other curves, planes, surfaces, volumes, and other objects in various dimensional spaces, together with vectors.

And using an angle, we can specify an amount of turn or change in direction.

So learning, using, or applying those ideas or math tools, we get to solve problems.

And this book can help. It can help learn them, and use them so that you can navigate to find solutions to problems. And in particular, it can help come up with answers to those **what**s and **how**s stated above. So it can help you grow and run your own math, and thus, can help achieve your solid math skill.

It is however, not a magic book giving you a math skill of high caliber overnight. And it can have many mistakes, too. There is no magic, and math is full of facts and ideas. And it is after all, not me and not your teacher but you who put together some of those facts and ideas, and understand it. Putting facts and ideas together, understanding it, and taking care of what you have learned, you grow your math. And this book can help.

This is a book of examples designed to help you grow your math, and assumes that you are a real beginner. This book requires though, time and effort, the amount of which need to be substantial, too, but will be worth it. That's because you want a substantial achievement, and will get it. And probably, you will get to see this book helping you get there much faster than expected. And then, you will get to see the way math runs.

In math, everything is an idea. So is a problem. And solving it, we put it many different ways. For instance, while expanding or reducing it, or modifying or converting it, we keep searching for the solution, approaching the solution, and eventually, can get there. So don't look for the solution outside the problem. The solution is inside the problem if the problem is properly made.

If it is not, no solution is the solution. And in fact, it is often the case a problem itself is the solution. We can put a problem in many different ways, and eventually, can end up with the solution.   How come then, is the solution no other than the problem?

For instance, the solution to $3232 \div 101$ is 32.   And we can put it this way:

$$3232 \div 101 = \frac{3232}{101} = \frac{32 \times 101}{101} = \frac{32}{1} = 32 \;\Rightarrow 3232 \div 101 = 32.$$

And we can get this, too: $32 \Rightarrow 3232 \div 101$.   How?

$$32 = \frac{32}{1} = \frac{32 \times 101}{101} = \frac{3232}{101} = 3232/101 = 3232 \div 101.   \text{Too easy?}$$

For another instance, the solution to $ax^2 + bx + c = 0$ is: $x = \frac{-b \pm \sqrt{b^2 - 4ac}}{2a}$, which is called the quadratic formula.   How come then, is the solution no other than the problem?

We can put it this way:

$$x = \frac{-b \pm \sqrt{b^2 - 4ac}}{2a} \Rightarrow 2ax = -b \pm \sqrt{b^2 - 4ac} \Rightarrow 2ax + b = \pm\sqrt{b^2 - 4ac}$$

$$\Rightarrow (2ax + b)^2 = b^2 - 4ac \Rightarrow 4a^2x^2 + 4abx + b^2 = b^2 - 4ac$$

$$\Rightarrow 4a^2x^2 + 4abx = -4ac \Rightarrow ax^2 + bx = -c \Rightarrow ax^2 + bx + c = 0.$$

And we can get this, too: $ax^2 + bx + c = 0 \Rightarrow x = \frac{-b \pm \sqrt{b^2 - 4ac}}{2a}$.   How?

$$ax^2 + bx + c = a(x^2 + \tfrac{b}{a} x) + c = a(x^2 + \tfrac{b}{a} x + \tfrac{b^2}{4a^2} - \tfrac{b^2}{4a^2}) + c = a(x^2 + \tfrac{b}{a} x + \tfrac{b^2}{4a^2}) - \tfrac{b^2}{4a} + c$$

$$= a(x + \tfrac{b}{2a})^2 - \tfrac{b^2 - 4ac}{4a} = 0 \Rightarrow a(x + \tfrac{b}{2a})^2 = \tfrac{b^2 - 4ac}{4a} \Rightarrow (x + \tfrac{b}{2a})^2 = \tfrac{b^2 - 4ac}{4a^2} \Rightarrow x + \tfrac{b}{2a} = \pm\sqrt{\tfrac{b^2 - 4ac}{4a^2}}$$

$$\Rightarrow x = -\tfrac{b}{2a} \pm \tfrac{\sqrt{b^2 - 4ac}}{2a} = \tfrac{-b \pm \sqrt{b^2 - 4ac}}{2a} \Rightarrow x = \tfrac{-b \pm \sqrt{b^2 - 4ac}}{2a}.$$

And we call the set of processes above, algebra.

So if a problem is well defined, that is, if it makes sense, we should be able to get it solved the way below:

A problem $\Rightarrow$ ... $\Rightarrow$ ... $\Rightarrow$ the solution, and thus: **the problem $\Rightarrow$ the solution**.

So solving a problem, we put it many different ways so that we can get to the solution.

And that's the way, math runs.

May your math run very well.

Seong R. Kim

B.S. Math. Michigan Tech. Univ.   M.S. Math. Rensselaer Polytechnic Institute

Notes:

This book is one of five books about some basics in elementary algebra, and covers equations often used in high schools and colleges or universities. And the equations are for curves called conic sections, often just called conics. There are five kinds in conics. And of the five, one is covered briefly here in this book, and is for circles.

So this book covers equations indicating circles, that is, equations for circles. And this book explains what such an equation is about, how it gets made, what it does or how it behaves, and what we can do with it or how to use it.   What then, is it for?

A circle is an idea in math, so it's a math idea, and is a tool in math. So it's a math tool. And we use it, solving problems, of course. So students need to get the idea.

And thus, this book helps you get the idea of a circle, that is, the concept of a math object called a circle, and see how to use it, because the book explains what it is and how it works, along with those stated above so that you can develop your own idea to make use of it, solving problems, of course.   What then, about the other conics?

They are covered in their individual books, too, which are as follows:

Algebra Examples Conics 1 Lines, which covers therefore, equations for lines, often called linear equations or equations of degree 1.

Algebra Examples Conics 2 Parabolas, which covers thus, equations for parabolas, often called quadratic equations or equations of degree 2.

Algebra Examples Conics 4 Ellipses, which is about equations for ellipses.

Algebra Examples Conics 5 Hyperbolas, explaining those for hyperbolas.

And each book is designed for those students who want to study calculus, want to major in science or engineering, or want to take IB courses in math, so each book covers materials in each category in such a depth and extent. And if you don't need that much, it will be sufficient to study the sections and the sets of examples bulleted in the table of contents.

Either way, the books will help you grab math ideas often used in real life as well as in math courses. The ideas are about lines, parabolas, circles, ellipses, hyperbolas, and their equations, so you will get to see what those curves and equations are about, how they work, and how to use them, and develop your own idea to make use of those, solving problems, of course.

In short, the books help you develop and grow your own idea to make use of math ideas, providing examples, showing all the steps and the ideas behind them, and explaining what the math ideas are about.

Contents

Note:

The drawings or graphs in this book are not exact, and are approximate or conceptual ones.

$\in$	"$a \in B$" means that a belongs to B. "$p, q,$ and $r \in W$" means that $p, q,$ and r belong to W.
$\Rightarrow$	"$A \Rightarrow B$." means that A implies B.
$\equiv$	$A \equiv B$ means that A and B are identical to each other.
$\neq$	$A \neq B$ means that A is not equal to B.
$|A|$	The magnitude of A. For instance, $|{-1}| = |1| = 1$.
$\therefore$	Therefore
$\Leftrightarrow$	"$A \Leftrightarrow B$" means "If A then B." and "If B then A." We can read $A \Leftrightarrow B$ as "A if and only if B." In such a case, we can say that $A = B$.
Δx and Δy	Suppose that (x_1, y_1) and (x_2, y_2) are two points in the x-y plane. Then, we get either of the two below. $\Delta x = x_2 - x_1$, and $\Delta y = y_2 - y_1$. $\Delta x = x_1 - x_2$, and $\Delta y = y_1 - y_2$.

Distance Formula

Suppose that d is the distance between two points (x_1, y_1) and (x_2, y_2) in the x-y plane. Then, we get: $d^2 = (\Delta x)^2 + (\Delta y)^2$.

₀. **What is a circle?**

A circle is in a plane, and is a set of points, from each of which, <u>the distance to a particular point is constant,</u> that is, the same. And the particular point is called the center of the circle, and the same distance is called the radius.

Like a line and a parabola, a circle is one of the basic curves called conic sections, of called conics, for short. So a circle is a conic.

And a circle is a simply closed curved-line segment that is symmetric and has infinitely many axes of symmetry, each of which is a diameter.

Fig. 0

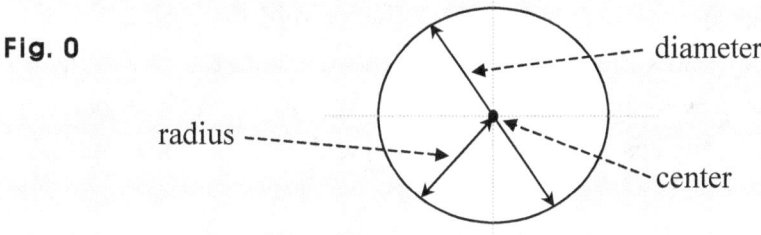

How then, can we get a circle?

As shown in the figure below, if cutting a right cone with a plane, we can get a cross section called a circle. So it's called a conic section, just called a conic, for short.

And if a cone is a right cone, the line connecting the vertex of the cone and the center of the base is perpendicular to the base, and thus, makes a right angle (90°) with the base.

The cross section in black below is a curve closed, and is a circle. So if the plane is parallel to the base, but does not include the vertex, the cross section is a circle.

Fig. 1

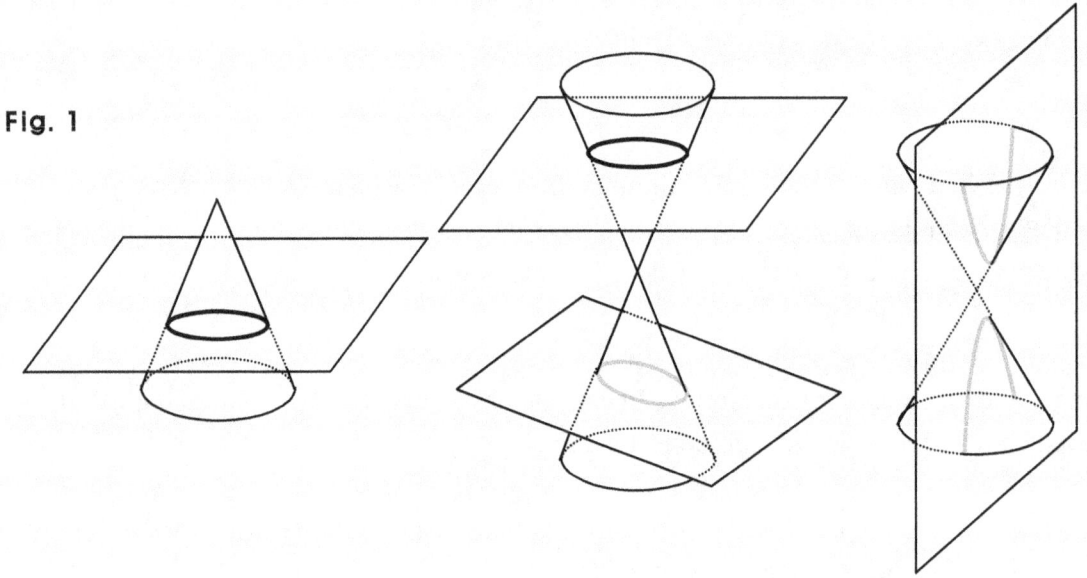

By the way, we have an object similar to a circle. Squeezing a circle, we get such an object, if you will. Such an object is called an ellipse, which is a conic, too, and is the closed curved line segment in gray in the figure above. So if the plane is neither parallel to the base, nor passing through the base, and does not include the vertex, the cross section is an ellipse.

If however, the plane is perpendicular to the base, and doesn't include the vertex, or if the plane is not parallel to the side of the cone (called the generator of the cone, too), and passes through the base, we get a cross section called a hyperbola.

And we can also explain a circle the way as follows.

Suppose that a point is moving along a curve in the *x-y* plane, and that the distance from the moving point to a particular point is constant. Then, the particular point is called the center, the distance is called the radius, and the curve is a circle.

So if a point (x, y) is in a circle, no matter where the point (x, y) may be, <u>the distance from the point (x, y) to the point called the center is constant</u>, that is, <u>the same</u>.

How then, can we explain or describe a particular circle?

Normally, in math, we put a circle in the *x-y* plain, and describe the circle specifying the center as well as the radius.

So for instance, if a circle is <u>centered at a point (5, 2)</u> in the *x-y* plane, and its <u>radius is 3</u>, we put the circle in a graph the way as follows:

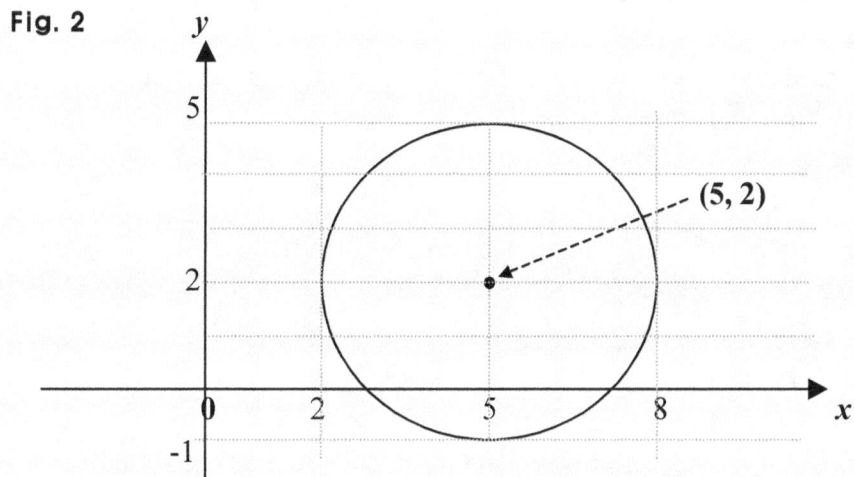

Fig. 2

And in math, making a circle, we define it. So defining a circle particular, we make the particular circle.   How then, can we define a particular circle?

As in the cases of other conics as a parabola, we can define a circle using an equation that indicates the circle. So we can define a particular circle producing the equation of it.

For instance, assuming *C* is the circle described above, we can define *C* the way below:

$$(x-5)^2 + (y-2)^2 = 3^2$$

And the equation above is in the standard form, and is a standard equation of a circle.

Putting the standard equation above in the general form, we just expand (or simplify) it. Then, we get: $x^2 + y^2 - 10x - 4y + 20 = 0$, which is a general equation of a circle.

And in general, putting a circle in the standard form, we can put it the way below:

$$(x - u)^2 + (y - v)^2 = r^2, \text{ where } r \neq 0.$$

What then, do we mean by u, v, and r?

The center is at (u, v), and the radius is r.

So just describing a particular circle, we often describe it specifying the radius. If however, we define a particular circle placed in the x-y plane, we want to specify the center and the radius.

And putting a circle in an equation in general, we can put it the way below:

$$x^2 + y^2 + ax + by + c = 0, \text{ where } \underline{a^2 + b^2 > 4c}.$$

And the equation above is in the <u>general form</u>, and is called a general equation of a circle.

What do we mean by, though, this: $a^2 + b^2 > 4c$?

It means that if the equation makes a circle, we get this: $a^2 + b^2 - 4c > 0$.
So if we get: $a^2 + b^2 \leq 4c$, the equation does <u>not</u> make a circle. Usually though, we just put it this way: $x^2 + y^2 + ax + by + c = 0$, assuming implicitly, $a^2 + b^2 - 4c > 0$.

For instance, $x^2 + y^2 - 2x - 4y + 6 = 0$ seems indicating a circle, but does not indicate it.

That's because we get: $a^2 + b^2 - 4c = (-2)^2 + (4)^2 - 4 \cdot 6 = 4 + 16 - 24 = -4 < 0$.
So we get: $a^2 + b^2 - 4c < 0 \Rightarrow a^2 + b^2 < 4c$.

And putting in fact, the equation in the standard form, we get:

$$x^2 + y^2 - 2x - 4y + 6 = x^2 - 2x + 1 - 1 + y^2 - 4y + 4 - 4 + 6 = (x-1)^2 + (y-2)^2 + 1 = 0$$
$$\Rightarrow (x-1)^2 + (y-2)^2 = -1,$$ which is impossible, and does not indicate a circle.

And we can give a name to an equation, too. For instance, assuming C is the equation of the circle above, we can put C the way as follows:

$$C(x, y) = x^2 + y^2 + ax + by + c = 0.$$ What then, about the standard equation?

Assuming S is the equation $(x-u)^2 + (y-v)^2 = r^2$, we can put S the way below, too:

$$S(x, y) = (x-u)^2 + (y-v)^2 - r^2 = 0,$$ where $r \neq 0$.   What if however, $r = 0$?

Then, the equation does not indicate a circle but just a point.   What point then, is it?

It is a point at (u, v).   It's because if $(x-u)^2 + (y-v)^2 = 0$, $(x-u)$ and $(y-v)$ both have to be 0 at the same time.   So we get: $x = u$ and $y = v$, and thus, we just get this: (u, v).

And we know that there is no circle of radius 0.   So if S indicates a circle, we get: $r \neq 0$.

And it is <u>not</u> the case where the equation $x^2 + y^2 = r^2$ is defined for x real and y real. ('x real' means every real number can be x.)

The equation is in fact, defined for $-r \leq x \leq r$, and $-r \leq y \leq r$.

That is, the domain is: $-r \leq x \leq r$, and the range is: $-r \leq y \leq r$.

That's simply because the circle is of radius r, and is centered at the origin $(0, 0)$.

And moving the circle above in the amount of u along the x-axis, and in the amount of v along the y-axis, we get another circle as follows: $(x-u)^2 + (y-v)^2 = r^2$.

What circle then, is it?

If we move the center of a circle, the circle moves the way the center gets moved.

So moving a circle, we move its center, keeping the radius intact, of course.
And in the case above, the center moves from the origin to the point (u, v).

So we can notice that the circle is: $(x - u)^2 + (y - v)^2 = r^2$, since the center is (u, v) and the radius is r. And it is <u>not</u> the case either, the equation is defined for x real and y real. What then, are the domain and the range?

It is defined for $-r + u \leq x \leq a + r$, and $-r + v \leq y \leq r + v$.

That is, the domain is: $u - r \leq x \leq u + r$, and the range is: $v - r \leq y \leq v + r$.

So if a circle is of radius r and is centered at (u, v), we can put it in the standard equation more specifically the way as follows:

$$(x - u)^2 + (y - v)^2 = r^2 \text{ where } r \neq 0, \ u - r \leq x \leq u + r, \text{ and } v - r \leq y \leq v + r.$$

Normally though, we just put it this way: $(x - u)^2 + (y - v)^2 = r^2$, assuming the others implicitly, of course

And a circle has some physical properties as follows.

To begin with, a circle can be said to be the simplest line segments curved and closed.

Suppose next, we have designated a particular point in the x-y plane.
Then, a circle is a set of all points from each of which the distance to the particular point is constant. In other words, all the individual distances from all the points in the circle to the particular point are the same. What then, is the particular point?

It is the center of the circle, of course.   What then, is the distance stated above?

It is the radius, of course. So a radius is a line segment connecting the center and a point in the circle. What then, is a line segment twice the radius?

It is a diameter, of course, and is passing through the center, and connecting two facing points in the circle.
How then, can we specify a circle? That is, what determines a particular circle?

The center and radius do so. Therefore, we can specify a circle by the center and radius.
A radius is the distance from the center of a circle to a point in the circle.
There are quite a few radii in a circle, then.
They all have the same length, of course, and there are in fact, infinitely many of those.
It can be thought that the center of a circle radiates the radii.

The center radiates the radius in every direction if you will.
That is, the center of a circle radiates the radii in all directions.
Besides, any line tangent to a circle is perpendicular to a line passing through the tangent point and the center of the circle. Thus, a line tangent to the circle is perpendicular to the line passing through the tangent point and including the radius of the circle.
So a line tangent to a circle is perpendicular to the radius that includes the tangent point.
And thus, a line tangent to a circle is perpendicular to any line parallel to that radius, too.

Fig. 3

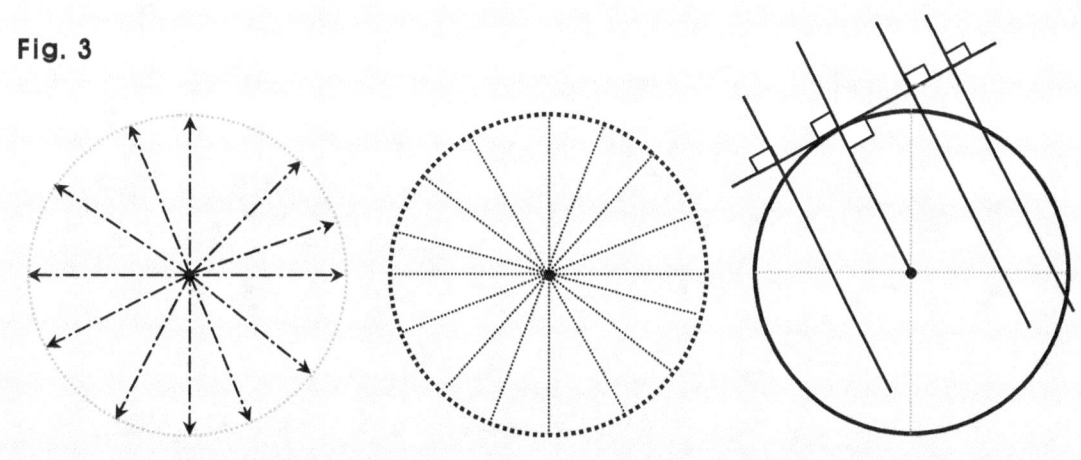

Examples 1 in Standard Forms

Label each circle below, and then find the equation of each of the circles.

Fig. 0

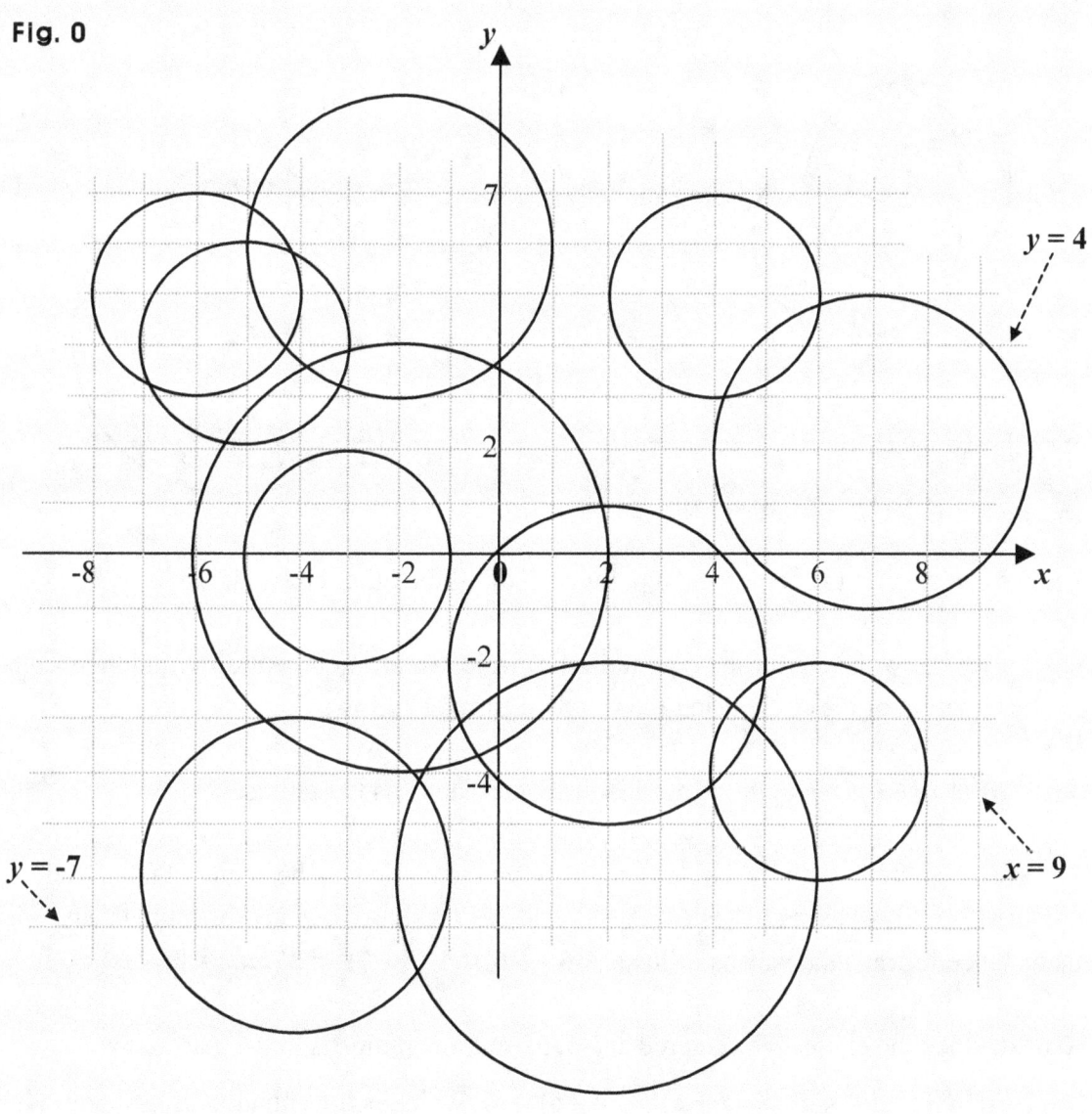

Suggestions or Solutions
To the Problems in the Examples

Beginning with the circles below, we can see first, that the circle *A* is centered at (-6, 5), and has a point (-6, 7) or (-4, 5).

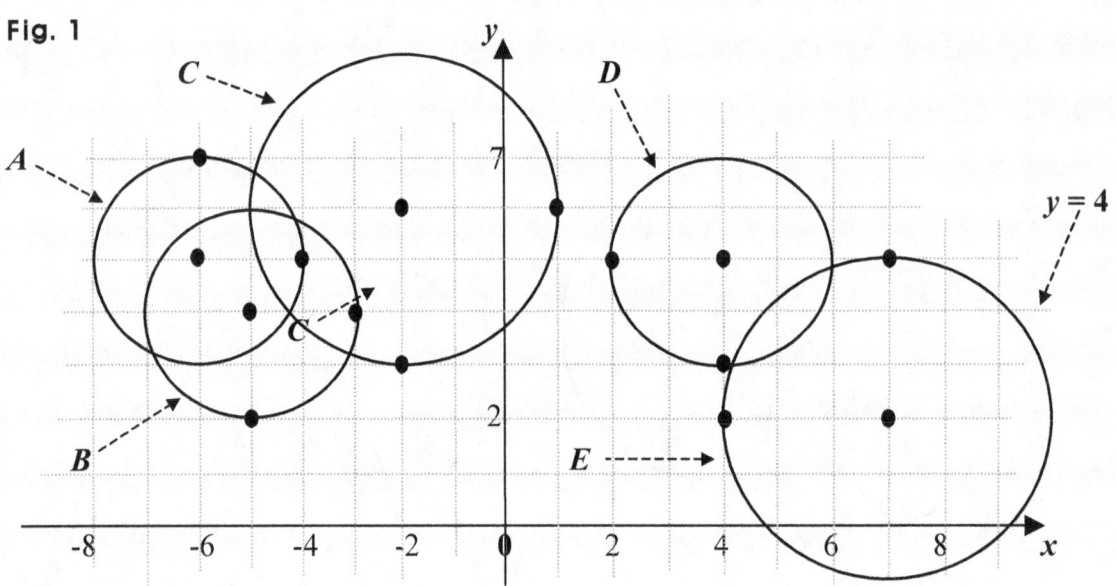

Fig. 1

And we can specify the center and a point each of all the other circles has.

Next, if we know the center and a point in the circle, we can get the radius, and then, can get the equation of the circle using the standard form as follows:

$(x - a)^2 + (y - b)^2 = r^2$, where (a, b) is the center, and r is the radius.

And thus, beginning with the circle *A*, we can see the circle has the center at (-6, 5) and a point at (-6, 7), so the radius is: $7 - 5 = 2$, and of course, we can get the radius using the distance formula the way as follows: $\{-6 - (-6)\}^2 + (5 - 7)^2 = 2^2 \Rightarrow r = 2$.

So next, since the circle *A* is centered at (-6, 5), and the radius is 2, we get:

$\{x - (-6)\}^2 + (y - 5)^2 = 2^2 \Rightarrow (x + 6)^2 + (y - 5)^2 = 4$, which is the equation of the circle *A*.

What if we use the other point (-4, 5) instead of the point (-6, 7)?

The point (-4, 5) is in the circle A, too, so we will get the same equation. Since the center is (-6, 5), the radius is: $-4 - (-6) = -4 + 6 = 2$, so we can get it the way as follows:

$$\{x - (-6)\}^2 + (y - 5)^2 = 2^2 \Rightarrow (x + 6)^2 + (y - 5)^2 = 4.$$

Note that <u>in math</u>, the same circles have the same center and the same radius. So even if the same radius, if the center is different, it's a different circle.

Next, moving on to the circle B, we can see that the center is (-5, 4) and it has a point at (-5, 2), so the radius is: $4 - 2 = 2$, and of course, we can get the radius using the distance formula the way as follows: $\{-5 - (-5)\}^2 + (4 - 2)^2 = 2^2 \Rightarrow r = 2$.

So next, since the circle B is centered at (-5, 4), and the radius is 2, we get:
$\{x - (-5)\}^2 + (y - 4)^2 = 2^2 \Rightarrow (x + 5)^2 + (y - 4)^2 = 4$, which is the equation of the circle B.

Next, looking at the circle C, we can see it has the center at (-2, 6) and a point at (-2, 3), so the radius is: $6 - 3 = 3$, and thus, we get:

$\{x - (-2)\}^2 + (y - 6)^2 = 3^2 \Rightarrow (x + 2)^2 + (y - 6)^2 = 9$, which is the equation of the circle C.

Next, examining the circle D, we can see it is centered at (4, 5) and has a point at (2, 5), so the radius is: $4 - 2 = 2$, and thus, we get:

$(x - 4)^2 + (y - 5)^2 = 2^2$, which is the equation of the circle D.

Next, examining the circle E, we can see it is centered at (7, 2) and has a point at (4, 2), so the radius is: $7 - 4 = 3$, and thus, we get:

$(x - 7)^2 + (y - 2)^2 = 3^2$, which is the equation of the circle E.

Next, moving on to the circles below, we can see that the circle F is centered at (-5, 6), and a point at (-7, 4) or (-3, 4).

Fig. 2

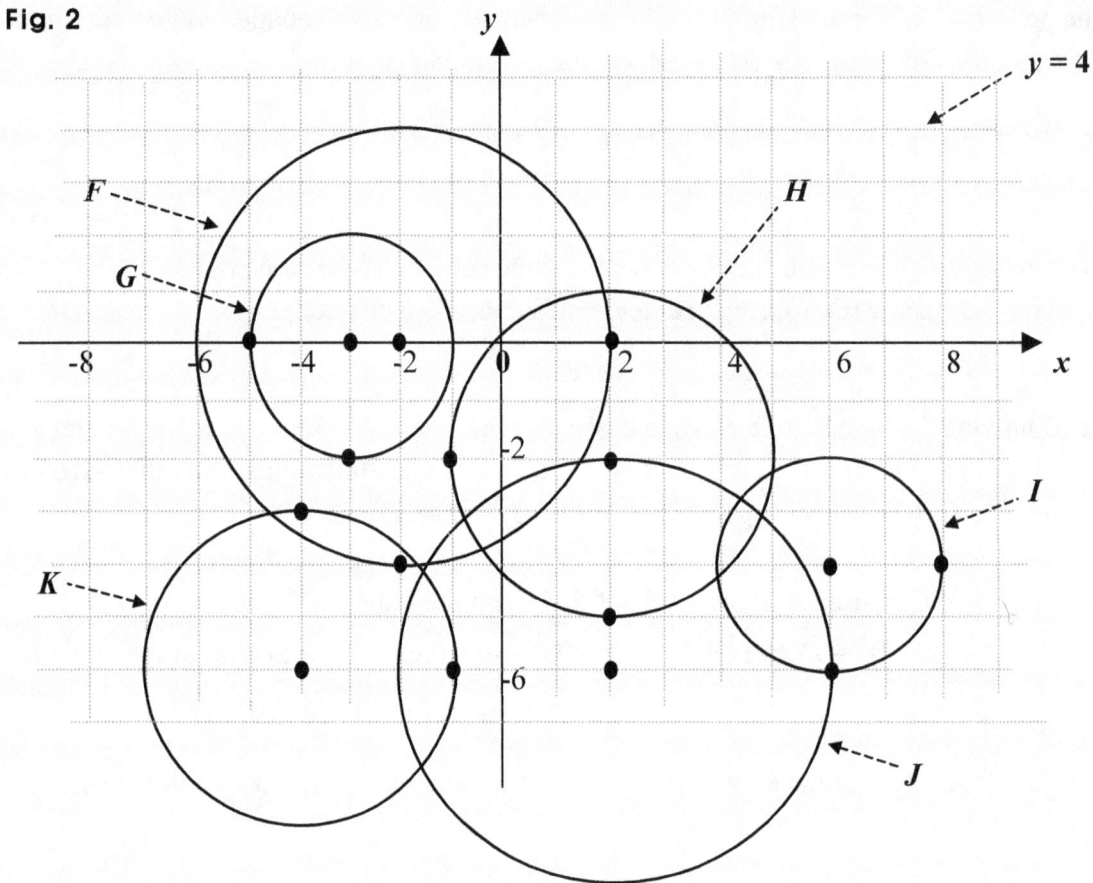

And knowing the center and a point in the circle, we can get the radius, and then, can get the equation of the circle using the standard form as follows:

$$(x - a)^2 + (y - b)^2 = r^2,$$ where (a, b) is the center, and r is the radius.

And thus, beginning with the circle **F**, we can see the circle has the center at (-2, 0) and a point at (2, 0), so the radius is: $2 - (-2) = 4$.

So next, since the circle **F** is centered at (-2, 0), and the radius is 2, we get:

$\{x - (-2)\}^2 + (y - 0)^2 = 4^2 \Rightarrow (x + 2)^2 + y^2 = 16$, which is the equation of the circle **F**.

Next, looking at the circle **G**, we can see it has the center at (-3, 0) and a point at (-5, 0), so the radius is: $-3 - (-5) = -3 + 5 = 2$, and thus, we get:

$\{x - (-3)\}^2 + (y - 0)^2 = 2^2 \Rightarrow (x + 3)^2 + y^2 = 4$, which is the equation of the circle **G**.

Next, examining the circle H, we can see it is centered at (2, -2) and has a point at (2, -5), the radius is: $|-5 - (-2)| = 3$, so we get:

$(x - 2)^2 + \{y - (-2)\}^2 = 2^2 \Rightarrow (x - 2)^2 + (y + 2)^2 = 4$, which is the equation of the circle H.

Next, examining the circle I, we can see it is centered at (6, -4) and has a point at (8, -4), so the radius is: $|6 - 8| = 2$, and thus, we get:

$(x - 6)^2 + \{y - (-4)\}^2 = 2^2 \Rightarrow (x - 6)^2 + (y + 4)^2 = 4$, which is the equation of the circle I.

Next, examining the circle J, we can see it is centered at (2, -6) and has a point at (2, -2), so the radius is: $|-6 - (-2)| = 4$, and thus, we get:

$(x - 2)^2 + \{y - (-6)\}^2 = 4^2 \Rightarrow (x - 2)^2 + (y + 6)^2 = 16$, which is the equation of the circle J.

Next, examining the circle K, since it is centered at (-4, -6) and has a point at (-1, -6), the radius is: $|-4 - (-1)| = 3$, so we get:

$\{x - (-4)\}^2 + \{y - (-6)\}^2 = 3^2 \Rightarrow (x + 4)^2 + (y + 6)^2 = 9$, which is the equation of K.

Examples 2 in Standard Forms

0. Specify the center and the radius of each circle below, and then, graph it.

0. $(x + 5)^2 + (y - 6)^2 = 4$ 1. $x^2 + (y - 2)^2 = 9$ 2. $(x - 5)^2 + (y - 3)^2 = 16$

3. $(x + 4)^2 + (y - 5)^2 = 12$ 4. $(x + 4)^2 + y^2 = 7$ 5. $(x + 4)^2 + (y + 3)^2 = 15$

6. $(x - \frac{9}{2})^2 + (y + 3)^2 = \frac{49}{4}$

Fig. 0

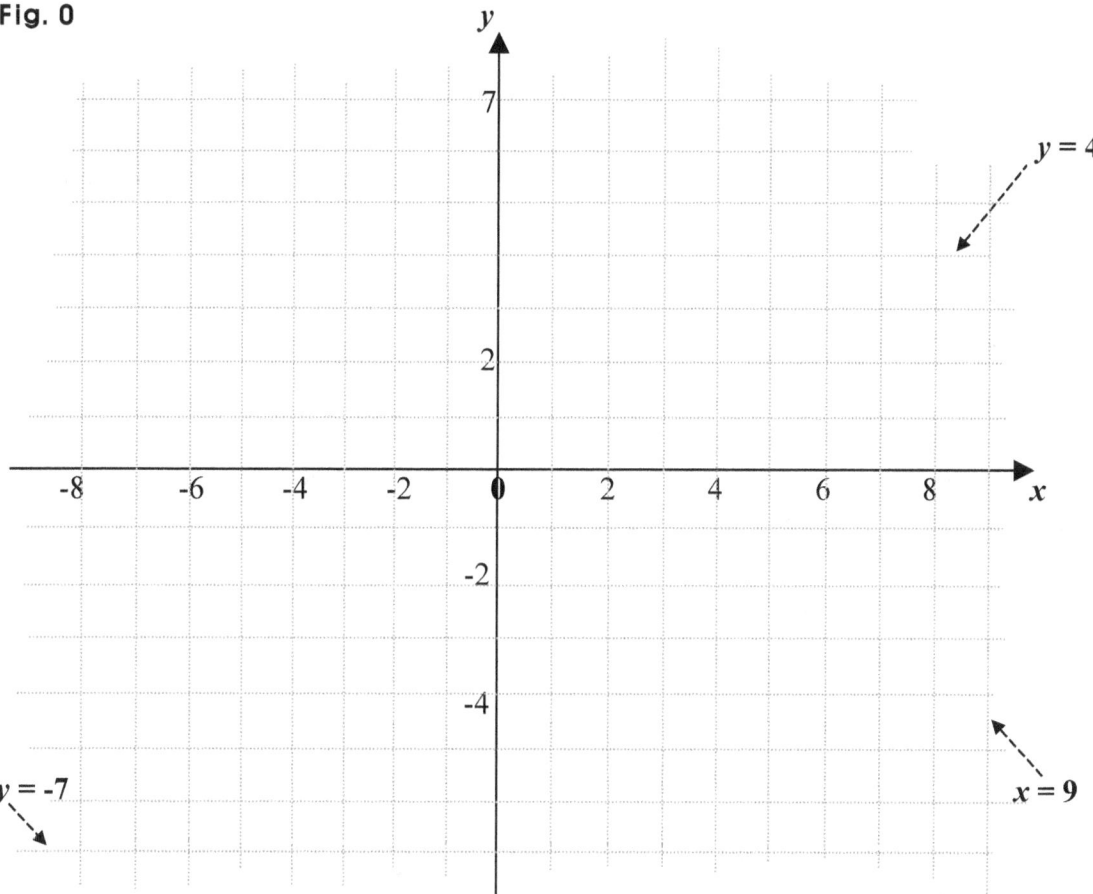

1. Find the center and the radius of each circle below:

0. $x^2 + 2x + y^2 + 2y + 1 = 0$

1. $x^2 - 3x + y^2 - 5y + 1 = 0$

2. $2x - x^2 + 3y - y^2 + 1 = 0$

3. $x^2 - 7x + y^2 + 2y + 1 = 0$

Suggestions or Solutions
To the Problems in the Example 0

Specify the center and the radius of each circle below, and then, graph it.

0. $(x + 5)^2 + (y - 6)^2 = 4$ 1. $x^2 + (y - 2)^2 = 9$ 2. $(x - 5)^2 + (y - 3)^2 = 16$

3. $(x + 4)^2 + (y - 5)^2 = 12$ 4. $(x + 4)^2 + y^2 = 7$ 5. $(x + 4)^2 + (y + 3)^2 = 15$

6. $(x - \frac{9}{2})^2 + (y + 3)^2 = \frac{49}{4}$

First, if we know the center and a point in the circle, we can get the radius, and then, can get the equation of the circle using the standard form as follows:

$(x - a)^2 + (y - b)^2 = r^2$, where (a, b) is the center, and r is the radius.

So to begin with, examining the circle $(x + 5)^2 + (y - 6)^2 = 4$, since $4 = 2^2$, we can see that the center is at (-6, 5), and the radius is 2.

Next, we can put the circle $x^2 + (y - 2)^2 = 9$ this way: $(x - 0)^2 + (y - 2)^2 = 3^2$, so we can see that the center is at (0, 2), and the radius is 3.

Next, examining the circle $(x - 5)^2 + (y - 3)^2 = 16$, since $16 = 4^2$, we can see that the center is at (5, 3), and the radius is 4.

Next, moving on to $(x + 4)^2 + (y - 5)^2 = 12$, we can put it the way below:

$\{x - (-4)\}^2 + (y - 5)^2 = (\sqrt{12})^2 = (2\sqrt{3})^2$. So the center is at (-4, 5), and the radius is $2\sqrt{3}$.

Next, moving on to $(x + 4)^2 + y^2 = 7$, we can put it this way: $\{x - (-4)\}^2 + (y - 0)^2 = 7$. So the center is at (-4, 0), and the radius is a square root of 7.

Moving next, on to $(x + 4)^2 + (y + 3)^2 = 15$, we can see that the center is at (-4, -3), and the radius is a square root of 15.

Next, moving on to $(x - \frac{9}{2})^2 + (y + 4)^2 = \frac{49}{4}$, we can put it the way below:

$(x - \frac{9}{2})^2 + \{y - (-4)\}^2 = \frac{49}{4} = (\frac{7}{2})^2$. So the center is at $(\frac{9}{2}, -4)$, and the radius is $\frac{7}{2}$.

And graphing the equations, we get:

Fig. 1

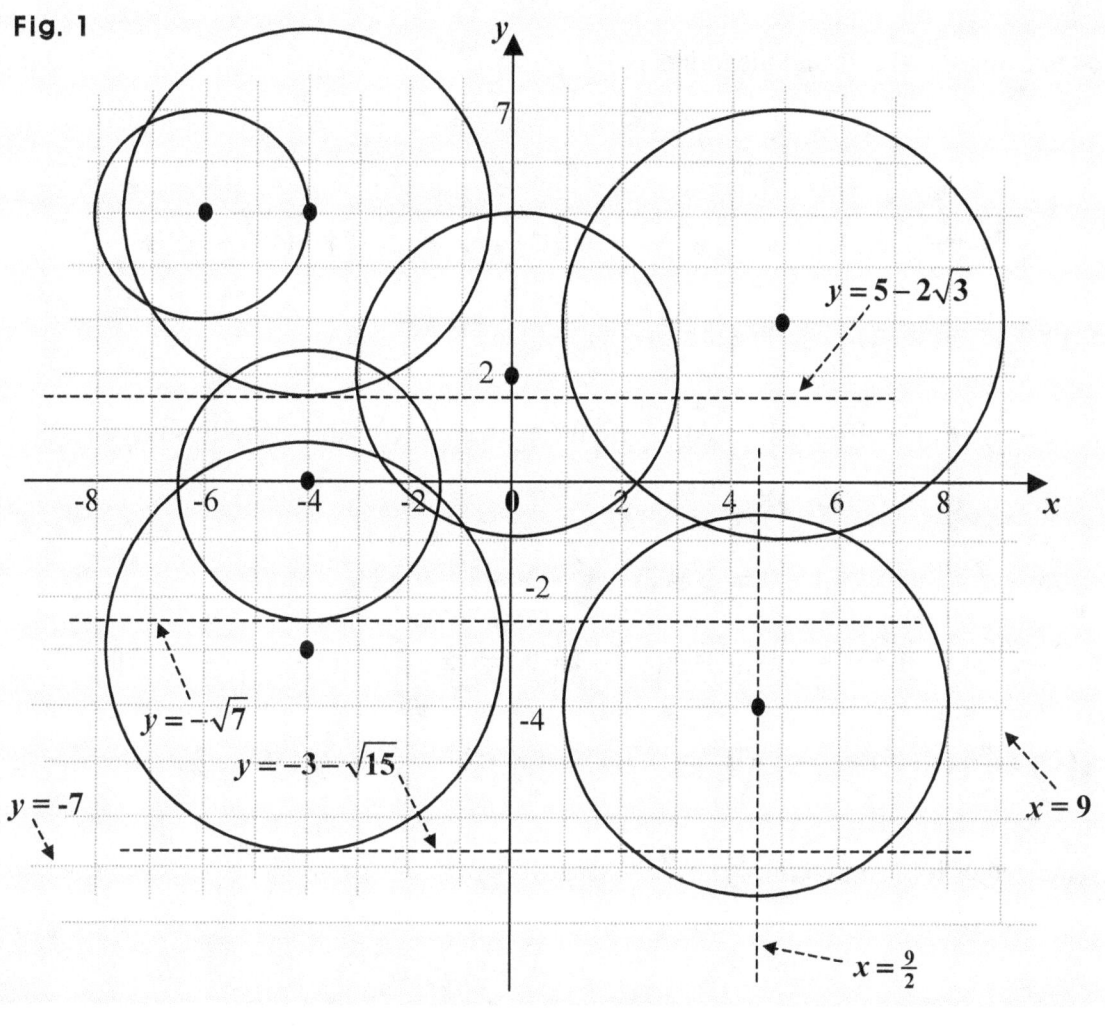

Suggestions or Solutions
To the Problems in the Example 1

0. $x^2 + 2x + y^2 + 2y + 1 = 0 \Rightarrow x^2 + 2x + 1 - 1 + y^2 + 2y + 1 = 0$

$\Rightarrow (x+1)^2 + (y+1)^2 - 1 = 0 \Rightarrow (x+1)^2 + (y+1)^2 = 1.$

So the center is (-1, -1) and the radius is 1.

1. $x^2 - 3x + y^2 - 5y + 1 = 0 \Rightarrow x^2 - 3x + (\frac{3}{2})^2 - (\frac{3}{2})^2 + y^2 - 5y + (\frac{5}{2})^2 - (\frac{5}{2})^2 + 1$

$\Rightarrow (x-\frac{3}{2})^2 + (y-\frac{5}{2})^2 - \frac{9}{4} - \frac{25}{4} + 1 = (x-\frac{3}{2})^2 + (y-\frac{5}{2})^2 - \frac{15}{2} = 0$

$\Rightarrow (x-\frac{3}{2})^2 + (y-\frac{5}{2})^2 = \frac{15}{2} = (\sqrt{\frac{15}{2}})^2.$ So the center is $(\frac{3}{2},\frac{5}{2})$, and the radius is $\sqrt{\frac{15}{2}}$.

2. $2x - x^2 + 3y - y^2 + 1 = 0 \Rightarrow x^2 - 2x + y^2 - 3y - 1 = 0$

$\Rightarrow x^2 - 2x + 1 - 1 + y^2 - 3y + (\frac{3}{2})^2 - (\frac{3}{2})^2 - 1 = 0 \Rightarrow (x-1)^2 + (y-\frac{3}{2})^2 - \frac{9}{4} - 1 = 0$

$\Rightarrow (x-1)^2 + (y-\frac{3}{2})^2 - \frac{9}{4} - 1 = 0 \Rightarrow (x-1)^2 + (y-\frac{3}{2})^2 = \frac{13}{4} = (\frac{\sqrt{13}}{2})^2.$

So the center is $(1,\frac{3}{2})$, and the radius is $\frac{\sqrt{13}}{2}$.

3. $x^2 - 7x + y^2 + 2y + 1 = 0 \Rightarrow x^2 - 7x + (\frac{7}{2})^2 - (\frac{7}{2})^2 + (y+1)^2 = 0$

$\Rightarrow (x-\frac{7}{2})^2 - (\frac{7}{2})^2 + (y+1)^2 = 0 \Rightarrow (x-\frac{7}{2})^2 + (y+1)^2 = (\frac{7}{2})^2.$

So the center is $(\frac{7}{2},-1)$, and the radius is $\frac{7}{2}$.

1. Equations for Circles 1

To begin with, putting a circle in an equation, we can get:

$$(x-5)^2 + (y-2)^2 = 9$$

What circle then, is it?

We can put it this way, too: $(x-5)^2 + (y-2)^2 = 3^2$.
And it is a circle <u>centered at (5, 2)</u>, and its <u>radius is 3</u>.
So we can put the circle in the *x-y* plane the way below:

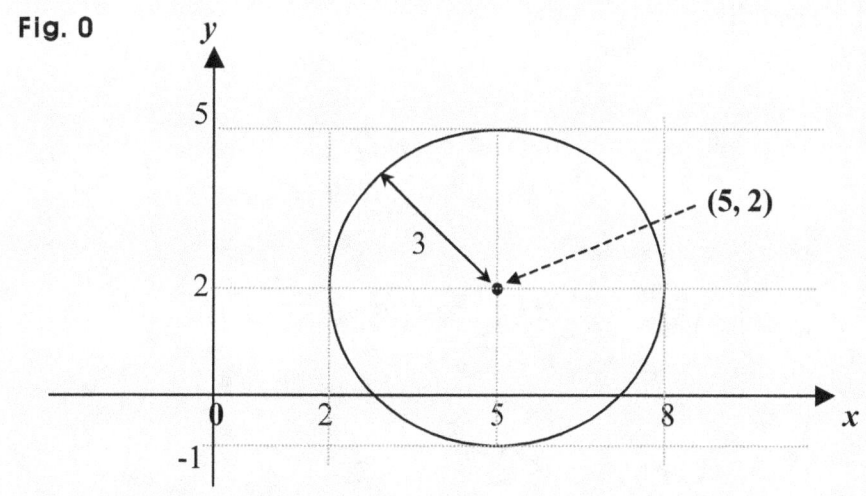

Fig. 0

What circle then, is the circle below?

$$x^2 + y^2 = 1$$

We can put it this way, too: $x^2 + y^2 = 1^2$, or $(x - 0)^2 + (y - 0)^2 = 1^2$.

So the circle is <u>centered at (0, 0)</u>, that is, at the origin, and the <u>radius is 1</u>.

And we call such a circle a unit circle, because the radius is 1, which is often called a unit, too. And we can put the unit circle in the *x-y* plane the way below:

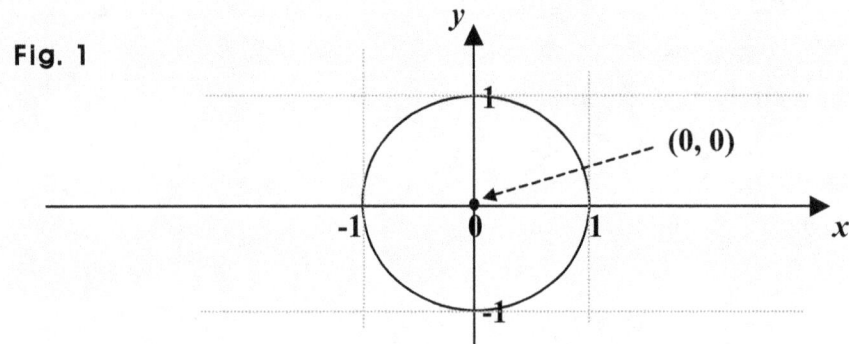

Fig. 1

What if the center is at (1, 1), and the radius is 1?

Then, the circle will be as follows: $(x - 1)^2 + (y - 1)^2 = 1^2$

And we can put the circle above in a graph the way below:

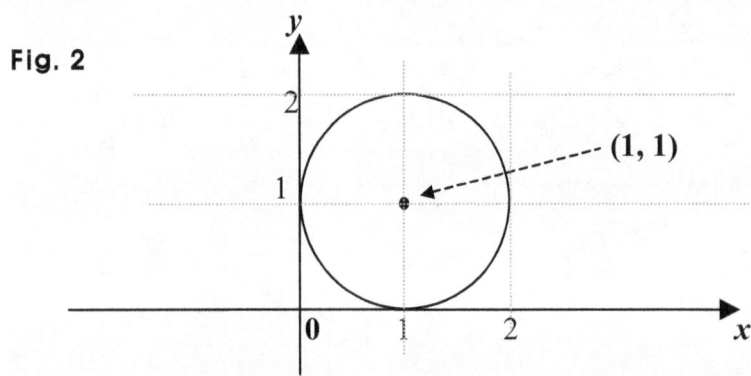

Fig. 2

How then, can we get such equations as above?

We know the fact that a circle is a set of points, from each of which, the distance to the point called the center is constant, that is, the same.   So?

So using the fact above, we can get the equation of a circle.

Thus, to begin with, suppose (u, v) is the center of a circle called C, a point $T(x, y)$ is an arbitrary point in the circle C, that is, a random point representing all the points in C, and the distance from T to the center is r.

Then, the radius is the distance, that is, r, because every point in the center is the distance away from the center. So we can put the three points T in the x-y plane the way as follows:

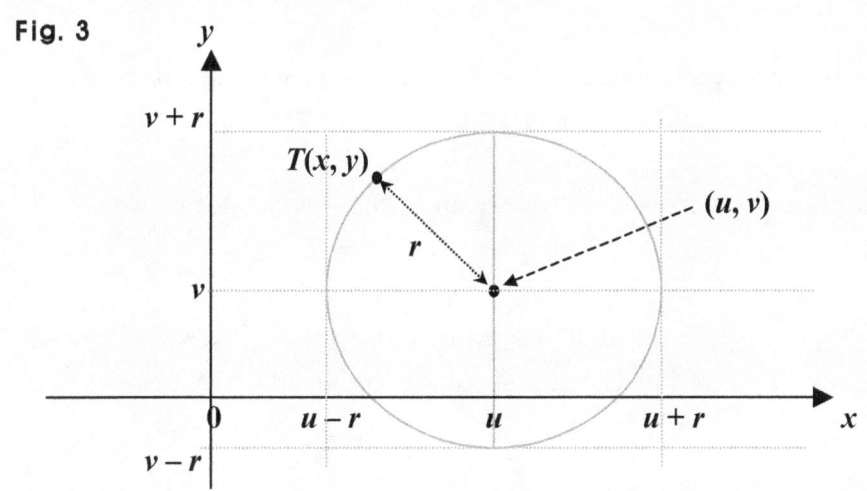

Fig. 3

Suppose next, the point T is moving along the circle C.
Then, the distance from $T(x, y)$ to the center is r. How come?

We know the fact that from any point in a circle, the sum of the two distances to two points called the foci is constant.   So?

So no matter where the point T may be in the circle C, the distance from the point T to the center (u, v) is always the same, and is r.   Using thus, the fact above, we can get the equation of E.   How then, can we use the fact?

We can use the distance formula, often called Pythagorean theorem.

What then, do we get?

We get an equation expressed in terms of the coordinates of the point T, that is, we get an equation in terms of x and y. And we call such an equation a connective equation.

So in this case, the connective equation is the equation that connects the two variables used as the coordinates of the arbitrary point $T(x, y)$ in the curve called the circle C.

How then, do we call the equation?

The equation explains every point in the circle C, that is, it indicates the circle C.
So the equation is called the equation of the circle C.

So let's now get the equation.   How then, can we apply the distance formula?

Assuming d is the distance between two points, Δx is the difference in x-coordinates, and Δy is the difference in y-coordinates, we can put the formula the way below:

$d^2 = (\Delta x)^2 + (\Delta y)^2.$

So using the distance formula, since the distance is r, we can get: $r^2 = (x - u)^2 + (y - v)^2.$

And the equation above is expressed in terms of the coordinates of the point $T(x, y)$, so it is an equation in terms of x and y. And we call such an equation a connective equation.

So in this case, the connective equation is the equation that connects the two variables used as the coordinates of the arbitrary point $T(x, y)$ in the curve called the circle C.

Thus, the equation of the circle C is: $(x - u)^2 + (y - v)^2 = r^2$, where $r \neq 0$.

And putting the circle C in the x-y plane, we get:

Fig. 4

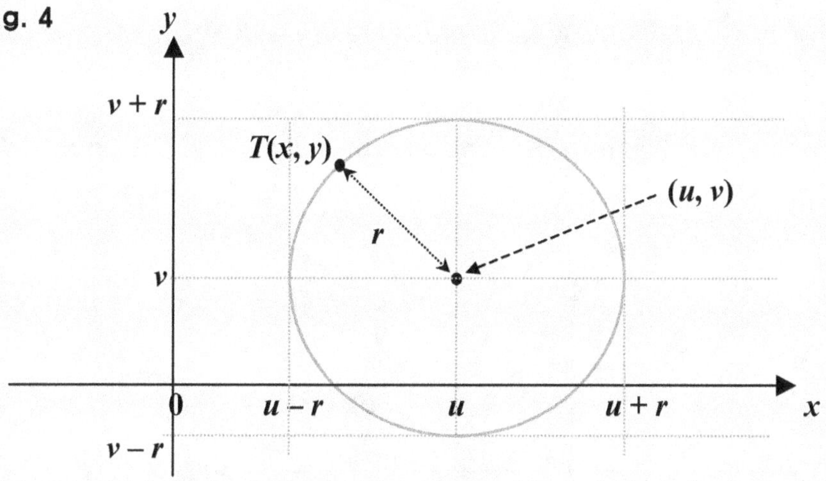

So we can now say that the circle C is centered at the point (u, v), and that the radius is r, that is, the diameter is $2r$.

And putting the circle C in its equation, we get: $(x - u)^2 + (y - v)^2 = r^2$, where $r \neq 0$.

What equation then, do we get if the center is at (-1, 3), and the radius is 2?

It is: $(x + 1)^2 + (y - 3)^2 = 2^2$.

So for another instance, if the center is at the origin, and the radius is r, the equation is: $x^2 + y^2 = r^2$.

And if a circle is a unit circle centered at (2, -1), the equation is: $(x - 2)^2 + (y + 1)^2 = 1^2$.

What then, about the domain and the range?

If the equation of a circle is: $(x - u)^2 + (y - v)^2 = r^2$, it is defined for $-r + u \leq x \leq a + r$, and $-r + v \leq y \leq r + v$.
That is, the domain is: $u - r \leq x \leq u + r$, and the range is: $v - r \leq y \leq v + r$.

What if the circle is: $x^2 + y^2 = r^2$?

Then, the domain is $-r \leq x \leq r$, that is, $|x| \leq r$, and also, the range is: $-r \leq y \leq r$ or $|y| \leq r$.

Fig. 5

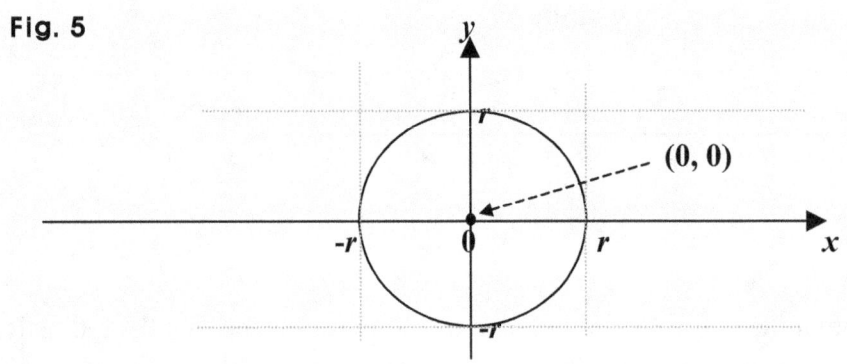

Now, putting threads together, we have four equations as follows.

The circle of radius 1 centered at the origin: $x^2 + y^2 = 1$.

A unit circle, a circle of radius 1 centered at (a, b): $(x - a)^2 + (y - b)^2 = 1$.

A circle of radius r centered at (a, b):

$(x - a)^2 + (y - b)^2 = r^2$ where a, b, and r are constant and $r > 0$.

The equation above is in the standard form.

And putting a circle in an equation in the general from, we get:

$x^2 + y^2 + ax + by + c = 0$ where $a^2 + b^2 > 4c$.

Let's now, put some circles in one graph.

In the figure below, a circle centered at **(a, b)** with the radius of **r** is dotted in black.

A unit circle centered at **(a, b)** is in gray.

And the unit circle centered at the origin is in black.

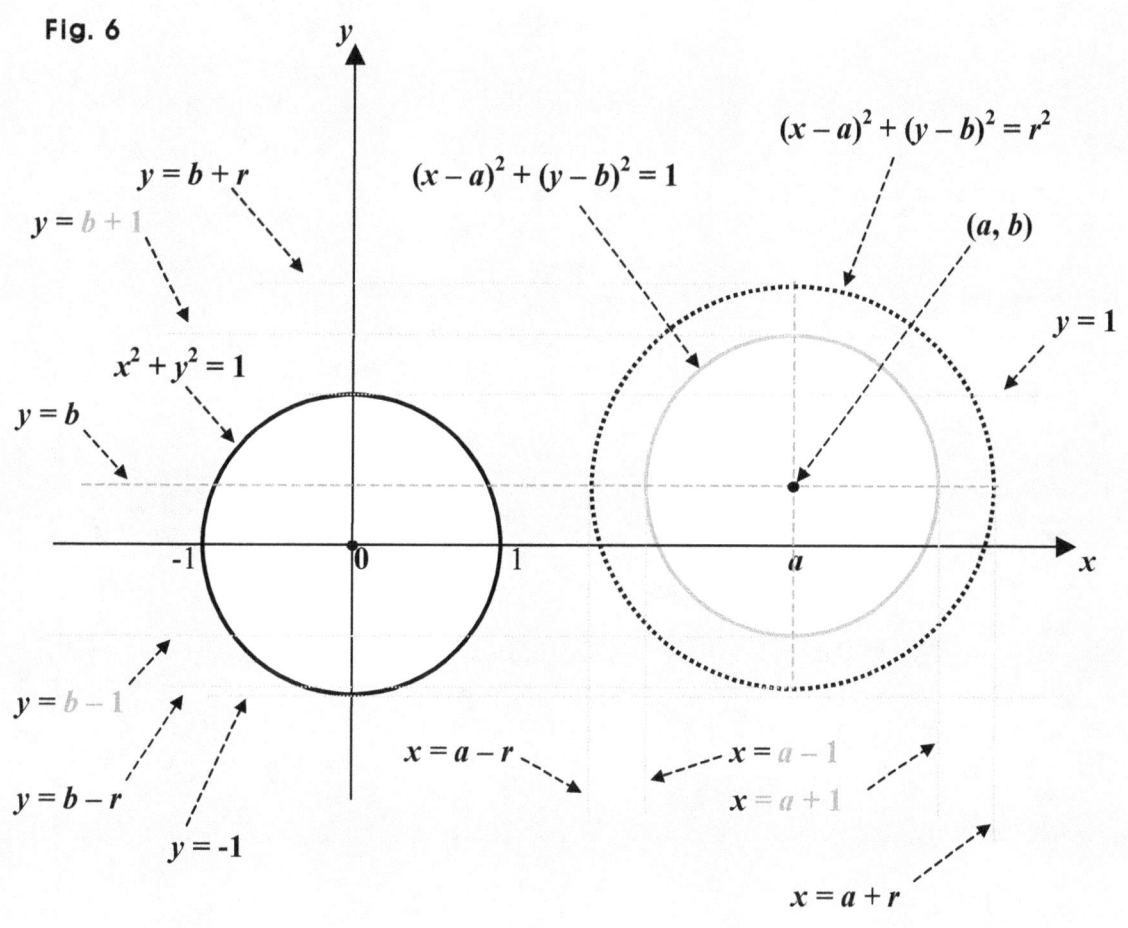

Fig. 6

And if **C** is the unit circle centered at the origin, moving the circle **C** by **a** along the **x**-axis, and by **b** along the **y**-axis, we get a unit circle centered at **(a, b)**.

And the two circles on the right are called concentric circles, since both have the same center.

Examples 1 in Circles

Put in a graph each of the equations in each example below:

0. $x^2 + y^2 = 1$, $x^2 + y^2 = 4$, $x^2 + y^2 = 2$,

and $x^2 + y^2 = c^2$ where c is constant, and $c > 0$.

1. $(x - 1)^2 + y^2 = 1$, $x^2 + (y - 1)^2 = 1$, $(u - 2)^2 + v^2 = 3$,   and $x^2 + (y + \frac{3}{2})^2 = 16$

2. $(s + 1)^2 + (t + 2)^2 = 1$, and $(x + \frac{1}{3})^2 + (y - \frac{2}{5})^2 = 3$

3   $4a^2 + 4b^2 = 1$, $9x^2 + 9(y - 1)^2 = 4$, and $7(x - 1)^2 + 7(y - 2)^2 = 3$

4. $(x - a)^2 + (y - b)^2 = c^2$ where a, b, and c are constant, and $c \neq 0$.

5. $d(x - a)^2 + d(y - b)^2 = c^2$ where a, b, c, and d are constant, $c \neq 0$, and $d > 0$.

6. $x^2 + y^2 - 4x - 6y + 9 = 0$

7. $t^2 + s^2 + 6t + 2s + 9 = 0$

8. $4x^2 + 4y^2 + 6x + 2y - \frac{13}{4} = 0$

9. $14x + 17y - 2x^2 - 2y^2 = -24$

A. $1.9u + 3.6v - \frac{1}{3}u^2 - \frac{1}{3}v^2 + 0.3 = 0$

Suggestions or Solutions
To the Problems in the Example 0

To begin with, the first equation is: $x^2 + y^2 = 1$.

The equation given indicates a unit circle centered at the origin **(0,0)**. So it is a circle of radius 1, and is the simplest and the most basic circle.

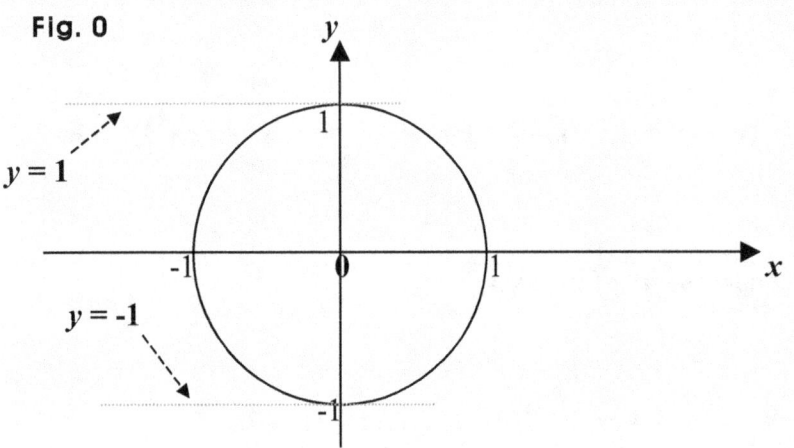

Fig. 0

And of the unit circle $x^2 + y^2 = 1$, changing the center or radius, or both, we can readily get the circle we want.

So for instance, moving the center to **(a, 0)**, we get:
$(x + a)^2 + y^2 = 1$, which is another circle of radius 1, which is a unit circle, too.

Moving the center to **(0, b)**, we get:
$x^2 + (y - b)^2 = 1$, which is another unit circle, too.

Moving the center to **(a, b)**, we get:
$(x - a)^2 + (y - b)^2 = 1$, which is a unit circle, also.

Changing the radius to **r**, and keeping the center intact, we get:
$x^2 + y^2 = r^2$, which is a circle centered at the origin **(0, 0)**.

Changing the radius to **r**, and moving the center to **(a, b)**, we get:
$(x - a)^2 + (y - b)^2 = r^2$, which is often called a standard equation for circles.

Next, the second equation is: $x^2 + y^2 = 4$.

We can put the equation above this way, too: $x^2 + y^2 = 2^2$.

The equation is called a connective equation connecting coordinates of any point that is a particular distance away from the origin **(0, 0)** in the *x-y* plane, and the particular distance is 2.

So the equation is the connective equation between the coordinates of an arbitrary point **(x, y)** representing all points in a curve, and 2 is the distance from the arbitrary point to the origin in the *x-y* plane.

So the curve is a circle where the center is the origin and the radius is 2.

Therefore, the equation given is the equation of the circle below.

Fig. 1

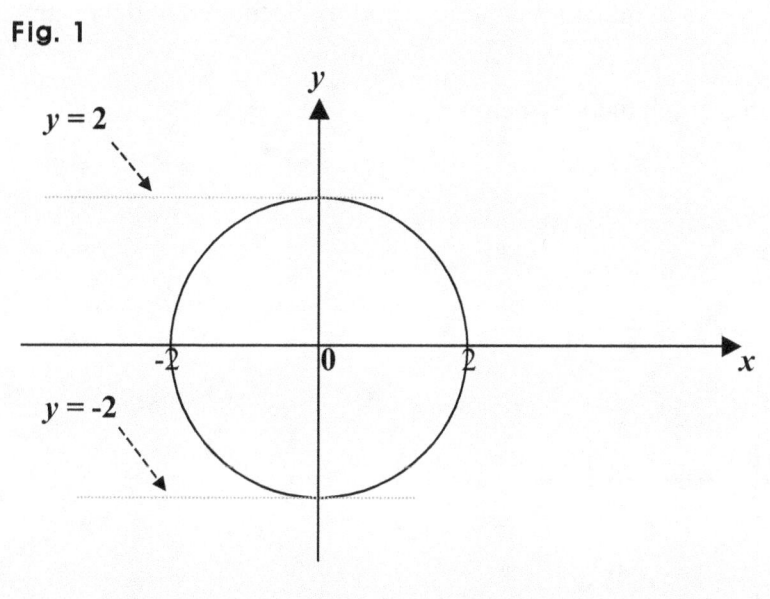

Next, the third equation is: $x^2 + y^2 = 2$.

We can put it this way, too: $x^2 + y^2 = (\sqrt{2})^2$.

So the equation indicates a circle where the center is the origin, and the radius is $\sqrt{2}$. And we can put the circle in a graph the way below:

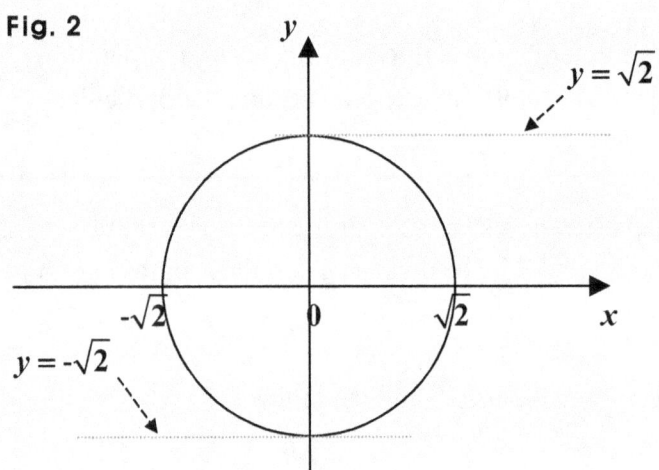

Fig. 2

And by the same token, $x^2 + y^2 = c^2$ indicates a circle of radius c centered at the origin.

And we can put the circle in a graph the way below:

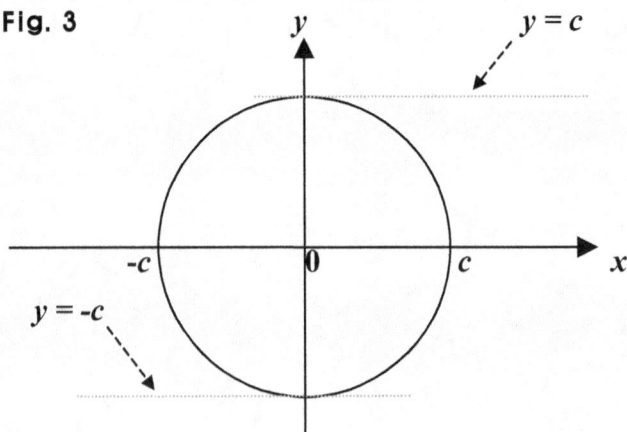

Fig. 3

Suggestions or Solutions
To the Problems in the Example 1

To begin with, the first equation is: $(x-1)^2 + y^2 = 1$.

It is the equation of a unit circle centered at **(1, 0)** in the *x-y* plane.

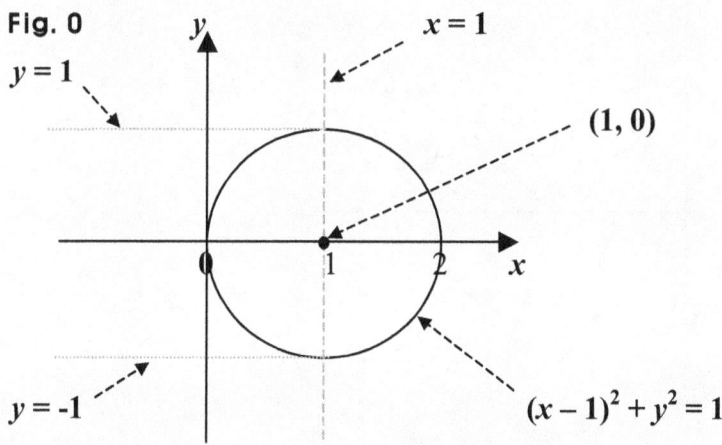

Fig. 0

$y = 1$

$x = 1$

(1, 0)

$y = -1$

$(x-1)^2 + y^2 = 1$

Next, the second equation is: $x^2 + (y-1)^2 = 1$.

And it is the equation of a unit circle centered at **(0, 1)**.
So we can put it in a graph the way below:

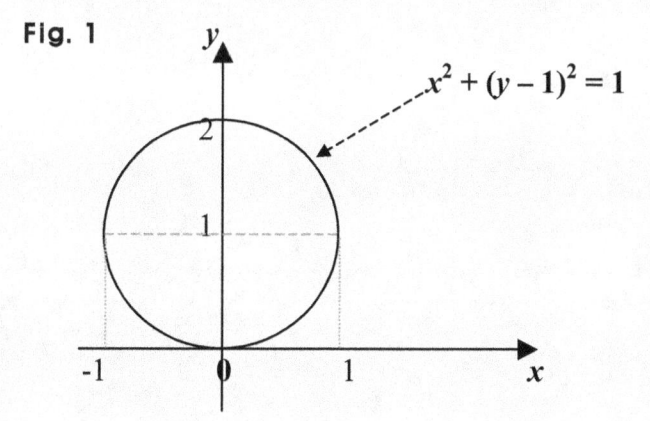

Fig. 1

$x^2 + (y-1)^2 = 1$

Next, the third equation is: $(u-2)^2 + v^2 = 3$.

And we can put the equation this way, too: $(u-2)^2 + v^2 = (\sqrt{3})^2$.

So it is a circle in the **u-v** plane, and is centered at **(-1, -2)**, and the radius is $\sqrt{3}$. And thus, we can put it in a graph the way blow:

Fig. 2

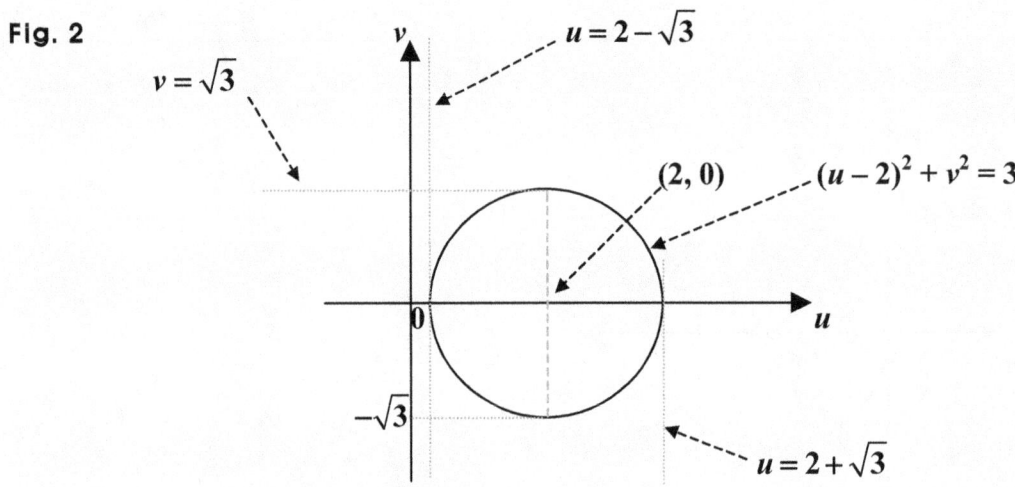

And next, the last equation is: $x^2 + (y+\frac{3}{2})^2 = 16$.

We can put the equation this way, too: $x^2 + (y+\frac{3}{2})^2 = 4^2$, so it is a circle where the center is $(0, -\frac{3}{2})$, and the radius is 4.   And thus, the graph is as follows:

Fig. 3

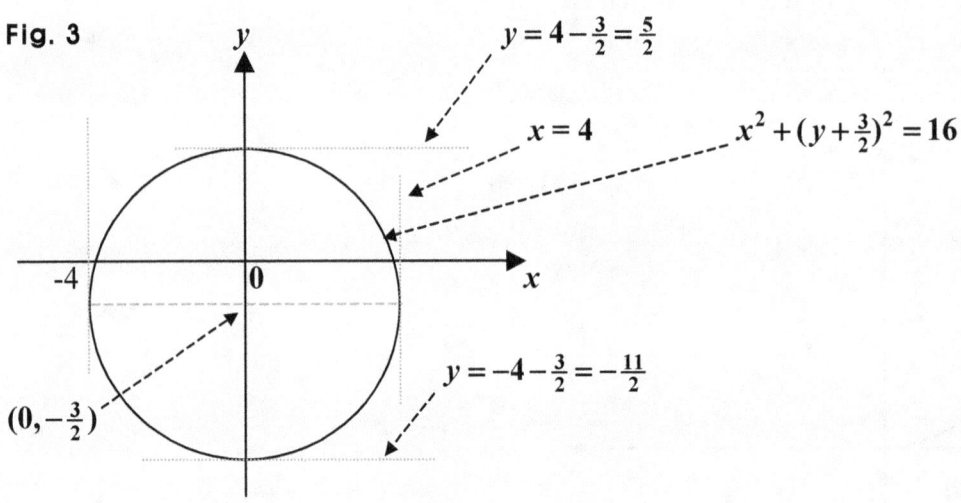

And as shown in the graph below, we can get the circle: $(x-1)^2 + y^2 = 1$, too, moving a unit circle centered at $(0, 0)$.   Moving the unit circle in the amount of 1 in the direction of the x-axis, we get the circle in black, which is the curve of the equation above.

So the circle in black below is of radius 1 and centered at $(1, 0)$, and thus, its equation is:

$(x-1)^2 + y^2 = 1$.

Fig. 4

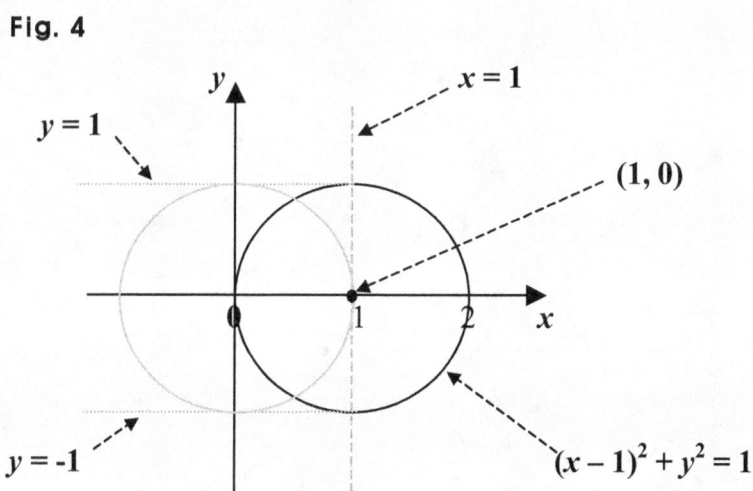

Let's take another look at the graph of the equation from another different perspective.

Suppose we keep intact the unit circle centered at the origin, and move the y-axis in the amount of -1 in the direction of the x-axis.

That is, we move only the y-axis to the left in the amount of 1.

Then, the origin gets moved in the same amount and direction, too.
And so do all the numbers in the x-axis.

Then, we get a new circle, and the new circle is a unit circle centered at $(1, 0)$.

Thus, the equation of the new circle is: $(x-1)^2 + y^2 = 1$.

And putting the ideas in graph, we can put them the way below:

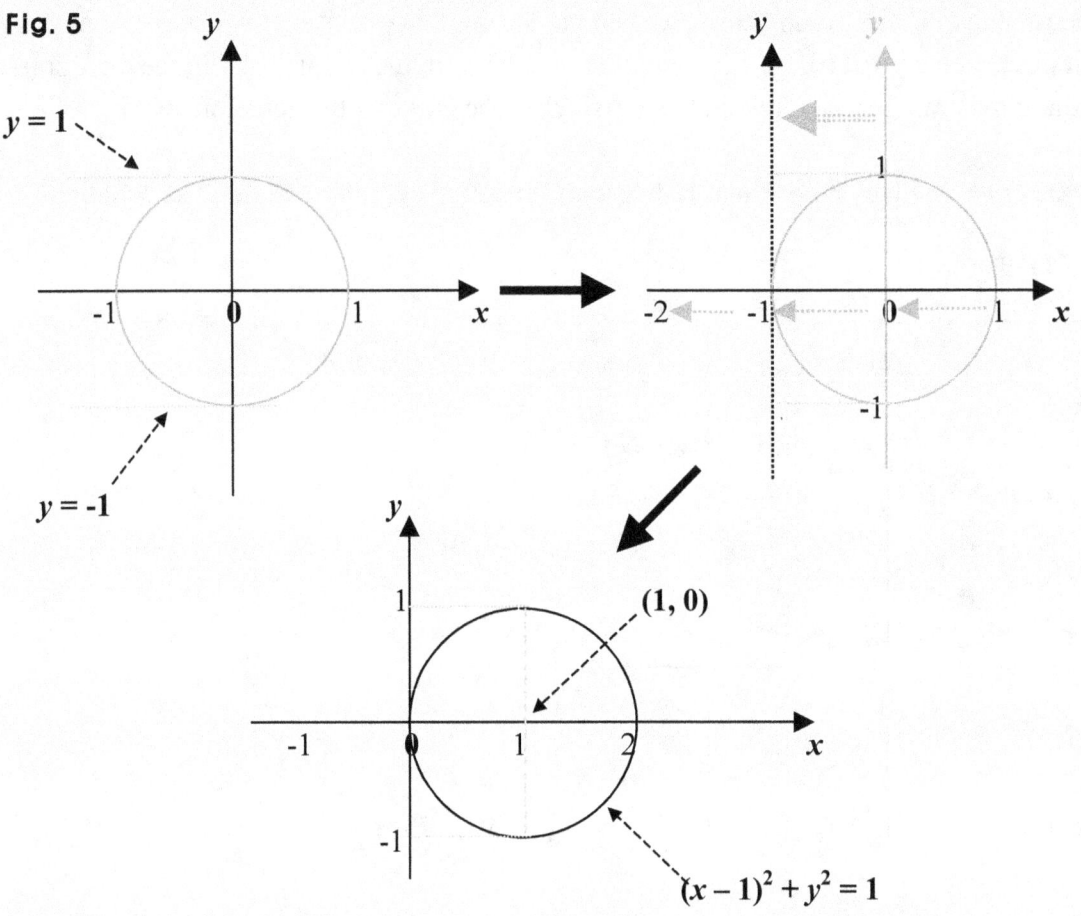

Fig. 5

Next, we can get the circle $x^2 + (y - 1)^2 = 1$, moving the unit circle centered at the origin in the amount of 1 in the direction of the y-axis only.

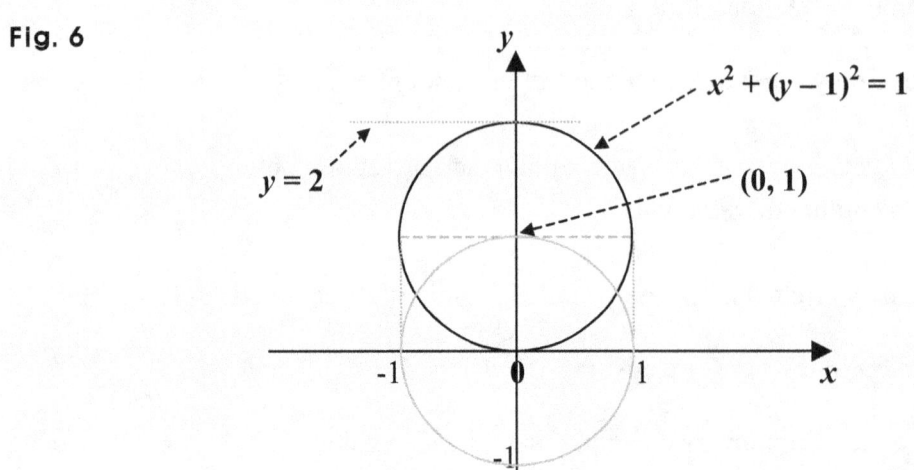

Fig. 6

Next, moving a circle $u^2 + v^2 = 3$ by 2 along the **u**-axis, we get: $(u - 2)^2 + v^2 = 3$.

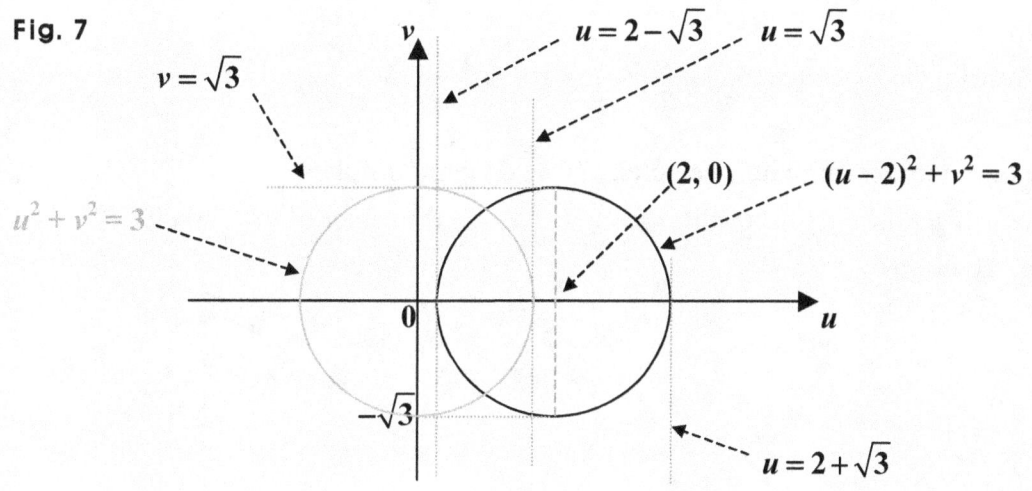

Fig. 7

And next, moving a circle $x^2 + y^2 = 16$ by $-\frac{3}{2}$ along the **y**-axis, that is, in the amount of $-\frac{3}{2}$ in the direction of the **y-axis,** we get: $x^2 + (y + \frac{3}{2})^2 = 16$.

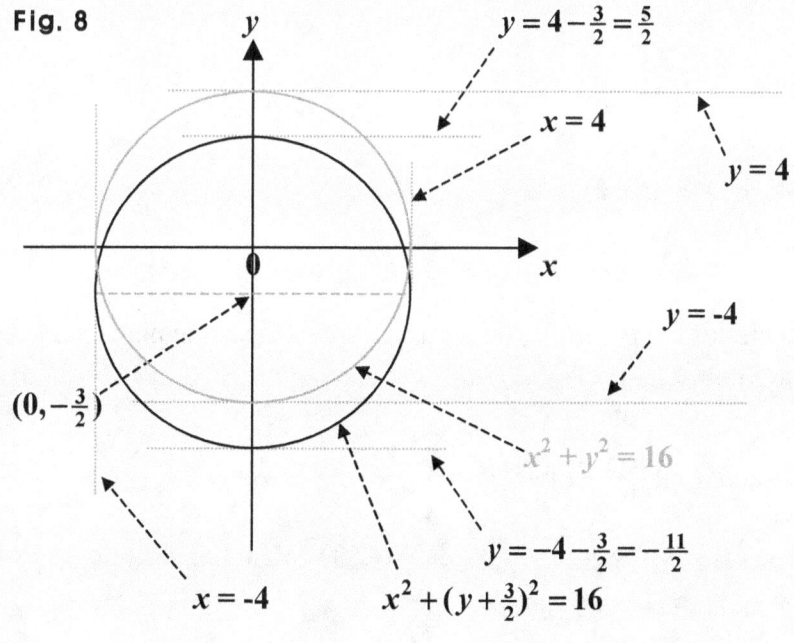

Fig. 8

Suggestions or Solutions
To the Problems in the Example 2

To begin with, the first equation is: $(s+1)^2 + (t+2)^2 = 1$.

It is the equation of a unit circle centered at **(-1, -2)** in the *s-t* plane.
Also, we can get the circle above if moving to **(-1, -2)** the center of the unit circle centered the origin.

Fig. 0

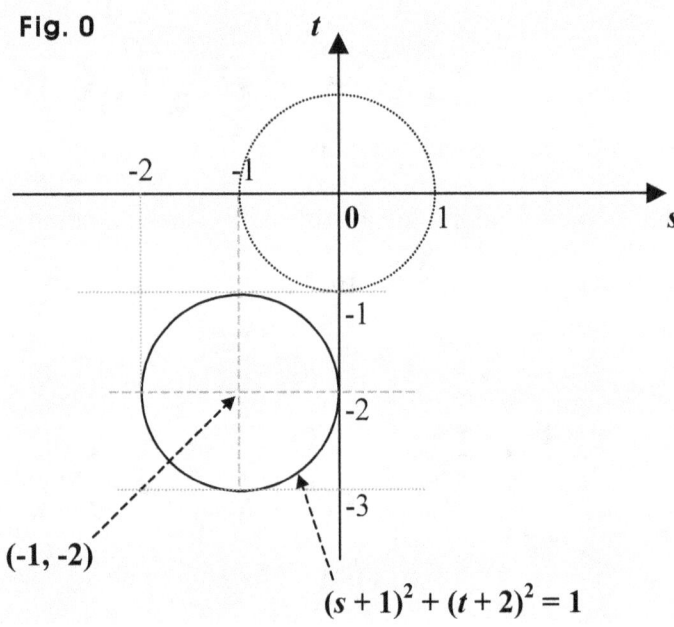

$(s+1)^2 + (t+2)^2 = 1$

By the way, in math, if two circles are the same, both circles are in the same plane, and have not only the same radii but the same centers, too.

And next, the second equation is: $(x+\frac{1}{3})^2 + (y-\frac{2}{5})^2 = 3$.

We can put the equation this way, too: $(x+\frac{1}{3})^2 + (y-\frac{2}{5})^2 = (\sqrt{3})^2$, which is a circle where the center is $(-\frac{1}{3}, \frac{2}{5})$, and the radius is $\sqrt{3}$.

And also, we can get the circle above moving a circle $x^2 + y^2 = 3$ by $-\frac{1}{3}$ in the direction of the x-axis and by $\frac{2}{5}$ in the direction of the y-axis. In the circle $x^2 + y^2 = 3$, the center is the origin, and the radius is $\sqrt{3}$.

So putting the two circles in a graph, we can put them the way below:

Fig. 1

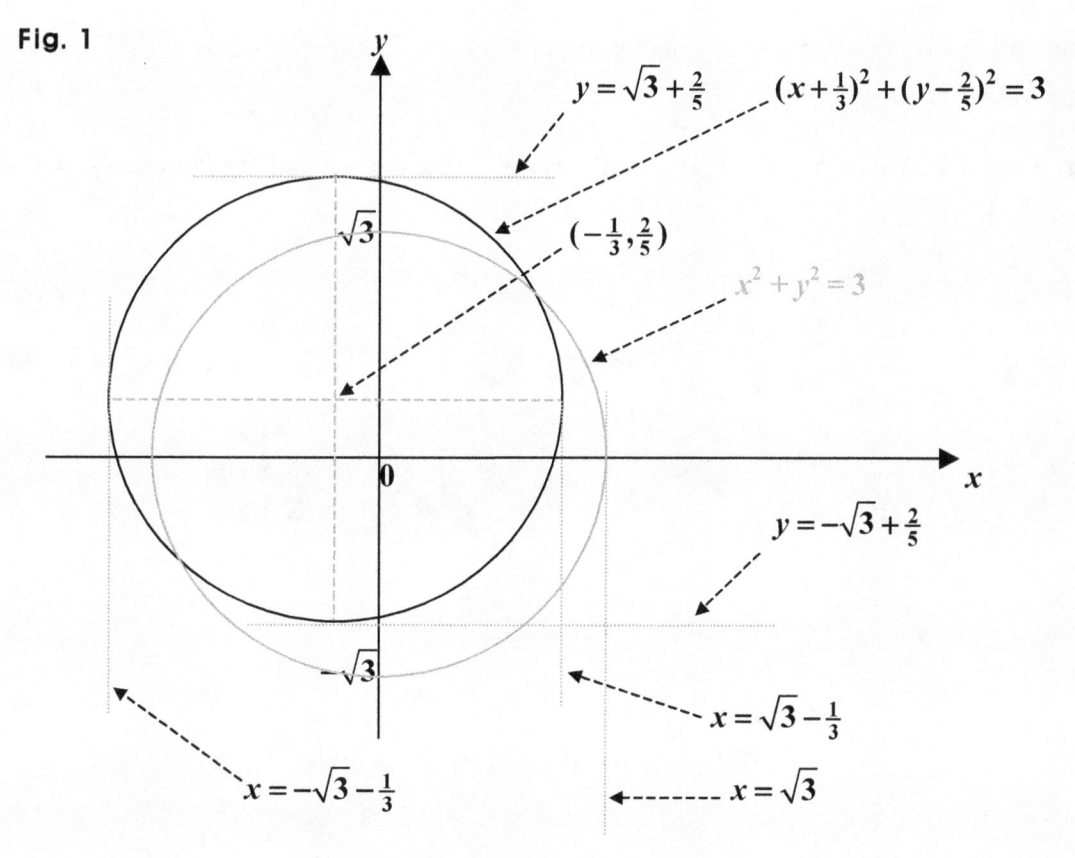

$y = \sqrt{3} + \frac{2}{5}$

$(x + \frac{1}{3})^2 + (y - \frac{2}{5})^2 = 3$

$(-\frac{1}{3}, \frac{2}{5})$

$x^2 + y^2 = 3$

$y = -\sqrt{3} + \frac{2}{5}$

$x = \sqrt{3} - \frac{1}{3}$

$x = -\sqrt{3} - \frac{1}{3}$

$x = \sqrt{3}$

Suggestions or Solutions
To the Problem in the Example 3

To begin with, the first equation is: $4a^2 + 4b^2 = 1.$

The equation indicates a circle in the **a-b** plane.

Converting the equation the way below, we can see the center and radius, which specify the circle, of course.

$4a^2 + 4b^2 = 1 \Rightarrow a^2 + b^2 = \frac{1}{4} = (\frac{1}{2})^2.$ So the center is the origin, and the radius is $\frac{1}{2}$.

Fig. 0

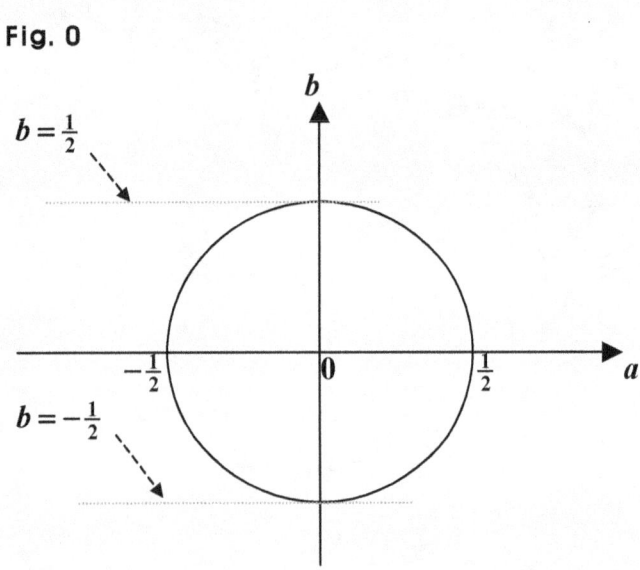

Next, the second equation is: $9x^2 + 9(y-1)^2 = 4.$

The equation is not in the standard form, and is not in the general form, either. It looks more like though, in the standard form. So it can be said to be in a semi standard form.

Thus, converting it into the complete standard equivalent, we get:

$9x^2 + 9(y-1)^2 = 4 \Rightarrow x^2 + (y-1)^2 = \frac{4}{9} = (\frac{2}{3})^2.$ So the center is **(0, 1)**, and the radius is $\frac{2}{3}$.

Also, translating a circle $9x^2 + 9y^2 = 4$, that is, $x^2 + y^2 = \frac{4}{9}$, by 1 in the direction of the y-axis, we can get the same circle, too, that is, the circle $9x^2 + 9(y - 1)^2 = 4$, which can be put this way: $x^2 + (y - 1)^2 = \frac{4}{9}$.

Fig. 1

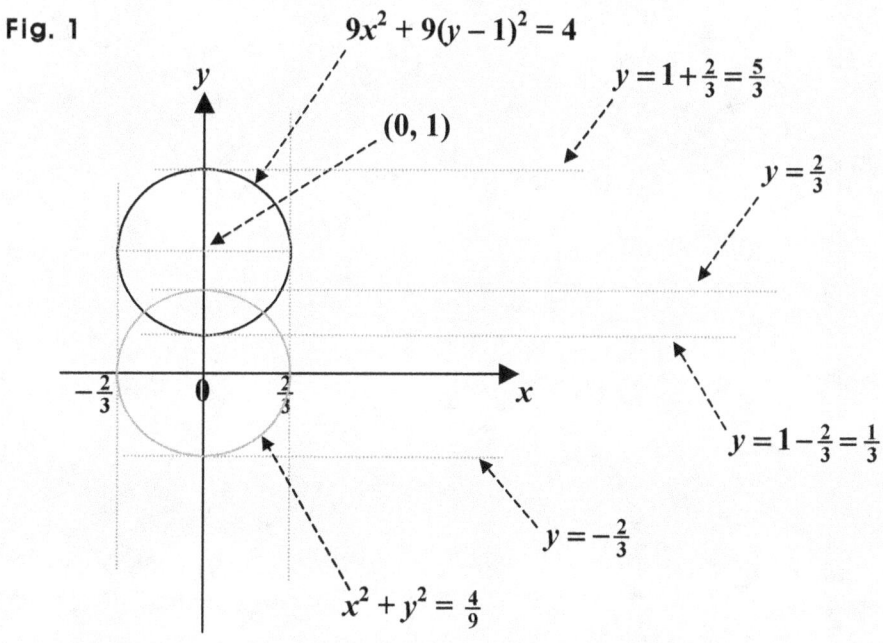

And next, the last equation is: $7(x - 1)^2 + 7(y - 2)^2 = 3$.

As the two equations above, this equation is not in the standard form, and is not in the general form, either. It looks more like the standard form, so it can be said to be quasi-standard form.   And putting the equation in the standard form, we get:

$$7(x - 1)^2 + 7(y - 2)^2 = 3 \Rightarrow (x - 1)^2 + (y - 2)^2 = \frac{3}{7} = \left(\sqrt{\frac{3}{7}}\right)^2$$

So the center is $(1, 2)$, and the radius is $\sqrt{\frac{3}{7}} \approx 0.655$. Also, we can get the same circle as the one above moving a circle $x^2 + y^2 = \frac{3}{7}$ by 1 in the direction of the x-axis and by 2 in the direction of the y-axis. In the circle $x^2 + y^2 = \frac{3}{7}$, the center is the origin, and the radius is $\sqrt{\frac{3}{7}}$. So putting them both in one graph, we can put them the way below:

40

Fig. 0

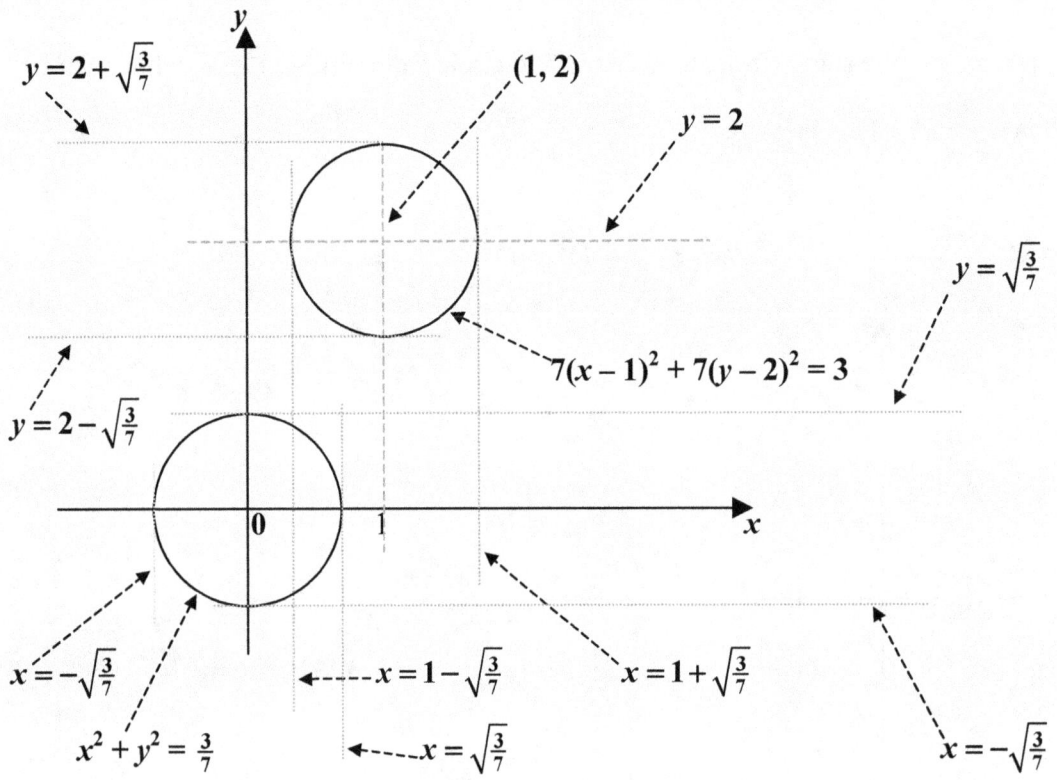

Suggestions or Solutions
To the Problem in the Example 4

Put in a graph the equation as follows: $(x - a)^2 + (y - b)^2 = c^2$ **where** a, b, **and** c **are constants and** $c \neq 0$.

Note that in $|x|$, the pair of the vertical bars can be called an absolute sign, and $|x|$ means the absolute value of x, and thus, is the magnitude of x. For instance, $|-2| = 2$.

We have: $(x - a)^2 + (y - b)^2 = c^2$. So the center is at (a, b), and the radius is $|c|$.

Depending on the signs of a and b, we can put the circle four different ways. Technically though, we have 5, because if a and b both are 0, the center is at the origin.

Fig. 0

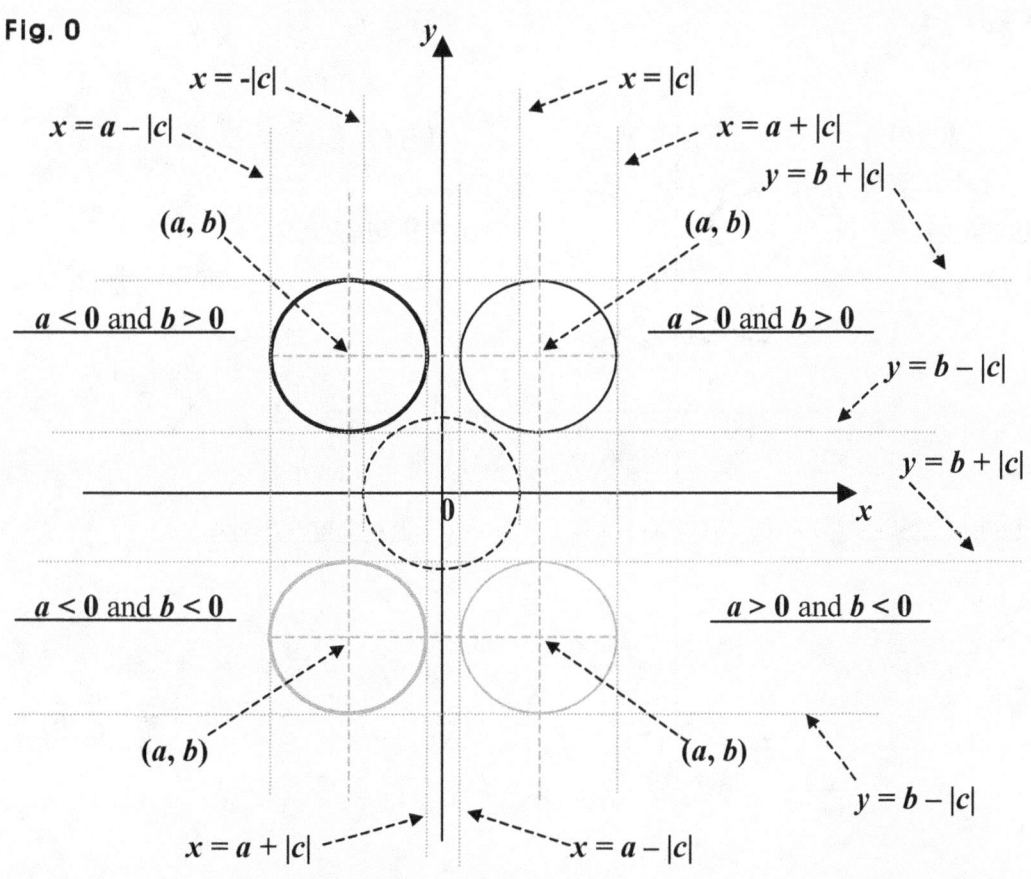

Suggestions or Solutions
To the Problem in the Example 5

Put in a graph the equation as follows: $d(x - a)^2 + d(y - b)^2 = c^2$ where a, b, c, and d are constants, $c \neq 0$, and $d > 0$.

Putting the equation in the standard form, we get: $(x - a)^2 + (y - b)^2 = \frac{c^2}{d} = (\frac{c}{\sqrt{d}})^2$.

So the center is at (a, b), and the radius is $|\frac{c}{\sqrt{d}}|$.

And we can have five different cases as below:

Fig. 0

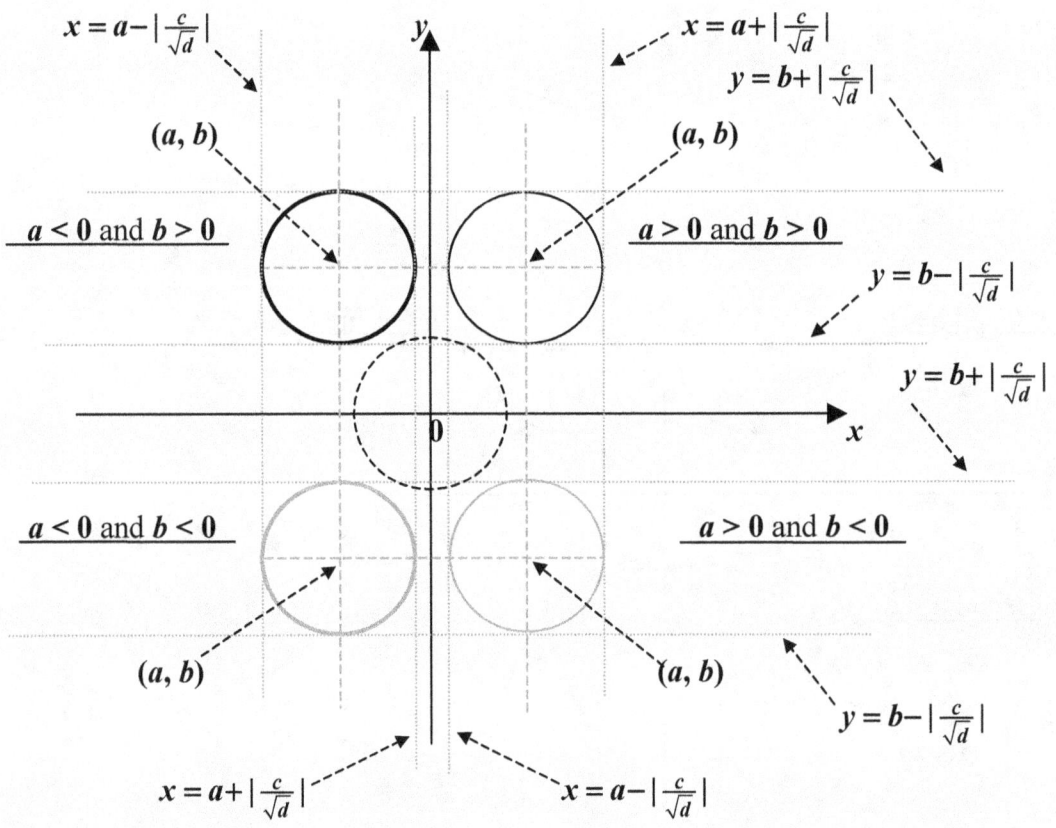

Suggestions or Solutions
To the Problem in the Example 6

Put in a graph the equation as follows: $x^2 + y^2 - 4x - 6y + 9 = 0.$

The equation indicates a circle, and is in the general form.
How do we know though, if it really does indicate a circle?

Putting the equation in the standard form, we can actually confirm that it indicates a circle. And if it is the case, we can quickly see the center and radius. So we can readily put the circle in a graph.

$$x^2 + y^2 - 4x - 6y + 9 = x^2 - 4x + y^2 - 6y + 9 = x^2 - 4x + 4 - 4 + y^2 - 6y + 9$$

$$= (x-2)^2 - 4 + (y-3)^2 = 0 \Rightarrow (x-2)^2 + (y-3)^2 = 4 = 2^2.$$

So it indicates a circle, and we can see that the center is **(2, 3)**, and the radius is **2**.

Fig. 0

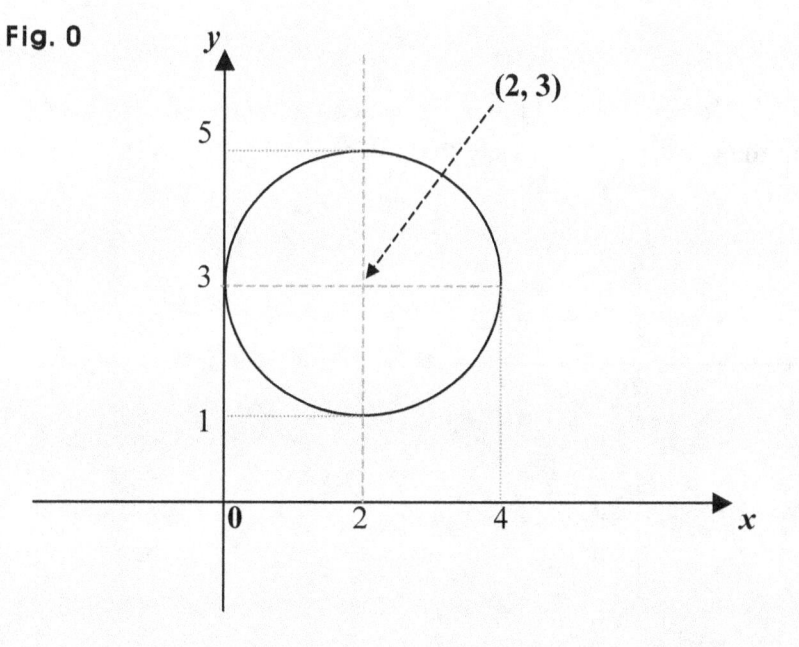

Suggestions or Solutions
To the Problem in the Example 7

Put in a graph the equation as follows: $t^2 + s^2 + 6t + 2s + 9 = 0$.

The equation represents a circle in the general form.
Putting the equation in the standard form, we can confirm that it indicates a circle. And if it is the case, we can readily see the center and the radius.

$$t^2 + s^2 + 6t + 2s + 9 = t^2 + 6t + s^2 + 2s + 9 = t^2 + 6t + 9 - 9 + s^2 + 2s + 1 - 1 + 9$$
$$= (t + 3)^2 + (s + 1)^2 - 9 - 1 + 9 = (t + 3)^2 + (s + 1)^2 - 1 = 0 \Rightarrow (t + 3)^2 + (s + 1)^2 = 1.$$

So it is a unit circle centered at **(-3, -1)** in the *t-s* plane.

Fig. 0

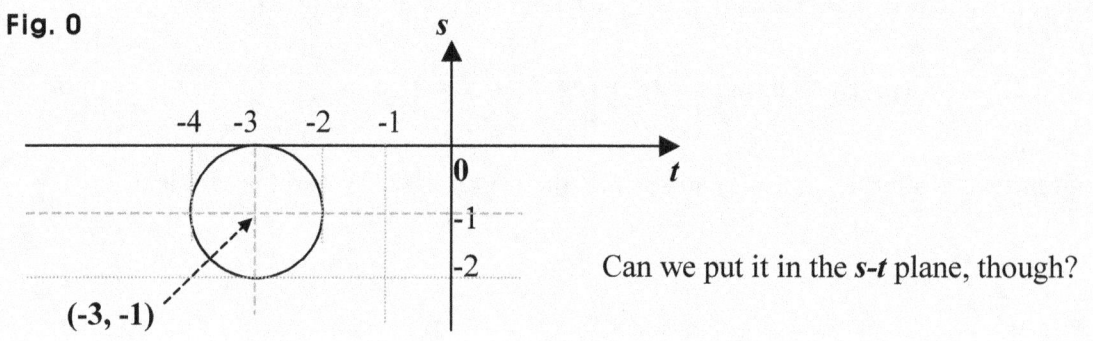

Can we put it in the *s-t* plane, though?

Of course, we can. We have: $(t + 3)^2 + (s + 1)^2 = 1$. So we can get: $(s + 1)^2 + (t + 3)^2 = 1$.
Putting thus, the circle in the *s-t* plane, we can say that the center is at **(-1, -3)**.

Fig. 1

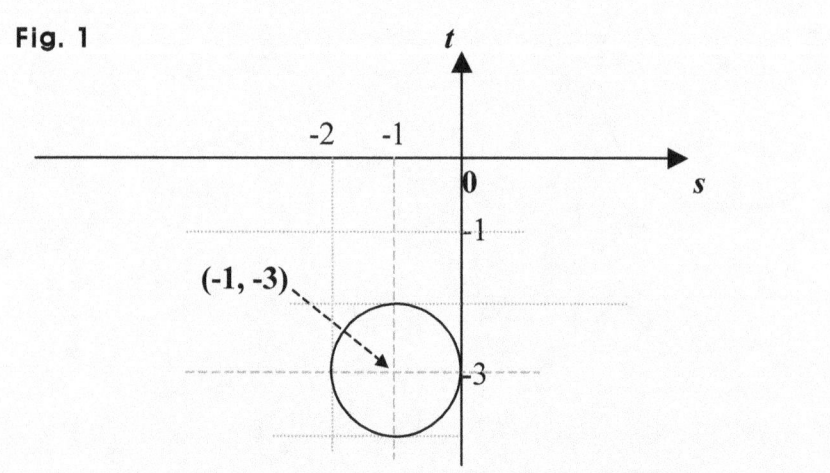

Suggestions or Solutions
To the Problem in the Example 8

Put in a graph the equation as follows: $4x^2 + 4y^2 + 6x + 2y - \frac{13}{4} = 0$.

$$4x^2 + 4y^2 + 6x + 2y - \tfrac{13}{4} = 0 \Rightarrow x^2 + y^2 + \tfrac{3}{2}x + \tfrac{1}{2}y - \tfrac{13}{16} = 0$$

$$\Rightarrow x^2 + \tfrac{3}{2}x + \tfrac{9}{16} - \tfrac{9}{16} + y^2 + \tfrac{1}{2}y + \tfrac{1}{16} - \tfrac{1}{16} - \tfrac{13}{16} = (x + \tfrac{3}{4})^2 + (y + \tfrac{1}{4})^2 - \tfrac{23}{16} = 0$$

$$\Rightarrow (x + \tfrac{3}{4})^2 + (y + \tfrac{1}{4})^2 = \tfrac{23}{16} = (\tfrac{\sqrt{23}}{4})^2$$

So it is a circle, where the center is at $(-\tfrac{3}{4}, -\tfrac{1}{4})$, and the radius $= \tfrac{\sqrt{23}}{4} \approx 1.199$.

Fig. 0

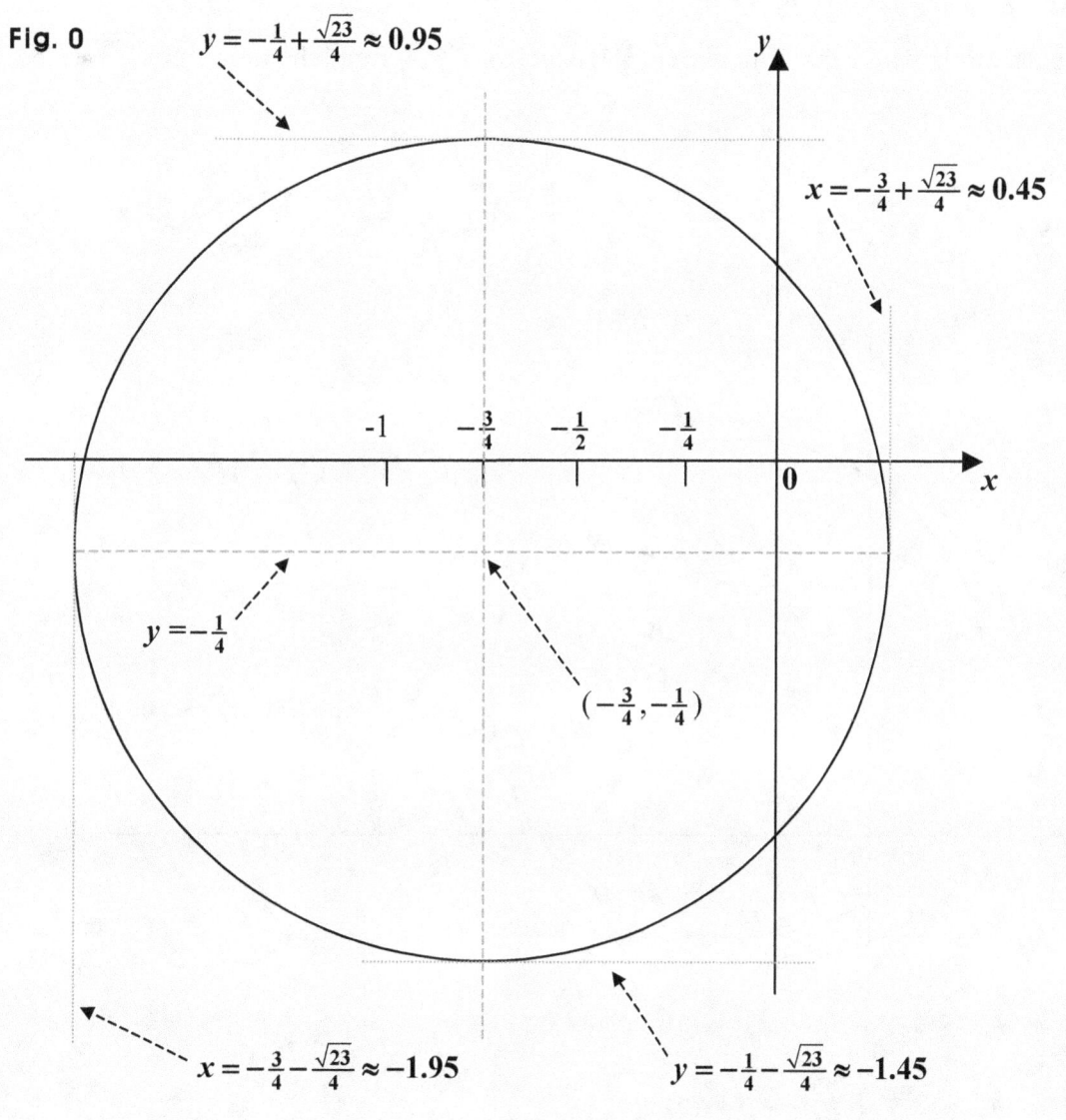

Suggestions or Solutions
To the Problem in the Example 9

Put in a graph the equation as follows: $14x + 17y - 2x^2 - 2y^2 = -24$.

$14x + 17y - 2x^2 - 2y^2 = -24 \Rightarrow 2x^2 + 2y^2 - 14x - 17y - 24 = 0.$

$\Rightarrow x^2 - 7x + y^2 - \frac{17}{2}y - 12 = 0 \Rightarrow x^2 - 7x + (\frac{7}{2})^2 - (\frac{7}{2})^2 + y^2 - \frac{17}{2}y + (\frac{17}{4})^2 - (\frac{17}{4})^2 - 12 = 0$

$\Rightarrow (x - \frac{7}{2})^2 + (y - \frac{17}{4})^2 - \frac{49}{4} - (\frac{17}{4})^2 - 12 = 0 \Rightarrow (x - \frac{7}{2})^2 + (y - \frac{17}{4})^2 = \frac{196+289+192}{16} = \frac{677}{16}$

$\Rightarrow (x - \frac{7}{2})^2 + (y - \frac{17}{4})^2 = \frac{677}{16} = (\frac{\sqrt{677}}{4})^2$

So it is a circle where the center is: $(\frac{7}{2}, \frac{17}{4})$, which is: $(\frac{14}{4}, \frac{17}{4})$, and the radius is: $\frac{\sqrt{677}}{4} \approx 6.5$.

Fig. 0

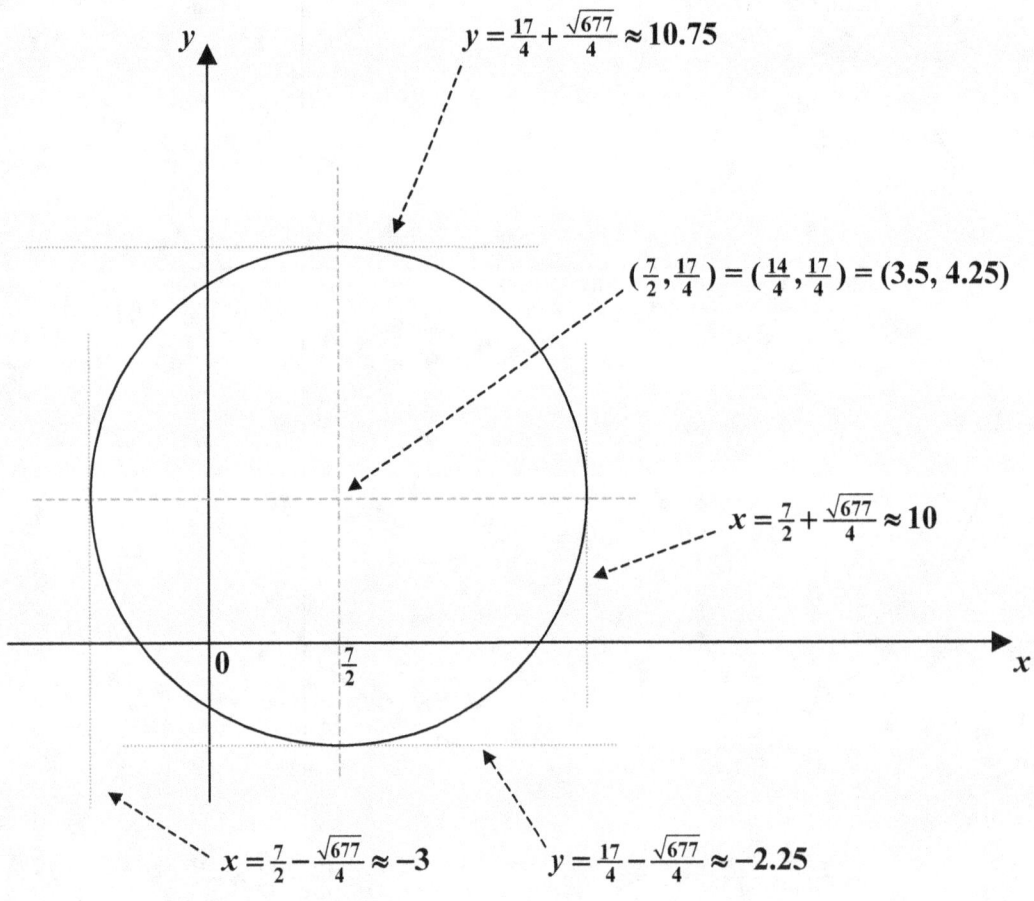

Suggestions or Solutions
To the Problem in the Example A

Put in a graph the equation as follows: $1.9u + 3.6v - \frac{1}{3}u^2 - \frac{1}{3}v^2 + 0.3 = 0.$

$1.9u + 3.6v - \frac{1}{3}u^2 - \frac{1}{3}v^2 + 0.3 = 0 \Rightarrow \frac{1}{3}u^2 + \frac{1}{3}v^2 - 1.9u - 3.6v - 0.3 = 0$

$\Rightarrow u^2 + v^2 - 5.7u - 10.8v - 0.9 = 0 \Rightarrow u^2 - 5.7u + v^2 - 10.8v - 0.9 = 0$

$\Rightarrow u^2 - 5.7u + \frac{5.7^2}{4} - \frac{5.7^2}{4} + v^2 - 10.8v + 5.4^2 - 5.4^2 - 0.9 = 0$

$\Rightarrow (u - \frac{5.7}{2})^2 + (v - 5.4)^2 - \frac{5.7^2}{4} - 5.4^2 - 0.9 = 0 \Rightarrow (u - \frac{5.7}{2})^2 + (v - 5.4)^2 = 38.1825$

So it is a circle, the center is: $(\frac{5.7}{2}, 5.4) = (2.85, 5.4)$, and the radius is: $\sqrt{38.1825} \approx 6.18$.

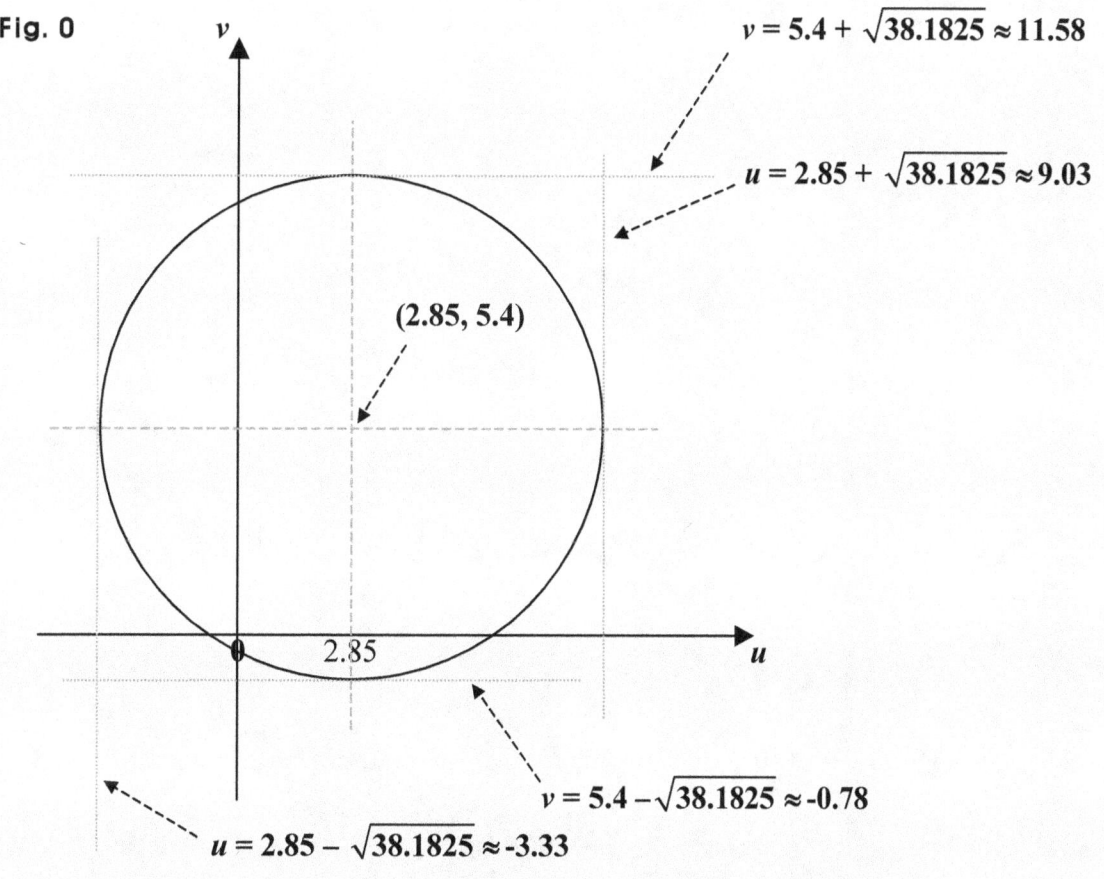

Fig. 0

$v = 5.4 + \sqrt{38.1825} \approx 11.58$

$u = 2.85 + \sqrt{38.1825} \approx 9.03$

(2.85, 5.4)

2.85

$v = 5.4 - \sqrt{38.1825} \approx -0.78$

$u = 2.85 - \sqrt{38.1825} \approx -3.33$

2. Equations for Circles 2

To begin with, putting a circle in an equation, we can get:

$$(x-1)^2 + (y-2)^2 = 3^2$$ What circle then, is it?

It is a circle of radius 3 centered at (1, 2).
So we can put the circle in the *x-y* plane the way below:

Fig. 0

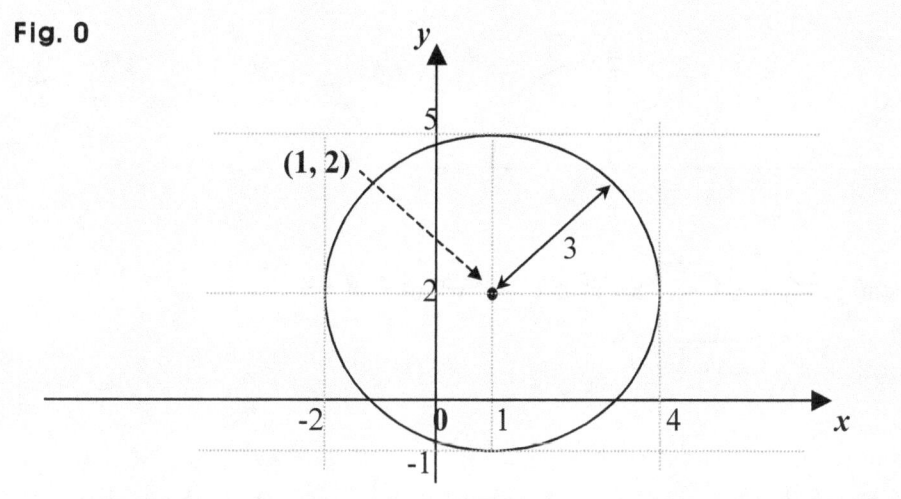

What circle then, is the circle as follows: $x^2 + y^2 = 1$?

It is a circle of radius 1 centered at (0, 0), that is, at the origin, and is called a unit circle.
And we can put the unit circle in the *x-y* plane the way below:

Fig. 1

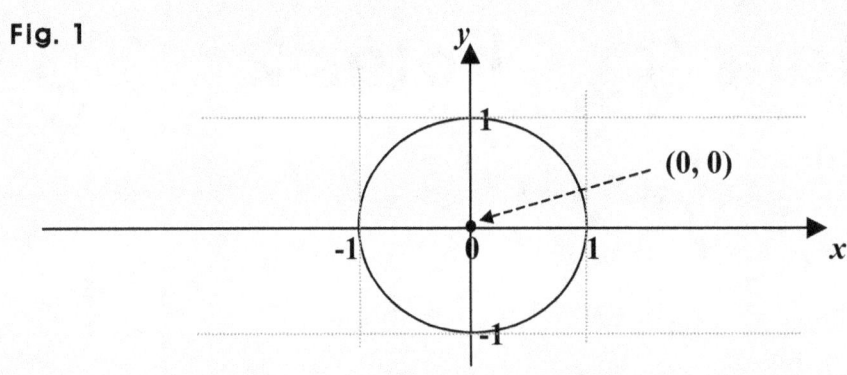

What then, about the equation as follows: $y = \sqrt{1-x^2}$?

It is not a circle but a half circle, and is the upper half of a circle where the radius is 1, and the center is at the origin. So we can put it in a graph the way below:

Fig. 2

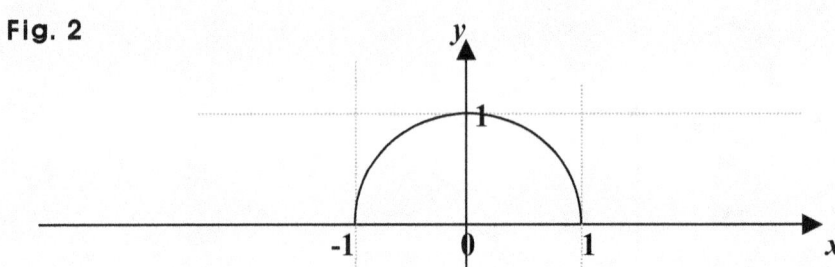

And in fact, we can get: $y = \sqrt{1-x^2} \Rightarrow y^2 = 1-x^2 \Rightarrow x^2 + y^2 = 1$.

What then, about this: $y = -\sqrt{1-x^2}$?

It is a half circle, too, but is this time, the lower half of the circle where the radius is 1, and the center is at the origin. So we can put it in a graph the way below:

Fig. 3

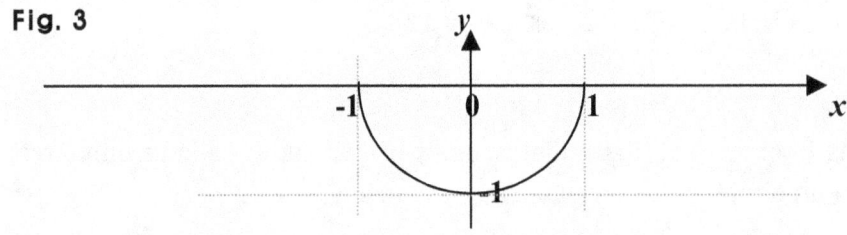

And we can get this, too, of course: $y = -\sqrt{1-x^2} \Rightarrow y^2 = 1 - x^2 \Rightarrow x^2 + y^2 = 1$.

What then, about this equation: $x = \sqrt{1-y^2}$?

It is a half circle, too, but is this time, the right half of the circle radius 1 centered at the origin. So we can put it in a graph the way below:

Fig. 4

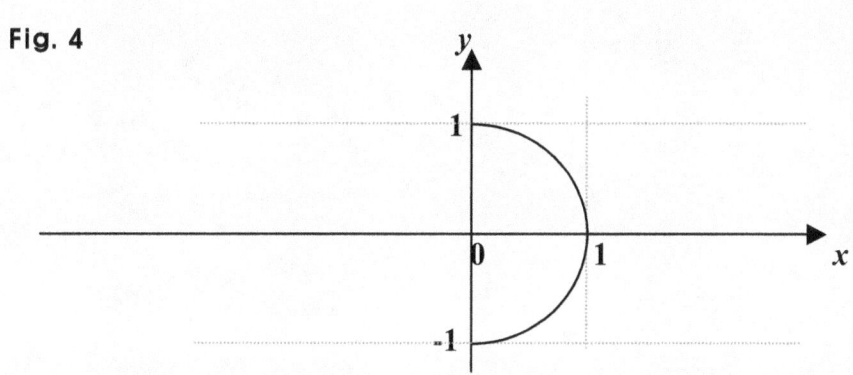

And we get this, too, of course: $x = \sqrt{1-y^2} \Rightarrow x^2 = 1 - y^2 \Rightarrow x^2 + y^2 = 1$.

And the left half is: $x = -\sqrt{1-y^2}$.

What then, about the equation as follows: $y = \sqrt{4-(x-1)^2}$?

It is a half circle, too, but is this time, the upper half of a circle where the radius is 2, and the center is at (1, 0).   So the graph is as follows:

Fig. 5

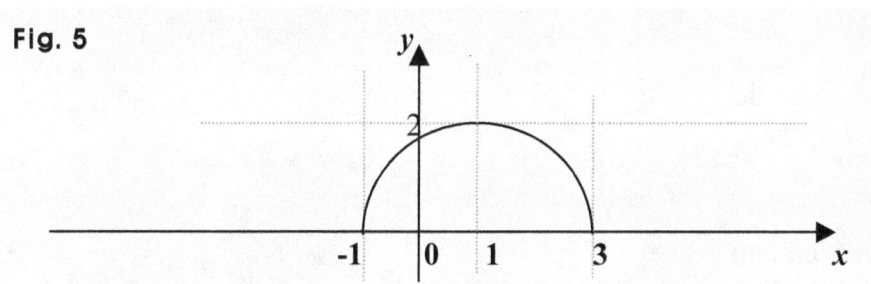

And we get this, of course: $y = \sqrt{4-(x-1)^2} \Rightarrow y^2 = 4-(x-1)^2 \Rightarrow (x-1)^2 + y^2 = 4 = 2^2$.

Similarly, the lower half is: $y = -\sqrt{4-(x-1)^2}$. And the graph is:

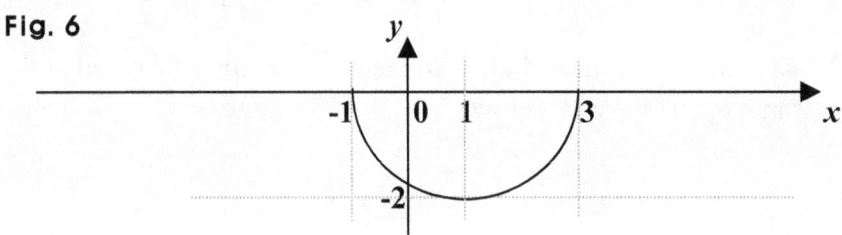

Fig. 6

What then, about the right half?

Its equation is not this: $x = \sqrt{4-(y-1)^2}$ but this: $x-1 = \sqrt{4-y^2}$.

In other words, it is: $x = \sqrt{4-y^2}+1$ or $x = 1+\sqrt{4-y^2}$.   And the graph is as below:

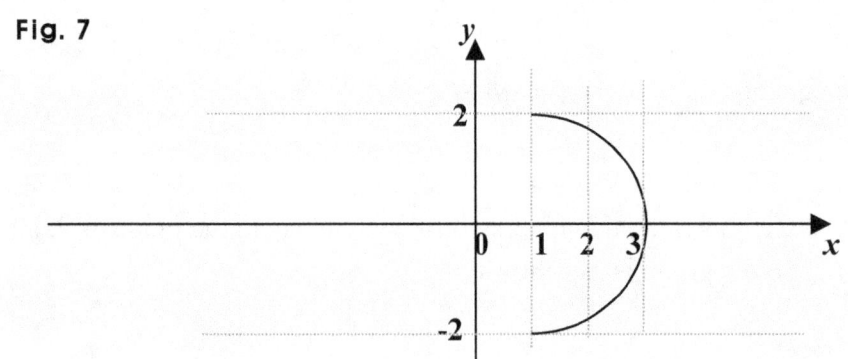

Fig. 7

And we get: $x-1 = \sqrt{4-y^2} \Rightarrow (x-1)^2 = 4-y^2 \Rightarrow (x-1)^2 + y^2 = 4 = 2^2$.

What then, about the left half?

Its equation is this: $x - 1 = -\sqrt{4 - y^2}$.

We can put it this way, too: $x = -\sqrt{4 - y^2} + 1$ or $x = 1 - \sqrt{4 - y^2}$.   And the graph is:

Fig. 8

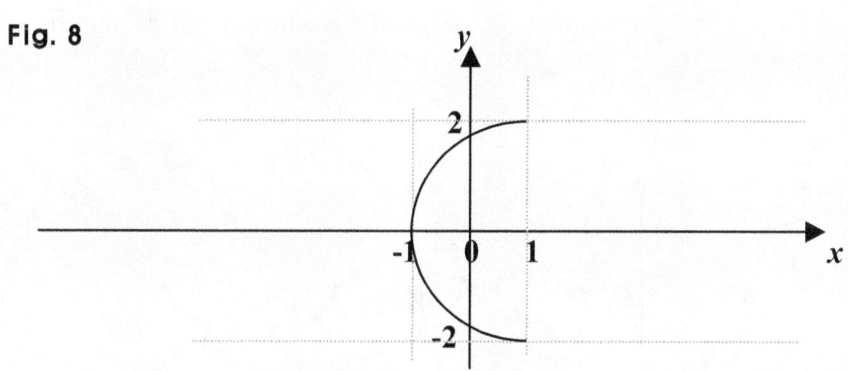

What about the upper half of a circle where the radius is 3, and the center is at (1, 2)?

Its equation is: $y - 2 = \sqrt{9 - (x - 1)^2}$.

And we can put it this way, too: $y = \sqrt{9 - (x - 1)^2} + 2$ or $y = 2 + \sqrt{9 - (x - 1)^2}$.
And the graph is:

Fig. 9

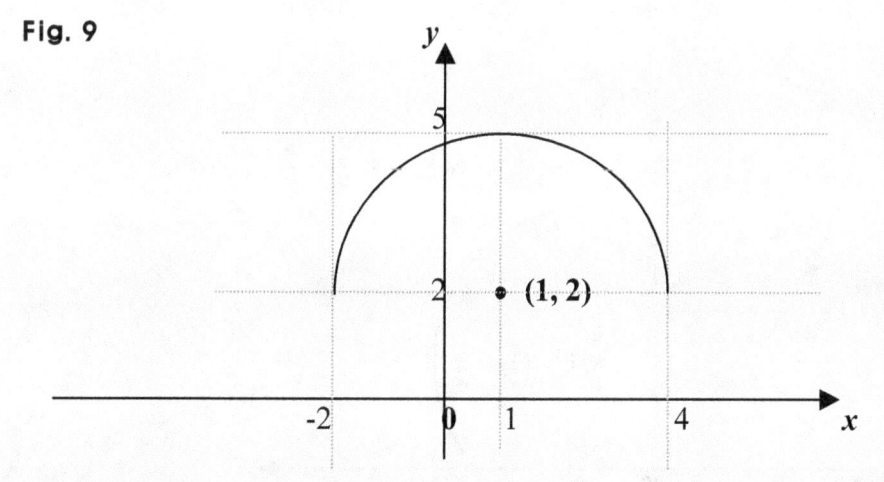

What then, about the lower half?

Its equation is: $y - 2 = -\sqrt{9 - (x-1)^2}$.

And we can put it this way, too: $y = -\sqrt{9 - (x-1)^2} + 2$ or $y = 2 - \sqrt{9 - (x-1)^2}$.

It can be put this way, too: $y = 2 - \sqrt{8 - x^2 + 2x}$ if what's inside the root is simplified.
And the graph is:

Fig. A

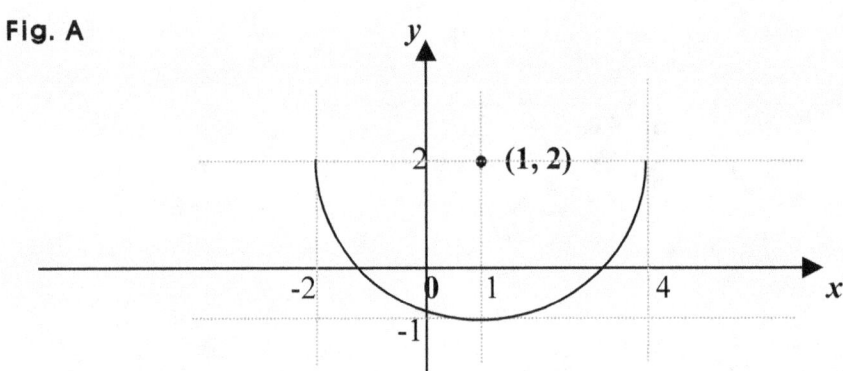

What about the right half?

Its equation is: $x - 1 = \sqrt{9 - (y-2)^2}$.

And we can put it this way, too: $x = \sqrt{9 - (y-2)^2} + 1$ or $x = 1 + \sqrt{9 - (y-2)^2}$.

Or it can be put this way, too: $x = 1 + \sqrt{5 - y^2 + 4y}$ if what's inside the root is simplified.
And the graph is:

Fig. B

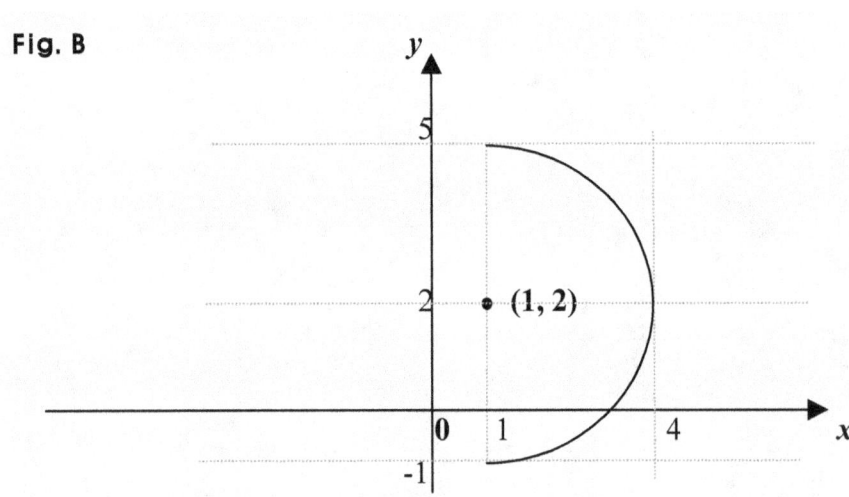

What about the left half?

Its equation is: $x - 1 = -\sqrt{9 - (y-2)^2}$.

And we can put it this way, too: $x = -\sqrt{9 - (y-2)^2} + 1$ or $x = 1 - \sqrt{9 - (y-2)^2}$.

It can be put this way, too: $x = 1 - \sqrt{5 - y^2 + 4y}$ if what's inside the root is simplified.

And the graph is:

Fig. C

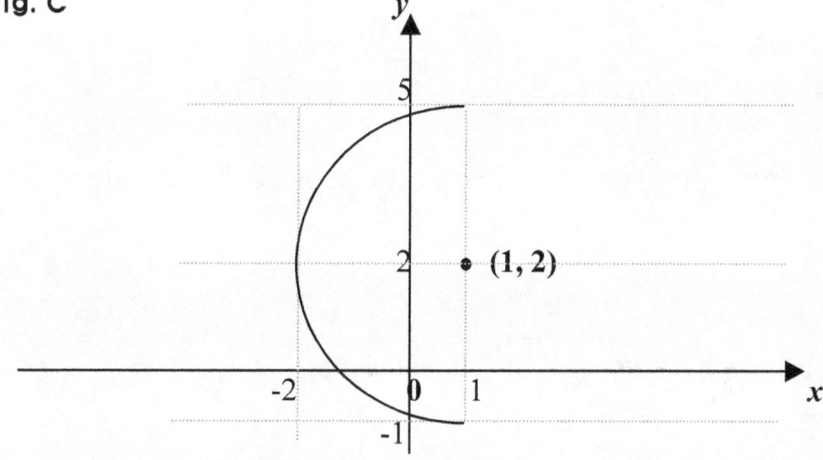

What then, about a quarter circle, that is, a quarter of a circle?

If it is the right upper quarter, and the whole circle is of radius 1, and is centered at the origin, the graph is as follows:

Fig. D

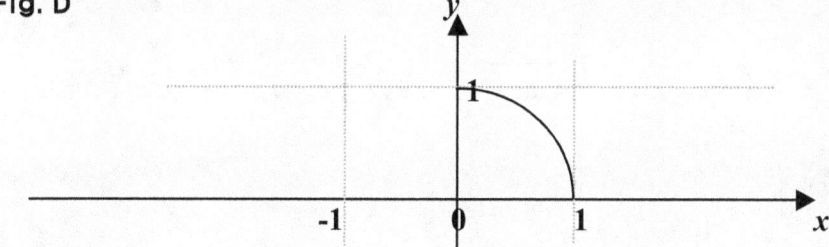

Then, we can see that the quarter is defined for $0 \leq x \leq 1$.

And the equation of the upper half is: $y = \sqrt{1-x^2}$.

So the upper quarter is: $y = \sqrt{1-x^2}$ for $0 \leq x \leq 1$, which is the domain.

And just producing: $y = \sqrt{1-x^2}$ without the domain, we mean the largest possible domain, so the domain is: $-1 \leq x \leq 1$.

So just showing this: $y = \sqrt{1-x^2}$, we mean this: $y = \sqrt{1-x^2}$ for $-1 \leq x \leq 1$.

And by the same token, the lower quarter is: $y = -\sqrt{1-x^2}$ for $0 \leq x \leq 1$.

What then, about the lower left quarter of a circle where the center is at (1, 2) and the radius is 3?

First, the lower half is: $y - 2 = -\sqrt{9-(x-1)^2}$. And the graph is as below:

Fig. E

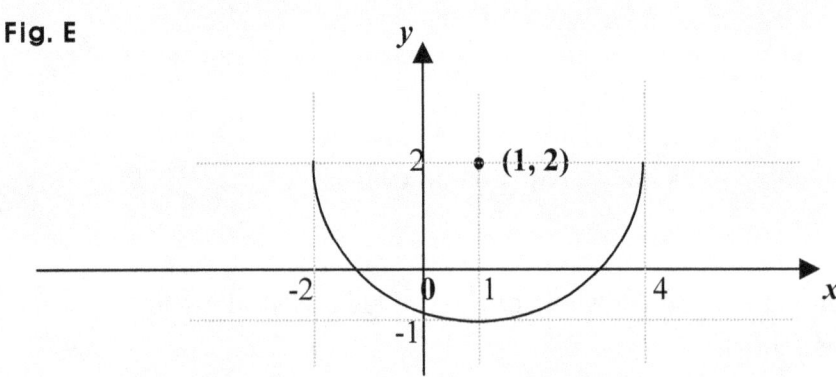

So the lower left quarter is: $y - 2 = -\sqrt{9-(x-1)^2}$ for $-2 \leq x \leq 1$.

And similarly, we can define many other parts of a circle.

Examples 2 in Circles

Doing algebra, we often get to solve equations that are in connection with circles.

Many are in fact, equations for circles whereas others are not, that is, they do not indicate any circle. And many look quite simple, but they only look simple.

Solve for x and y in each equation below.

0. $x^2 + y^2 = 0$.

1. $ax^2 + ay^2 = 0$ where a is a constant.

2. $ax^2 + ay^2 = a$ where a is a constant.

3. $ax^2 + ay^2 = a^2$ where a is a constant.

4. $ax^2 + ay^2 = a^3$ where a is constant.

5. $x^2 + (y - c)^2 = 0$ where c is constant.

6. $\frac{x^2}{d} + \frac{(y-c)^2}{d} = 0$ where c and d are constants.

7. $(x - a)^2 + (y - b)^2 = 0$ where a and b are constant.

8. $w(x - u)^2 + w(y - v)^2 = 0$ where u, v, and w are constant.

Suggestions or Solutions
To the Problem in the Example 0

Solve for x and y the equation as follows: $x^2 + y^2 = 0$.

To begin with, neither of x^2 and y^2 can be negative.

Next, if $x \neq 0$, we get: $x^2 > 0$, and if $y \neq 0$, we get: $y^2 > 0$.

So anyway, if $x \neq 0$, we get: $x^2 \neq 0$, and also, if $y \neq 0$, we get: $y^2 \neq 0$.

So if either of x and y is not 0, we get: $x^2 + y^2 \neq 0$.

Next, if $x = 0$, we get: $x^2 = 0$, and if $y = 0$, we get: $y^2 = 0$.

That is, if $x = 0$, and $y = 0$, we get: $x^2 = 0$, and $y^2 = 0$.

So if $x = 0$, and $y = 0$, we get: $x^2 + y^2 = 0$.

And we know if either of x and y is not 0, we get: $x^2 + y^2 \neq 0$.

So only if $x = 0$, and $y = 0$, we get: $x^2 + y^2 = 0$.

In other words, if $x^2 + y^2 = 0$, we get: $x = 0$, and $y = 0$.

And thus, the solution is: $x = 0$, and $y = 0$.

And in this case, we say that if and only if $x = 0$, and $y = 0$, we get: $x^2 + y^2 = 0$, and also, we say that if and only if $x^2 + y^2 = 0$, we get: $x = 0$, and $y = 0$.

And in math, symbolically, we put it this way: $(x = 0, \text{ and } y = 0) \Leftrightarrow x^2 + y^2 = 0$.

And also, we can put it this way, too: $x^2 + y^2 = 0 \Leftrightarrow (x = 0, \text{ and } y = 0)$.

And it is not rare to see similar cases when we do algebra.

For instance, we can have a situation where $(x - 2)^2 + (y^2 + 3y - 4)^2 = 0$.

Then, we need to have: $x - 2 = 0$, and $y^2 + 3y - 4 = 0$.

So we get: $x = 2$, and also, get: $y + 3y - 4 = (y - 1)(y + 4) = 0 \Rightarrow y = 1$ or -4.

Thus, we get: $x = 2$, and $y = 1$ or -4, which means: <u>$x = 2$ and $y = 1$</u>, or <u>$x = 2$ and $y = $ -4</u>.

In sum, we have: $A^2 + B^2 = 0 \Leftrightarrow A = 0$ and $B = 0$.   In other words, we have:

$A^2 + B^2 = 0 \Rightarrow A = 0$ and $B = 0$, and also, $(A = 0$ and $B = 0) \Rightarrow A^2 + B^2 = 0$.

And more generally: $A^2 + B^2 + C^2 + \ldots = 0 \Leftrightarrow A = 0, B = 0, C = 0, \ldots$

If m and n are positive even numbers, we get: $A^m + B^n = 0 \Leftrightarrow A = 0$ and $B = 0$.

And if p, q, and r are positive and even, we get:

$A^p + B^q + C^r + \ldots = 0 \Leftrightarrow A = 0, B = 0, C = 0, \ldots$

Suggestions or Solutions
To the Problem in the Example 1

Solve for x and y the equation as follows: $ax^2 + ay^2 = 0$ where a is a constant.

We can put it this way: $ax^2 + ay^2 = 0 \Rightarrow a(x^2 + y^2) = 0$.

So to begin with, if $a = 0$, then x and y both can take any real number.

Next, if $a \neq 0$, we get: $a(x^2 + y^2) = 0 \Rightarrow x^2 + y^2 = 0$, which means x and y both are 0.

Therefore, the solution is:

If $a = 0$, x and y are any real numbers.
If $a \neq 0$, we get: $x = y = 0$.

(Note that the material below is for advanced students.)

Now, what if $ax^2 + by^2 = 0$, where a and b are constant?

To begin with, if $a = 0$, and $b = 0$, then x and y both can take any real number.

Next, if $a = 0$, and $b \neq 0$, then x can take any real number, but $y = 0$.

Next, if $a \neq 0$, and $b = 0$, then $x = 0$, but y can take any real number.

Next, if $a \neq 0$, and $b \neq 0$, then x and y both are 0. Are they 0 only? Are they?

It's not quite the case, because we can have a case where $\frac{a}{b} = -1$.
In other words, we can have a situation where $a + b = 0$, and both a and b are not 0.
Then, we can get: $a = -b$.   So what?

So we get: $ax^2 + by^2 = ax^2 - ay^2 = a(x^2 - y^2) = 0 \Rightarrow x^2 - y^2 = 0$ since $a \neq 0$.
Thus, we get: $x^2 - y^2 = 0 \Rightarrow (x + y)(x - y) = 0 \Rightarrow y = x$ or $y = -x$.

What then, do we mean by this: $y = x$ or $y = -x$?

Both are lines, that is, each of the two equations indicates a line in the x-y plane. So what?

The two lines $y = x$ and $y = -x$ both pass through the origin.

And we have a fact that each point in a line has a pair of coordinates. One is the x-coordinate, and the other is the y-coordinate.

So we can say that:

• At each point in the line, $y = x$, the x-coordinate is the *value* of x, and the y-coordinate is the *value* of y, and they satisfy the equation, $y = x$, and thus, satisfy: $ax^2 + by^2 = 0$.

And the same is true for the line $y = -x$, too. So we can say that:

• At each point in the line, $y = -x$, the x-coordinate is the value of x, and the y-coordinate is the value of y, and they satisfy the equation, $y = -x$, and thus, satisfy: $ax^2 + by^2 = 0$.

So the coordinates of each of all the points in the two lines $y = x$ and $y = -x$ are respectively the values of x and y in $ax^2 + by^2 = 0$.   Therefore, the solution is that:

• If $a = 0$, and $b = 0$, then x and y both can take any real number.

• If $a = 0$, and $b \neq 0$, then x can take any real number, but $y = 0$.

• If $a \neq 0$, and $b = 0$, then $x = 0$, but y can take any real number.

• And if $a \neq 0$, and $b \neq 0$, then:

If $a + b \neq 0$, then x and y both are 0.

If $a + b = 0$, then the coordinates of each of all the points in two lines $y = x$ and $y = -x$ are respectively the values of x and y in $ax^2 + by^2 = 0$.

So more specifically, if $a \neq 0$, $b \neq 0$, and $a + b = 0$, the values of x and y are as follows:

...(-2, -2), ... (-0.3, -0.3), ... (0, 0), (0.1, 0.1), ... $(\frac{1}{3}, \frac{1}{3})$, ... (1, 1), ... (3, 3), ...

...(-2, 2), ... (-0.3, 0.3), ... (-0.1, 0.1), ... $(-\frac{1}{3}, \frac{1}{3})$, ... (1, -1), ... (3, -3), ...

That is to say that:

... $x = -2$ and $y = -2$. ... $x = -0.3$ and $y = -0.3$. ... $x = 0$ and $y = 0$. ...

... $x = 0.1$ and $y = 0.1$. ... $x = \frac{1}{3}$ and $y = \frac{1}{3}$. ... $x = 1$ and $y = 1$. ...

... $x = -2$ and $y = 2$. ... $x = -0.3$ and $y = 0.3$. ... $x = -0.1$ and $y = 0.1$. ...

... $x = -\frac{1}{3}$ and $y = \frac{1}{3}$. ... $x = 1$ and $y = -1$. ... $x = 3$ and $y = -3$. ...

Suggestions or Solutions
To the Problem in the Example 2

Solve for x and y the equation as follows: $ax^2 + ay^2 = a$ where a is a constant.

We can put it this way: $ax^2 + ay^2 = a \Rightarrow a(x^2 + y^2) = a$.

So to begin with, if $a = 0$, then x and y both can take any real numbers.

Next, if $a \neq 0$, we get: $a(x^2 + y^2) = a \Rightarrow x^2 + y^2 = 1$.

Then, what do we mean by this: $x^2 + y^2 = 1$?

It is a circle, and is the circle of radius 1 centered at the origin.
That is, the equation indicates a unit circle centered at the origin in the *x-y* plane.

Fig. 0

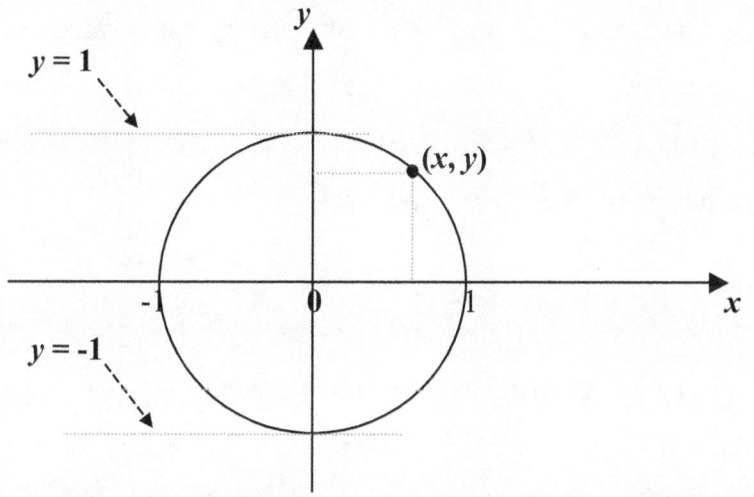

So what?

Just about the same story as in the example 1 above
Each point in a circle has a pair of coordinates.
One is the *x*-coordinate, and the other is the *y*-coordinate.

So at each and every point in the circle, $x^2 + y^2 = 1$, the x-coordinate is the value of x in $ax^2 + ay^2 = a$, and the y-coordinate is the value of y in $ax^2 + ay^2 = a$.

Thus, the coordinates of each of all the points in the circle, $x^2 + y^2 = 1$, are respectively the values of x and y in the equation $ax^2 + ay^2 = a$. For instance:

If we choose 1 for x, then we get: $x^2 + y^2 = 1 \Rightarrow 1 + y^2 = 1$.
So we have to choose 0 for y.

Next, if $\frac{1}{2}$ or $-\frac{1}{2}$ is chosen for x, we get: $(\frac{1}{2})^2 + y^2 = 1$.

So next, we want to get the solution to the equation, $(\frac{1}{2})^2 + y^2 = 1$, that is, the value of y.

Then, we get: $(\frac{1}{2})^2 + y^2 = 1 \Rightarrow y^2 = 1 - \frac{1}{4} = \frac{3}{4} \Rightarrow y = \pm\frac{\sqrt{3}}{2}$.

So we get: $x = \pm\frac{1}{2} \Rightarrow y = \pm\frac{\sqrt{3}}{2}$.

Thus, one solution is: $x = \frac{1}{2}$ and $y = \frac{\sqrt{3}}{2}$. And another is: $x = -\frac{1}{2}$ and $y = -\frac{\sqrt{3}}{2}$.

Suppose this time, that we choose $\frac{\pm\sqrt{2}}{2}$ for x.

Then, we want to get the solution to the equation, $(\frac{\pm\sqrt{2}}{2})^2 + y^2 = 1$, that is, the value of y.

Then, we get: $(\frac{\pm\sqrt{2}}{2})^2 + y^2 = 1 \Rightarrow (\frac{\sqrt{2}}{2})^2 + y^2 = 1 \Rightarrow \frac{2}{4} + y^2 = 1 \Rightarrow y^2 = \frac{1}{2} \Rightarrow y = \pm\sqrt{\frac{1}{2}}$.

Meanwhile, we have $\frac{1}{2} = \frac{2}{4}$. So we get: $y = \pm\sqrt{\frac{1}{2}} = \pm\sqrt{\frac{2}{4}} = \pm\frac{\sqrt{2}}{2} = \frac{\pm\sqrt{2}}{2}$.

Thus, y has to be $\frac{\pm\sqrt{2}}{2}$, too.

So another solution is: $x = \frac{\sqrt{2}}{2}$ and $y = \frac{\sqrt{2}}{2}$. And another is: $x = -\frac{\sqrt{2}}{2}$ and $y = -\frac{\sqrt{2}}{2}$.

Therefore, the entire solution is that:

If $a = 0$, then x and y both can take any real numbers.

If $a \neq 0$, then the coordinates of each of all the points in the circle, $x^2 + y^2 = 1$, are respectively the values of x and y in the equation $ax^2 + ay^2 = a$. And note that in this case, we have: $|x| \leq 1$, and $|y| \leq 1$, that is, $-1 \leq x \leq 1$, and $-1 \leq y \leq 1$.

Suggestions or Solutions
To the Problem in the Example 3

Solve for x and y the equation as follows: $ax^2 + ay^2 = a^2$ where a is a constant.

We can put it this way, too: $ax^2 + ay^2 = a^2 \Rightarrow a(x^2 + y^2) = a^2$.

So to begin with, if $a = 0$, then x and y can be any real numbers.

And next, if $a > 0$, we get: $a(x^2 + y^2) = a^2 \Rightarrow x^2 + y^2 = a$.   So what?

The equation represents a circle centered at the origin with the radius of $\sqrt{a}$.

Therefore, in $ax^2 + ay^2 = a^2$, the values of x and y are respectively the coordinate values at each of all the points in the circle. That is, the x-coordinate at every point in the circle is the value of x in $ax^2 + ay^2 = a^2$, and the y-coordinate is the value of y.
And note that in this case, we have:

$|x| \leq \sqrt{a}$, and $|y| \leq \sqrt{a}$, that is, $-\sqrt{a} \leq x \leq \sqrt{a}$, and $-\sqrt{a} \leq y \leq \sqrt{a}$.

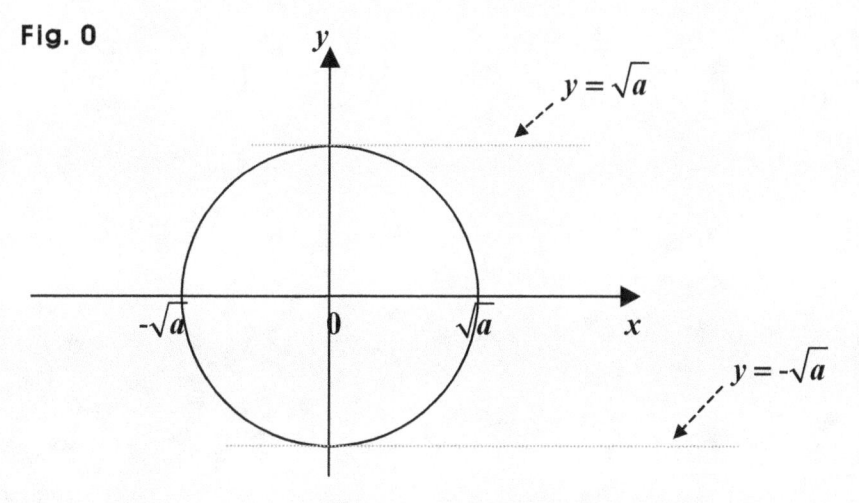

Fig. 0

That's not it though, because we have another case where $a < 0$.

If $a < 0$, we get: $x^2 + y^2 = a < 0$, which however, is not possible, and thus, we have no numbers for x and y.   Why not?

That's because $x^2 + y^2$ cannot be < 0, since any real number squared is ≥ 0.

Therefore, the solution to $ax^2 + ay^2 = a^2$ is that:

First, if $a = 0$, then x and y can be any real numbers.

Next, if $a > 0$, then x is the x-coordinate, and y is the y-coordinate of each of all the points in the circle $x^2 + y^2 = a$.

And next, if $a < 0$, there are no values for x and y.

Suggestions or Solutions
To the Problem in the Example 4

Solve for x and y the equation as follows: $ax^2 + ay^2 = a^3$ where a is a constant.

We can have: $ax^2 + ay^2 = a^3 \Rightarrow a(x^2 + y^2) = a^3$.

So first, if $a = 0$, then x and y can take any real numbers.

Next, if $a \neq 0$, we get: $a(x^2 + y^2) = a^3 \Rightarrow x^2 + y^2 = a^2$.

We know $x^2 + y^2 = a^2$ is an equation of a circle of radius a centered at the origin.

So the coordinate values of each of all the points in the circle above can be respectively the values of x and y.   And in this case, note that we have:

$|x| \leq |a|$, and $|y| \leq |a|$, that is, $-|a| \leq x \leq |a|$, and $-|a| \leq y \leq |a|$.

Why not just a but $|a|$, though?

In the problem definition, a is said to be just a constant.
That is, it is not the case where a has to be positive or 0 only.

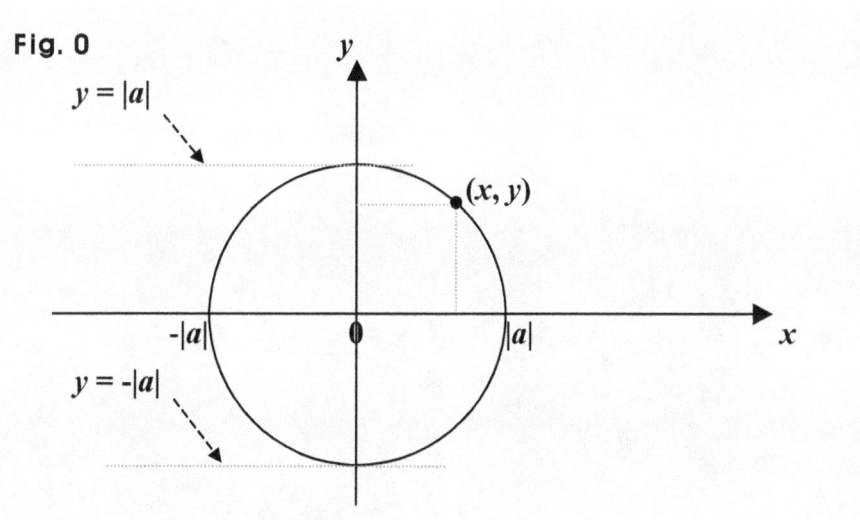

Fig. 0

So what if a is negative?

Even if a is negative, the equation of the circle above still holds since $a^2 > 0$, and thus, there is no problem with it.

Therefore, the solution to $ax^2 + ay^2 = a^3$ is that:

First, if $a = 0$, then x and y can be any real numbers.

Next, if $a \neq 0$, the value of x is the x-coordinate, and the value of y is the y-coordinate of each of all the points in the circle $x^2 + y^2 = a^2$.

Suggestions or Solutions
To the Problem in the Example 5

Solve for x and y the equation as follows: $x^2 + (y - c)^2 = 0$ where c is constant.

To begin with, if $x^2 + (y - c)^2 = 0$, both x^2 and $(y - c)^2$ have to be 0 at the same time.

Therefore, we get: $x = 0$, and $y - c = 0 \Rightarrow y = c$.

Since there is no circle of radius 0, the equation does not indicate any circle at all.

What if $x^2 + a(y - c)^2 = 0$ where a and c are constant?

Then, x^2 and $a(y - c)^2$ both have to be 0 at the same time.

Thus, we need to take care of a case where $a = 0$, and another case where $a \neq 0$.

Then, we get:

First, if $a \neq 0$, we get: $x^2 = 0 \Rightarrow x = 0$, and $(y - c)^2 = 0 \Rightarrow y - c = 0 \Rightarrow y = c$.
So we get: $x = 0$ and $y = c$.

And next, if $a = 0$, we get: $x^2 = 0 \Rightarrow x = 0$, and $(y - c)^2$ can be any real number.
So x can be 0 only, but y can be any real number. How come y can be any real number?

If $(y - c)^2$ can be any real number, then $y - c$ can be any real number, too.
By the same token, if $y - c$ can be any real number, then y can be any real number, also.
What about c, though?

It is just a constant, so no matter what value c may get, it is fixed, and thus, y can get any value since $y - c$ can be any real number. Therefore, the solution to $x^2 + a(y - c)^2 = 0$ is:

If $a \neq 0$, we get: $x = 0$ and $y = c$.

If $a = 0$, then x can be 0 only, but y can be any real number.

Suggestions or Solutions
To the Problem in the Example 6

Solve for x and y the equation $\frac{x^2}{d} + \frac{(y-c)^2}{d} = 0$ where c and d are constants.

First of all, there is no division by 0.
So we need to have: $d \neq 0$ since d is the denominator in the equation given.

Next, multiplying by d both sides of the equation, we get: $x^2 + (y-c)^2 = 0$.

So we get: $x^2 + (y-c)^2 = 0 \Rightarrow x = 0$, and $y - c = 0$, that is, $y = c$.

Therefore, the solution is: $x = 0$, and $y = c$.

What if we have: $\frac{x^2}{d} + \frac{(y-c)^2}{d} = d$?

Then, multiplying by d both sides of the equation, we get: $x^2 + (y-c)^2 = d^2$, which is an equation of a circle, and the circle has a radius of d, and is centered at $(0, c)$.

So if $d > 0$, the value of x is the x-coordinate, and that of y is the y-coordinate of each of all the points in the circle above. In this case, too, however, we may want to note that:

$|x| \leq |d|$, and $|y - c| \leq |d|$, that is, $-d \leq x \leq d$, and $-d \leq y - c \leq d$.
In other words: $-d \leq x \leq d$, and $c - d \leq y \leq c + d$.

Fig. 0

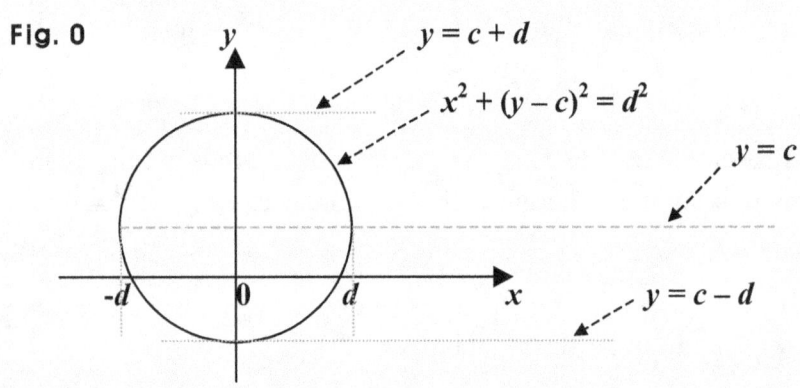

What if $d < 0$?

Even if d is negative, we still get a circle, which is the same as the one above.

That's because the negative sign gets canceled out.

Suppose for instance, $d = -2$.

Then, we get: $\frac{x^2}{d} + \frac{(y-c)^2}{d} = d \Rightarrow \frac{x^2}{-2} + \frac{(y-c)^2}{-2} = -2 \Rightarrow x^2 + (y-c)^2 = (-2)^2 = 4 = 2^2$

$\Rightarrow x^2 + (y-c)^2 = 2^2$, which is the equation of a circle of radius 2 centered at $(0, c)$.

Suggestions or Solutions
To the Problem in the Example 7

Solve for x and y the equation $(x - a)^2 + (y - b)^2 = 0$ where a and b are constant.

To begin with, if $(x - a)^2 + (y - b)^2 = 0$, we get: $x - a = 0$, and $y - b = 0$ at the same time.

Therefore, the solution is that $x = a$ and $y = b$.

What if we want to solve for x, y, and z the equation as follows?

$(x - a)^2 + (y - b)^2 + (z - c)^2 = 0$ where a, b, and c are constant.

First of all, in the equation given, x, y, z, a, b, and c can have real numbers only.

So $x - a$, $y - b$, and $z - c$ are real, too, and thus, neither of $(x - a)^2$, $(y - b)^2$, and $(z - c)^2$ can be negative.

So first, if $x - a = 0$, we get: $(x - a)^2 = 0$, if $y - b = 0$, we get: $(y - b)^2 = 0$, and if $z - c = 0$, we get: $(z - c)^2 = 0$.

Next, if $x - a \neq 0$, we get: $(x - a)^2 > 0$, if $y - b \neq 0$, we get: $(y - b)^2 > 0$, and if $z - c \neq 0$, we get: $(z - c)^2 > 0$.

Thus, we can get three cases as follows:

• If either of $(x - a)$, $(y - b)$, and $(z - c)$ is 0, and the others are not, then either of $(x - a)^2$, $(y - b)^2$, and $(z - c)^2$ is 0, and the others are positive, so we get:

$(x - a)^2 + (y - b)^2 + (z - c)^2 > 0.$

• If $(x - a)$, $(y - b)$, and $(z - c)$ are all non-zero, then all of $(x - a)^2$, $(y - b)^2$, and $(z - c)^2$ are positive, so we get: $(x - a)^2 + (y - b)^2 + (z - c)^2 > 0$.

• If $(x - a)$, $(y - b)$, and $(z - c)$ all are 0, then all of $(x - a)^2$, $(y - b)^2$, and $(z - c)^2$ are 0, too, so we get: $(x - a)^2 + (y - b)^2 + (z - c)^2 = 0$.

Thus, $(x - a)$, $(y - b)$, and $(z - c)$ all have to be 0 at the same time.

In other words, unless $(x - a)$, $(y - b)$, and $(z - c)$ all are 0 at the same time, we get: $(x - a)^2 + (y - b)^2 + (z - c)^2 > 0$.

Thus, we need to have $x - a = 0$, $y - b = 0$, and $z - c = 0$.

Therefore, the solution is: $x = a$, $y = b$, and $z = c$.

Suggestions or Solutions
To the Problem in the Example 8

Solve for x and y the equation as follows:

$w(x - u)^2 + w(y - v)^2 = 0$ **where u, v, and w are constant.**

We can have: $w(x - u)^2 + w(y - v)^2 = 0 \Rightarrow w\{(x - u)^2 + (y - v)^2\} = 0$.

So first, if $w = 0$, then $(x - u)$ and $(y - v)$ can be any real numbers

Thus, if $(x - u)$ is any real number, we can say that ($x = u +$ any real number) is any real number.

So x can be any real number.

And the same is true for y, too.

Thus, x and y can be any real number.

Next, if $w \neq 0$, then we get:

$w\{(x - u)^2 + (y - v)^2\} = 0 \Rightarrow (x - u)^2 + (y - v)^2 = 0$

So we get: $x - u = 0$ and $y - v = 0$, and thus, we get: $x = u$ and $y = v$.

Therefore, the solution to $w(x - u)^2 + w(y - v)^2 = 0$ is:

First, if $w = 0$, then x and y are any real numbers.

Next, if $w \neq 0$, we get: $x = u$, and $y = v$.

₃.Equations for Circles 3

In math, making a circle or defining a circle, we put it in an equation, and produce the equation of the circle.   What then, do we need making or defining a circle?

We need the radius and the center. What then, do we do with those?

We can get the equation of the circle putting them into a template. And the template is called the standard equation for circles. So we can call it a formula, too.

And the template is: $(x - a)^2 + (y - b)^2 = r^2$, where (a, b) is the center, and r is the radius.

So for instance, if (-2, 3) is the center, and 2 is the radius, we can define the circle, and the circle is: $(x + 2)^2 + (y - 3)^2 = 2^2$.

So given the center and the radius, we can quickly define the circle, that is, produce the equation of it using the template above.

What if we are given only either of the two?

Given the center only, we can come up with infinitely many circles that share the center. And we call those circles concentric circles.
So given the center only, we cannot find the circle.
Next, given the radius only, we can come up with infinitely many circles, too, and each circle has a different center. So given the radius only, we cannot find the circle, either.

What if this time, given a particular point, can we find a particular or unique circle passing through the particular point?

No, we can't. That's because we can have more than one circle that can pass through the particular point. We can have in fact, infinitely many of those.

Fig. 0

What if this time, we are given two particular points?

Then again, we can construct infinitely many circles that can share the two points.

Fig. 1

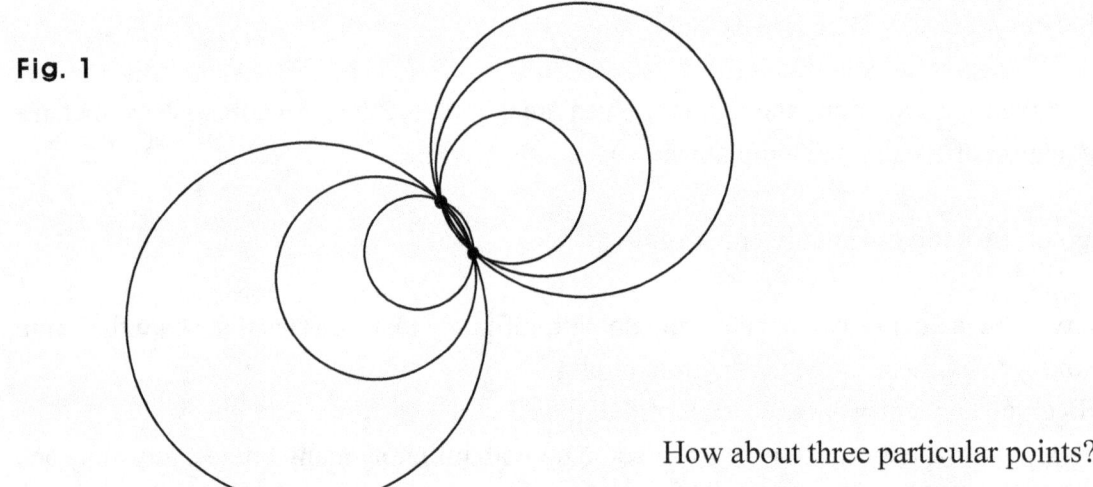

How about three particular points?

Then, we are likely to construct a particular circle.   Why not definitely but likely?

That's because if the three points are in a line, we are not able to make a circle.

Fig. 2

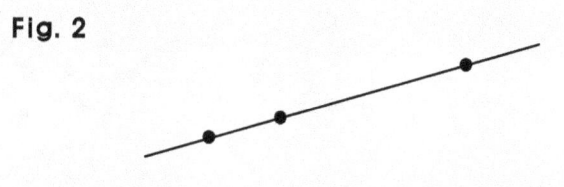

So the three points are not in a line if a circle includes the three.

Fig. 3

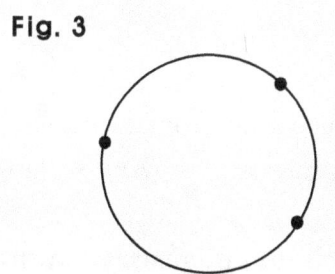

Having three points not in a line, we can make (define) a particular circle.
That is, only one circle passes though the three points. How then, can we get the circle?

We can easily get it using the equation as follows: $x^2 + y^2 + ax + by + c = 0$.   How?

Putting each of the three points into the equation, we get an equation for *a*, *b*, and *c*.
So we get a system of three equations for *a*, *b*, and *c*.
And solving the system, we can get the equation of the circle.

What then, about the standard equation: $(x - a)^2 + (y - b)^2 = c^2$?

We can still use it in the case, too, where three points not in a line are given.

So let's now find, for instance, a circle that has three points: **(1, 1)**, **(2, 3)**, and **(4, 2)** using the standard equation.   Putting the three points in a graph, we get:

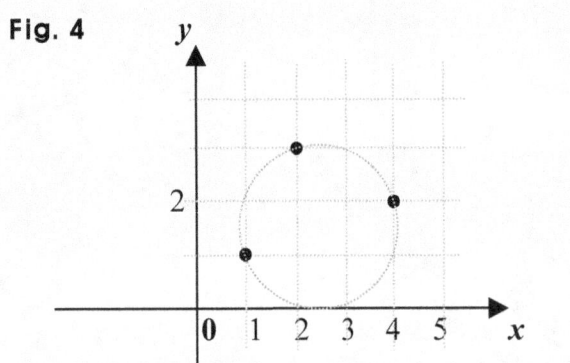

Fig. 4

We will find the center first, and then, get the radius.

The three points are in a circle, so each point is the same distance away from the center. We can use the distance formula to find the center, and the formula is: $d^2 = (\Delta x)^2 + (\Delta y)^2$.

In the formula above, d is the distance between two points, Δx is the difference between x-coordinates, and Δy is the difference between y-coordinates.

So assuming first, the center is **(s, t)**, since the distance from each point to the center is the same, we can get two equations the way as follows:

One is: $(a-1)^2 + (b-1)^2 = (a-2)^2 + (b-3)^2$ if we use two points **(1, 1)** and **(2, 3)**.
And the other is: $(s-1)^2 + (t-1)^2 = (s-4)^2 + (t-2)^2$ if we use **(1, 1)** and **(4, 2)**.

So we now have a system of two equations for s and t.
The equations above look complicated, but can be simplified by some factorizations.

We have a factorization identity: $u^2 - v^2 = (u+v)(u-v)$.   So we get:

$(s-1)^2 + (t-1)^2 = (s-2)^2 + (t-3)^2 \Rightarrow (s-1)^2 - (s-2)^2 = (t-3)^2 - (t-1)^2$
$\Rightarrow \{(s-1) + (s-2)\}\{(s-1) - (s-2)\} = \{(t-3) + (t-1)\}\{(t-3) - (t-1)\}$
$\Rightarrow (2s-3)\cdot 1 = (2t-4)\cdot(-2) \Rightarrow 2s - 3 = -4t + 8 \Rightarrow 2s + 4t = 11$.

$(s-1)^2 + (t-1)^2 = (s-4)^2 + (t-2)^2 \Rightarrow (s-1)^2 - (s-4)^2 = (t-2)^2 - (t-1)^2$

$\Rightarrow \{(s-1) + (s-4)\}\{(s-1) - (s-4)\} = \{(t-2) + (t-1)\}\{(t-2) - (t-1)\}$

$\Rightarrow (2s-5)\cdot 3 = (2t-3)\cdot(-1) \Rightarrow 6s - 15 = -2t + 3 \Rightarrow 6s + 2t = 18 \Rightarrow 3s + t = 9.$

Thus, we get a new system simplified as follows: $2s + 4t = 11$, and $3s + t = 9$.
Solving the system above, we can find the center (s, t).
Then, we can directly put the center into the standard form: $(x - s)^2 + (y - t)^2 = c^2$.

So next, we want to get the radius, which is c in this case.
We don't really have to find c, though.   How come?

We have only to find c^2, since the equation $(x - a)^2 + (y - b)^2 = c^2$ has c^2 and not c.

And the actual solution to the system is: $s = 2.5$, and $t = 1.5$.   So we get:

$c^2 = (s-1)^2 + (t-1)^2 = (2.5-1)^2 + (1.5-1)^2 = 1.5^2 + 0.5^2 = (\tfrac{3}{2})^2 + (\tfrac{1}{2})^2 = \tfrac{9+1}{4} = \tfrac{10}{4} = \tfrac{5}{2}.$

Therefore, the circle is: $(x - \tfrac{3}{2})^2 + (y - \tfrac{1}{2})^2 = \tfrac{5}{2}.$

Anyway, finding c, we get: $c^2 = \tfrac{10}{4} \Rightarrow c = \pm\tfrac{\sqrt{10}}{2}$, and c is a radius, so we get: $c = \tfrac{\sqrt{10}}{2}.$

• Suppose this time, two points are given, one is the center, and the other is in a circle.

Then, the center is given, so we can take advantage of the standard equation as follows:
$(x - a)^2 + (y - b)^2 = c^2$, where (a, b) is the center, and c is the radius.

First, putting the center into the form, we get an equation where c is the only unknown.
So next, to find c, we can put the other point into the equation.

So for instance, if the center is $(1, 2)$, putting it first into the standard equation, we get:
$(x - a)^2 + (y - b)^2 = c^2 \Rightarrow (x - 1)^2 + (y - 2)^2 = c^2.$

And next, if the point given is $(2, 2)$, putting it into the equation above, we get:
$(x - 1)^2 + (y - 2)^2 = c^2 \Rightarrow (2 - 1)^2 + (2 - 2)^2 = c^2 \Rightarrow 1 = c^2.$

So the circle is: $(x-1)^2 + (y-2)^2 = 1$, which is a unit circle centered at **(1, 2)**.

• Suppose this time, we have two points and a radius, but the center is not given.

Then, we get to find not one circle particular but two particular circles.

That is because in a plane, we can have not one but two particular points, each of which is the same distance away from the two points given, and the same distance is the radius.

For instance, we can have two points C_1 and C_2 in the same plane, and each of C_1 and C_2 is the same distance away from the other two points as shown below.

Fig. 5

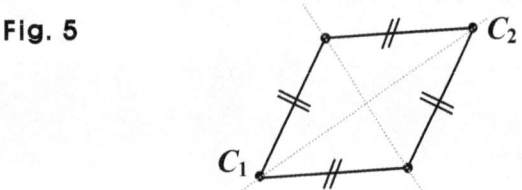

So C_1 and C_2 are the centers of two particular circles, which have the same radii.

Fig. 6

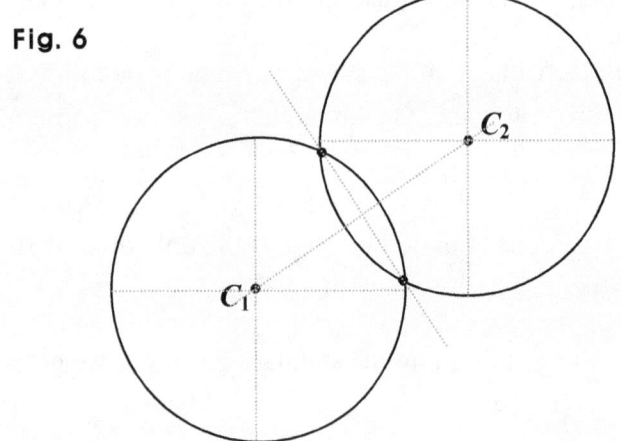

Then, we may want to begin with finding the two centers.
How do we find the two centers, though?

We can use the idea of the center of a circle. What's the idea?

The idea is that the center is the same distance away from every point in the circle, and that the same distance is the radius. The two centers can be thus, found by the distance formula. So assuming for instance, the two points are $(1, 2)$ and $(3, 4)$, and the radius is 5, we get a system of two equations for a and b as follows:

One is: $(a - 1)^2 + (b - 2)^2 = 5^2$, and the other is: $(a - 3)^2 + (b - 4)^2 = 5^2$.

And we can get a simpler version of the system using a factorization.
We have a factorization identity: $u^2 - v^2 = (u + v)(u - v)$.   So we get:

$(a - 1)^2 + (b - 2)^2 = (a - 3)^2 + (b - 4)^2 \Rightarrow (a - 1)^2 - (a - 3)^2 = (b - 4)^2 - (b - 2)^2$

$\Rightarrow \{(a - 1) + (a - 3)\}\{(a - 1) - (a - 3)\} = \{(b - 4) + (b - 2)\}\{(b - 4) - (b - 2)\}$

$\Rightarrow (2a - 4)2 = (2b - 6)(-2) \Rightarrow 4a - 8 = -4b + 12 \Rightarrow 4a + 4b = 20 \Rightarrow a + b = 5$.

What about the other of the two equations in the system?

We can use either of $(a - 1)^2 + (b - 2)^2 = 5^2$ and $(a - 3)^2 + (b - 4)^2 = 5^2$.

Choosing for instance, the first of the two above, we get a simpler system where:
$a + b = 5$, and $(a - 1)^2 + (b - 2)^2 = 5^2$.

Then, to begin with, we get: $a + b = 5 \Rightarrow b = 5 - a$.

So we get: $(a - 1)^2 + (b - 2)^2 = 5^2 \Rightarrow (a - 1)^2 + (5 - a - 2)^2 = 5^2$.

Meanwhile, $(a - 1)^2 + (5 - a - 2)^2 = (a - 1)^2 + (3 - a)^2 = (a - 1)^2 + (a - 3)^2$. How come?

$(x - c)^2 = (-c + x)^2 = \{(-1)(c - x)\}^2 = (-1)^2(c - x)^2 = 1(c - x)^2 = (c - x)^2$.

In other words, $(x - c)^2 = \{-(x - c)\}^2 = (-x + c)^2 = (c - x)^2$.   For instance, $7^2 = (-7)^2 = 7^2$.

So next, we get: $(a - 1)^2 + (a - 3)^2 = a^2 - 2a + 1 + a^2 - 6a + 9 = 2a^2 - 8a + 10$.

Since we began with $(a-1)^2 + (b-2)^2 = 5^2$, we now get: $2a^2 - 8a + 10 = 5^2$.

Thus, we get: $2a^2 - 8a + 10 = 5^2 \Rightarrow 2a^2 - 8a - 15 = 0$.

Using the quadratic formula, we can get the values of a. Don't remember the formula?

Assuming: $ax^2 + bx + c = 0$, we get: $x = \frac{-b \pm \sqrt{b^2 - 4ac}}{2a}$, which is the formula.

So using the quadratic formula, we get:

$a = \frac{8 \pm \sqrt{64 - 4 \cdot 2 \cdot (-15)}}{2 \cdot 2} = \frac{8 \pm \sqrt{184}}{4} = \frac{8 \pm \sqrt{4 \cdot 46}}{4} = \frac{8 \pm 2\sqrt{46}}{4} = \frac{4 \pm \sqrt{46}}{2}$. Such a long calculation, isn't it?

We have a simpler version, which is: $x = \frac{-k \pm \sqrt{k^2 - ac}}{a}$, where $b = 2k$ in the formula.

So using the simpler version, we get: $a = \frac{4 \pm \sqrt{16 - 2 \cdot (-15)}}{2} = \frac{4 \pm \sqrt{46}}{2}$.

What if we can't remember the formula at all at the exam?

Then, put the equation into the complete square. We have: $2a^2 - 8a - 15 = 0$. So we get:

$2a^2 - 8a - 15 = 2(a^2 - 4a) - 15 = 2(a^2 - 4a + 4 - 4) - 15 = 2\{(a-2)^2 - 4\} - 15$

$= 2(a-2)^2 - 8 - 15 = 2(a-2)^2 - 23 = 0 \Rightarrow (a-2)^2 = \frac{23}{2} \Rightarrow a - 2 = \pm\sqrt{\frac{23}{2}} \Rightarrow a = 2 \pm \sqrt{\frac{23}{2}}$.

How come we get: $2 \pm \sqrt{\frac{23}{2}} = \frac{4 \pm \sqrt{46}}{2}$?

$2 \pm \sqrt{\frac{23}{2}} = \frac{4}{2} \pm \sqrt{\frac{46}{4}} = \frac{4}{2} \pm \frac{\sqrt{46}}{2} = \frac{4 \pm \sqrt{46}}{2}$.

Next, we have: $b = 5 - a$. So we get: $b = 5 - a = 5 - \frac{4 \pm \sqrt{46}}{2} = \frac{6 \mp \sqrt{46}}{2}$. How come $\mp$?

Doing it separately, we get:

When $a = \frac{4+\sqrt{46}}{2}$, we get: $b = 5 - a = 5 - \frac{4+\sqrt{46}}{2} = \frac{6-\sqrt{46}}{2}$.

When $a = \frac{4-\sqrt{46}}{2}$, we get: $b = 5 - a = 5 - \frac{4-\sqrt{46}}{2} = \frac{6+\sqrt{46}}{2}$.

Now, putting threads together, we get: $a = \frac{4\pm\sqrt{46}}{2}$, and $b = \frac{6\mp\sqrt{46}}{2}$.

So one center is $(\frac{4+\sqrt{46}}{2}, \frac{6-\sqrt{46}}{2})$, and the other is $(\frac{4-\sqrt{46}}{2}, \frac{6+\sqrt{46}}{2})$.

For reference, the centers are approximately **(5.39, -0.39)** and **(-1.39, 6.39)**.

Next, putting each center into the standard form directly, and setting:

$(a_1, b_1) = (\frac{4+\sqrt{46}}{2}, \frac{6-\sqrt{46}}{2})$, and $(a_1, b_1) = (\frac{4+\sqrt{46}}{2}, \frac{6-\sqrt{46}}{2})$, we get:

$(x - a_1)^2 + (y - b_1)^2 = 5^2$, and $(x - a_2)^2 + (y - b_2)^2 = 5^2$, since the radius is 5.

• Now, what if we are given information not enough to determine a particular circle?

For instance, we are given only two points in the particular circle, and of course, neither of the two is the center. And for another instance, we are given the center only.

Then, we will have to find a group of circles. That is, the group of circles is the solution.

What do we mean by a group of circles, though?

Suppose we are given the center only, and for instance, the center is **(1, -2)**.

Then, using the standard form $(x - a)^2 + (y - b)^2 = c^2$, we can get: $(x - 1)^2 + (y + 2)^2 = c^2$.

So the solution can be such a group of concentric circles as follows:
$(x - 1)^2 + (y - 2)^2 = c^2$ where c is a positive integer.

Concentric circles share the same center.

The equation above indicates each of all circles where the centers are **(1, 2)**, but the radii are different from each other. That is, the radii are 0.1, 0.15, 1, 1.9, 2, 3, 4, etc.

Fig. 7

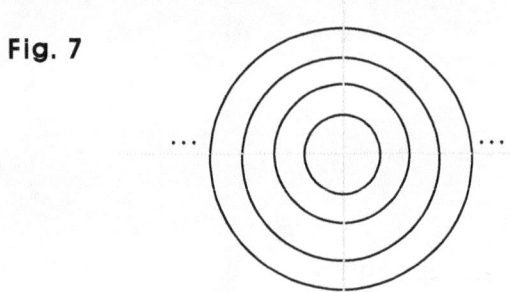

That's not it, of course.

So for another instance, the solution can also be a group of circles where the centers are in a line, or a group of circles tangent to each other.

Fig. 8

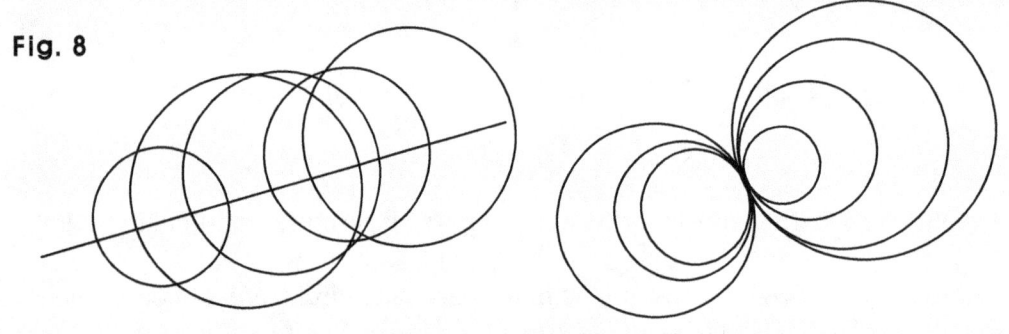

Doing a problem with finding a particular circle, we should be able to find information that can determine the particular circle.

Sometimes, the problem is so straightforward that all necessary information is given to us directly.

For instance, three points in a particular circle are specified as in the first example above.

Some other times, all the information is given in implicit manner.

In such cases, we should be able to extract or deduce such information from the problem description. That is, we need to track down the information on the center and radius.

Examples 3 in Circles

A circle is one of curved line segments simply closed. It is in fact, the simplest of all line segments closed.   How come though, it is that simple?

Either curved or not, a line segment is a collection of points in a plane.
Every point in a line segment called a circle is the same distance away from a particular point called the center, and the same distance is called the radius.

So saying a circle, we mean the radius and the center.
And a circle is that simple.   When it comes to a problem though, it is not that simple.

Working with a curve, we put it in an equation if we can. So working with a circle, we work with the equation of it. And putting it in a graph, we can work with it more readily since we can actually see it. And the same is true for any other curves, too, of course. How then, do we put a circle in an equation?

We can do it by means of the definition for circles, which is no other than the distance formula, called Pythagorean theorem, too. So using the formula, we get an equation called a standard equation. And the standard equation is: $(x - a) + (y - b) = c^2$.

In the equation above, (a, b) is the center, and c is the radius, of course.
So plugging in the center and the radius, we get the equation, so we can call it a template.

And simplifying the standard equation, we get an equivalent equation called a general equation, which takes this form: $x^2 + y^2 + Ax + By + C = 0$, which is called therefore, the general equation of a circle.

Now, doing the examples in this set and subsequent sets, we get to find a lot more about the idea called a circle.

Find the circle in each case below assume the circle is in the *x-y* plane.

0. Two points **(2, 1)** and **(-3, 2)** are endpoints of a diameter of the circle.

1. Three points **(2, 1)**, **(1, 3)**, and **(-3, 2)** are in the circle.

2. A point **(2, 1)** is in the circle centered at **(-3, 4)**.

3. A point **(2, 1)** is in the circle where the radius is 2.

Suggestions or Solutions
To the Problem in the Example 0

Two points (2, 1) and (-3, 2) are endpoints of a diameter of the circle.

To begin with, the center is $(\frac{2+(-3)}{2}, \frac{1+2}{2}) = (-\frac{1}{2}, \frac{3}{2})$.

Next, assuming d is the diameter, and r is the radius, we get:

$d^2 = \{2 - (-3)\}^2 + (1 - 2)^2 = 25 + 1 = 26 \Rightarrow d^2 = 26 \Rightarrow r = \frac{d}{2} \Rightarrow r^2 = \frac{d^2}{4} = \frac{26}{4} = \frac{13}{2}$.

Therefore, the circle is $(x + \frac{1}{2})^2 + (y - \frac{3}{2})^2 = \frac{13}{2}$.

If not quite sure of the idea behind the processes above, follow the steps below:

Given the center and the radius in a circle, we can find the circle, of course.
In this problem however, they are not given, are they?

Of course, they are, but hidden somewhere in the problem.
So we've got to find them. Where can we find them, though?

They are in a diameter.
What diameter? Are there many diameters in a circle?

We can have two interpretations of a diameter in a circle.
One is a line segment connecting two points facing each other in a circle.
The other is the length of such a line segment.

Usually, saying *the* diameter, we mean the length, and saying just a diameter though, we mean such a line segment.
So a circle has infinitely many diameters, each of which has the same length, of course.

And the same is true for a radius, too.

We can have therefore, two interpretations of a radius in a circle.

One is a line segment connecting two points, one of the two is the center, and the other is a point in the circle.
The other interpretation is the length of such a line segment.

So usually, saying *the* radius, we mean the length, and saying just a radius however, we mean such a line segment.
Thus, a circle has infinitely many radii, each of which has the same length, of course.

Now, in this problem, a diameter is given, so we may want to begin with it.

A diameter has the center and radius, and can be specified by its two endpoints.

The distance between the two endpoints is the diameter, which is twice the radius.

So the center is the midpoint between the two endpoints, **(2, 1)** and **(-3, 2)**, and the radius is half the diameter, of course. So what should we find first?

We may want to find the midpoint first, which is the center.
So suppose now, that the midpoint is **(s, t)**.

Then, we get: $s = \frac{2+(-3)}{2} = -\frac{1}{2}$, and $t = \frac{1+2}{2} = \frac{3}{2}$. So the center is $(-\frac{1}{2}, \frac{3}{2})$.
So next, what should we move on to?

It is the radius, which is half the diameter, of course. So what do we want to begin with?

We want to begin with the two points **(2, 1)** and **(-3, 2)**, since both are the endpoints of a diameter of the circle.

Thus, the diameter is the distance between **(2, 1)** and **(-3, 2)**. So first, assuming the diameter is *d*, we get: $d^2 = (\Delta x)^2 + (\Delta y)^2 = \{2 - (-3)\}^2 + (1 - 2)^2 = 25 + 1 = 26 \Rightarrow d^2 = 26$.

Next, assuming the radius is r, we get: $r = \frac{d}{2} \Rightarrow r^2 = \frac{d^2}{4} = \frac{26}{4} = \frac{13}{2}$.   Why r^2, though?

We are going to use the standard form, $(x - a)^2 + (y - b)^2 = r^2$, in which we have r^2.

Now that we have the center and radius, we can use the standard form to get the circle.

Thus, we get: $\{x - (-\frac{1}{2})\}^2 + (y - \frac{3}{2})^2 = r^2 = \frac{13}{2}$, so the circle is $(x + \frac{1}{2})^2 + (y - \frac{3}{2})^2 = \frac{13}{2}$.

In short:

To begin with, the center is $(\frac{2+(-3)}{2}, \frac{1+2}{2}) = (-\frac{1}{2}, \frac{3}{2})$.

Next, assuming d is the diameter, and r is the radius, we get:

$d^2 = \{2 - (-3)\}^2 + (1 - 2)^2 = 25 + 1 = 26 \Rightarrow d^2 = 26 \Rightarrow r = \frac{d}{2} \Rightarrow r^2 = \frac{d^2}{4} = \frac{26}{4} = \frac{13}{2}$.

Therefore, the circle is $(x + \frac{1}{2})^2 + (y - \frac{3}{2})^2 = \frac{13}{2}$.

Also, even if given two points only, we can still find a particular circle if one of the two is the center, and the other is in the circle. How?

We can find the radius, and then, the circle.   How can we find the radius?

We can find it using the distance formula.   How come?

Every point in a circle is the radius away from the center, so if we are given two points, one is the center, and the other is in the circle, then we can find the radius using the distance formula, since the distance between the two is the radius.

Suggestions or Solutions
To the Problem in the Example 1

Three points (2, 1), (1, 3), and (-3, 2) are in the circle.

Suppose the circle is: $x^2 + y^2 + ax + by + c = 0$ where a, b, and c are constant. Then, we get: $(2, 1) \Rightarrow 4 + 1 + 2a + b + c = 0$, $(1, 3) \Rightarrow 1 + 9 + a + 3b + c = 0$, and $(-3, 2) \Rightarrow 9 + 4 - 3a + 2b + c = 0$.

So we get: $2a + b + c + 5 - (a + 3b + c + 10) = 0 \Rightarrow a - 2b - 5 = 0$, and $a + 3b + c + 10 - (-3a + 2b + c + 13) \Rightarrow 4a + b - 3 = 0$.

Thus, we get: $4(a - 2b - 5) - (4a + b - 3) = 0 \Rightarrow -9b - 17 = 0 \Rightarrow b = -\frac{17}{9} \Rightarrow$ $4a + b - 3 = 4a - \frac{17}{9} - 3 = 0 \Rightarrow 4a = \frac{44}{9} \Rightarrow a = \frac{11}{9}$.

So we get: $2a + b + c + 5 = 2 \cdot \frac{11}{9} - \frac{17}{9} + c + 5 = c - \frac{5}{9} + 5 = c + \frac{40}{9} = 0 \Rightarrow c = -\frac{40}{9}$.

Therefore, the circle is: $x^2 + y^2 + \frac{11}{9}x - \frac{17}{9}y - \frac{40}{9} = 0$.

If not quite sure of the idea behind the processes above, follow the steps below:

Three particular points in a plane can determine a circle if the three are not in a line, of course. So given three points not in a line, we should be able to find a particular circle. How do we know though, if the three points are not in a line?

We can construct a line using two of the three, and then, can see if the other is in the line by plugging it into the line.

So first, using **(2, 1)** and **(1, 3)**, we can get the slope, which is: $\frac{3-1}{1-2} = -2$, and thus, the line is as follows: $y - 1 = -2(x - 2)$.

(If not sure of how to get a line this way, refer to **CONICS 1**.)

Next, putting **(-3, 2)** into the line above, we get:

On the left hand side, **2 – 1 = 1**, and on the right hand side **-2(–3 – 2) = 10**.

We can see that both sides are not equal, so the three points are not in a line.
So we can find the circle passing through the three. How?

We can find it using the general form: $x^2 + y^2 + ax + by + c = 0$ where *a*, *b*, and *c* are constant. (We can use the standard form, too, of course.)

Putting each of the three points into the general form above, we get an equation for the three constants *a*, *b*, and *c*, and therefore, we can set up a system of three equations for the three constants.

First, putting **(2, 1)** into the form, we get: $4 + 1 + 2a + b + c = 0 \Rightarrow 2a + b + c + 5 = 0$.

Next, putting **(1, 3)** into it, we get: $1 + 9 + a + 3b + c = 0 \Rightarrow a + 3b + c + 10 = 0$.

Finally, putting **(-3, 2)** into it, we get: $9 + 4 - 3a + 2b + c = 0 \Rightarrow -3a + 2b + c + 13 = 0$.

Then, eliminating *c*, we get:

First, $2a + b + c + 5 - (a + 3b + c + 10) = 0 \Rightarrow a - 2b - 5 = 0$.

Next, $a + 3b + c + 10 - (-3a + 2b + c + 13) \Rightarrow 4a + b - 3 = 0$.

Eliminating *a*, we get: $4(a - 2b - 5) - (4a + b - 3) = 0 \Rightarrow -9b - 17 = 0 \Rightarrow b = -\frac{17}{9}$.

So next, we get: $b = -\frac{17}{9} \Rightarrow 4a + b - 3 = 4a - \frac{17}{9} - 3 = 0 \Rightarrow 4a = \frac{44}{9} \Rightarrow a = \frac{11}{9}$.

Thus, we get: $2a + b + c + 5 = 2 \cdot \frac{11}{9} - \frac{17}{9} + c + 5 = c - \frac{5}{9} + 5 = c + \frac{40}{9} = 0 \Rightarrow c = -\frac{40}{9}$.

So the circle is: $x^2 + y^2 + \frac{11}{9}x - \frac{17}{9}y - \frac{40}{9} = 0$, where we can't see however, the center and the radius.

So let's put the circle in the standard form. We don't have to for this problem, of course. Just checking though

Putting the equation above into a pair of complete squares, we get the standard version.

So first, for x, we get: $x^2 + \frac{11}{9}x = x^2 + \frac{11}{9}x + (\frac{11}{2\cdot9})^2 - (\frac{11}{2\cdot9})^2 = (x + \frac{11}{18})^2 - (\frac{11}{18})^2$.

And next, for y, we get: $y^2 - \frac{17}{9}y = y^2 - \frac{17}{9}y + (\frac{17}{2\cdot9})^2 - (\frac{17}{2\cdot9})^2 = (y - \frac{17}{18})^2 - (\frac{17}{18})^2$.

So we get:

$$x^2 + y^2 + \frac{11}{9}x - \frac{17}{9}y - \frac{40}{9} = 0 \Rightarrow (x + \frac{11}{18})^2 - (\frac{11}{18})^2 + (y - \frac{17}{18})^2 - (\frac{17}{18})^2 - \frac{40}{9} = 0$$

$$\Rightarrow (x + \frac{11}{18})^2 + (y - \frac{17}{18})^2 - (\frac{11}{18})^2 - (\frac{17}{18})^2 - \frac{2\cdot40}{2\cdot9} = 0 \Rightarrow (x + \frac{11}{18})^2 + (y - \frac{17}{18})^2 - \frac{11^2 + 17^2 + 80\cdot18}{18^2} = 0$$

$$\Rightarrow (x + \frac{11}{18})^2 + (y - \frac{17}{18})^2 = \frac{11^2 + 17^2 + 80\cdot18}{18^2} = \frac{1850}{18^2} = \frac{5^2\cdot74}{18^2} = \frac{1850}{324} \Rightarrow (x + \frac{11}{18})^2 + (y - \frac{17}{18})^2 = \frac{1850}{324},$$

which indicates a circle of radius $\frac{5\sqrt{74}}{18}$ centered at $(-\frac{11}{18}, \frac{17}{18})$.

In short:

Suppose the circle is: $x^2 + y^2 + ax + by + c = 0$ where a, b, and c are constant.
Then, we get: $(2, 1) \Rightarrow 4 + 1 + 2a + b + c = 0$, $(1, 3) \Rightarrow 1 + 9 + a + 3b + c = 0$, and
$(-3, 2) \Rightarrow 9 + 4 - 3a + 2b + c = 0$.

So we get: $2a + b + c + 5 - (a + 3b + c + 10) = 0 \Rightarrow a - 2b - 5 = 0$, and
$a + 3b + c + 10 - (-3a + 2b + c + 13) \Rightarrow 4a + b - 3 = 0$.

Thus, we get: $4(a - 2b - 5) - (4a + b - 3) = 0 \Rightarrow -9b - 17 = 0 \Rightarrow b = -\frac{17}{9} \Rightarrow$
$4a + b - 3 = 4a - \frac{17}{9} - 3 = 0 \Rightarrow 4a = \frac{44}{9} \Rightarrow a = \frac{11}{9}$.

So we get: $2a + b + c + 5 = 2\cdot\frac{11}{9} - \frac{17}{9} + c + 5 = c - \frac{5}{9} + 5 = c + \frac{40}{9} = 0 \Rightarrow c = -\frac{40}{9}$.

Therefore, the circle is: $x^2 + y^2 + \frac{11}{9}x - \frac{17}{9}y - \frac{40}{9} = 0$.

Suggestions or Solutions
To the Problem in the Example 2

A point (2, 1) is in the circle centered at (-3, 4).

Suppose the circle is $(x - a) + (y - b) = c^2$, where (a, b) is the center and c is the radius. The center is at **(-3, 4**, and the point **(2, 1)** is in the circle.
So we get: $(\Delta x)^2 + (\Delta y)^2 = c^2 \Rightarrow (-3 - 2)^2 + (4 - 1)^2 = 25 + 9 = 34 = c^2$.

Therefore, the circle is: $(x + 3)^2 + (y - 4)^2 = 34$.

If not quite sure of the idea behind the processes above, follow the steps below:

Having the center and the radius, we can get a particular circle.
We are given a point **(2, 1)** in the circle, together with the center, which is at **(-3, 4)**.
Where then, is the radius?

It is between the center and the point given.
Every point in a circle is the same distance away from the center of the circle, and the same distance is the radius. So what's the radius?

The radius in the circle we want is the distance from the center **(-3, 4)** to the point **(2, 1)**.

We can get the distance by the distance formula, often called Pythagorean theorem, too.

And next, putting the center and the radius into the template, the standard form where:
$(x - a) + (y - b) = c^2$, where **(a, b)** is the center and c is the radius, we can get the circle.

So first, applying the distance formula to the center and the point, we get:
$(\Delta x)^2 + (\Delta y)^2 = c^2 \Rightarrow (-3 - 2)^2 + (4 - 1)^2 = 25 + 9 = 34 = c^2$, which is the radius squared.

Therefore, the circle we are after is: $(x + 3)^2 + (y - 4)^2 = 34$, where the radius is $\sqrt{34}$.

Putting it in the general form, we just expand the version standard above, so we get:

$(x + 3)^2 + (y - 4)^2 = x^2 + 6x + 9 + y^2 - 8y + 16 = x^2 + y^2 + 6x - 8y + 25 = 34$
$\Rightarrow x^2 + y^2 + 6x - 8y - 9 = 0$.

In short:

Suppose the circle is $(x - a) + (y - b) = c^2$, where (a, b) is the center and c is the radius.

The center is at $(-3, 4$, and the point $(2, 1)$ is in the circle.

So we get: $(\Delta x)^2 + (\Delta y)^2 = c^2 \Rightarrow (-3 - 2)^2 + (4 - 1)^2 = 25 + 9 = 34 = c^2$.

Therefore, the circle is: $(x + 3)^2 + (y - 4)^2 = 34$.

Suggestions or Solutions
To the Problem in the Example 3

A point (2, 1) is in the circle where the radius is 2.

Since the radius is 2, we can put the circle in such an equation as follows:
$(x - a)^2 + (y - b)^2 = 2^2$ where (a, b) is the center.

Then, since the point **(2, 1)** is in the circle, we get:
$(2 - a)^2 + (1 - b)^2 = 2^2$, which is the same as $(a - 2)^2 + (b - 1)^2 = 2^2$.

Therefore, the solution to this problem is every circle that satisfies an equation below:
$(x - a)^2 + (y - b)^2 = 2^2$ where a and b satisfy $(a - 2)^2 + (b - 1)^2 = 2^2$.

If not quite sure of the idea behind the processes above, follow the steps below:

Given the center and the radius, we may want to use the template (standard form) below:
$(x - a)^2 + (y - b)^2 = c^2$, where (a, b) is the center, and c is the radius.

The radius is given, and is 2, but the center is not given.   Where then, is the center?

Every point in a circle is the radius away from the center, so the center of the circle we are after is 2 away from the point **(2, 1)**.

Is there only one point though, that is 2 away from the point **(2, 1)**? Can we have only one point the distance from which to the point **(2, 1)** is 2?

No, it's not the case.
There can be many points that can be the center of the circle we are after, and thus, the solution to this problem can be more than one circle.   How come?

There are a lot of things can be 2 away from the point. We can have in fact, many things that can be the same distance away from a particular point in a plane. Many lines can be the same distance away from a point in a plane.

And the same is true for points, too. So there can be many points the same distance away from a particular point in a plane. That is in fact, what a circle is about, isn't it?

So many circles can have the same radii, and also, can pass through the same point. That is, many circles of a particular radius can share a particular point.

For instance, many circles of radius 1 can pass through a point (1, 2).

Fig. 0

All the circles above have the same radii, and share the same point.

Now, in this problem, the particular radius is 2, and the particular point is **(2, 1)**.

So all circles of radius 2 share the point **(2, 1)**.
That is, all circles of radius 2 meet altogether at **(2, 1)**.

In fact, infinitely many can be such, so the solution is a group of infinitely many circles. Thus, we want to produce an equation representing a group of circles.

So let's put them in an equation using the template: $(x - a)^2 + (y - b)^2 = c^2$.

First of all, we know that the radius is 2, so c is 2 in the template (standard) above.

So we get: $(x - a)^2 + (y - b)^2 = 2^2$, for now.

Next, the point **(2, 1)** is in the circle, so putting the point into the equation above, we get such an equation as follows: $(2 - a)^2 + (1 - b)^2 = 2^2$.   What equation is it, though?

Suppose now, C is the circle we are after. Then, the circle C is: $(x - a)^2 + (y - b)^2 = 2^2$.

And (a, b) is the center of the circle C, and is a point, too, of course.

Also, a and b are assumed to be constants that can take a pair of real numbers at a time.

So every point (a, b) satisfying the equation $(2 - a)^2 + (1 - b)^2 = 2^2$ can be the center of the circle C.

Then, the equation above is the connective expression between the coordinates of (a, b), which represents all points that can be the center of C. Where are all those points, then?

They are in a circle $(x - 2)^2 + (y - 1)^2 = 2^2$, and are all the points in the circle, so each of all the points in the circle is the center of the circle C. How come?

The connective expression $(2 - a)^2 + (1 - b)^2 = 2^2$ is the same as $(a - 2)^2 + (b - 1)^2 = 2^2$.

So taking for variables, a and b in the expression above, the expression can be taken for an equation of a circle in an a-b plane, where the a-axis is perpendicular to the b-axis as in the case where the x-y plane is set up.

Thus, the equation $(a - 2)^2 + (b - 1)^2 = 2^2$ indicates a circle where the center is the point $(2, 1)$ and the radius is 2 in the a-b plane. So (a, b) can be an arbitrary point in the circle.

Now, putting in the x-y plane, the circle $(a - 2)^2 + (b - 1)^2 = 2^2$, we just replace a with x, and b with y.
Then, we get: $(x - 2)^2 + (y - 1)^2 = 2^2$, which is the circle where every point is the center of the circle C.   How come?

Suppose now, D is the circle $(x - 2)^2 + (y - 1)^2 = 2^2$, where the center is a point $(2, 1)$.
Suppose also, A is a circle where the radius is 2 and the center is a point in the circle D.

Then, A passes through the center of D, which is the point $(2, 1)$.   How come?

The distance from every point in **D** to the center of **D** is 2 since the radius of **D** is 2.

The same is true for the circle **A**, too, and the center of **A** is in **D**. So the center of **D** is in **A**, and thus, **A** passes through the center of **D**, which is the point **(2, 1)**.

And the same is true for each of all the other points in the circle **D**, too.

So each point in the circle **D** can be a center of a circle of radius 2 centered at **(2, 1)**.

If two circles share the same radius, and one of the two passes through the center of the other, the center of the one is in the other, and the other passes through the center of the one, and has its center in the one. So both are in the same situation.

Fig. 1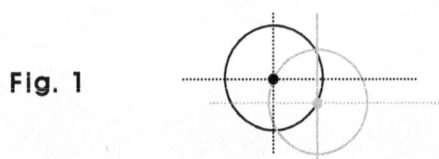

Thus, we get a group of circles, each of which passes through the point **(2, 1)**, and has a radius of 2, and a center that is one of all the points in the circle **D**.

So each point **(a, b)** satisfying the expression $(a - 2)^2 + (b - 1)^2 = 2^2$ is the center of each of all the circles in the group.

We know a circle of radius 2 centered at **(a, b)** is indicated by $(x - a)^2 + (y - b)^2 = 2^2$.
Therefore, the solution is every circle that satisfies the equation as follows:

$(x - a)^2 + (y - b)^2 = 2^2$ where *a* and *b* satisfy $(a - 2)^2 + (b - 1)^2 = 2^2$.

So "$(x - a)^2 + (y - b)^2 = 2^2$ where *a* and *b* satisfy $(a - 2)^2 + (b - 1)^2 = 2^2$." can represent a group of circles, and for each pair of values of *a* and *b*, $(x - a)^2 + (y - b)^2 = 2^2$ indicates a circle of radius 2 centered at **(a, b)**.

And putting in a graph some of the circles in the group, we can get:

Fig. 2

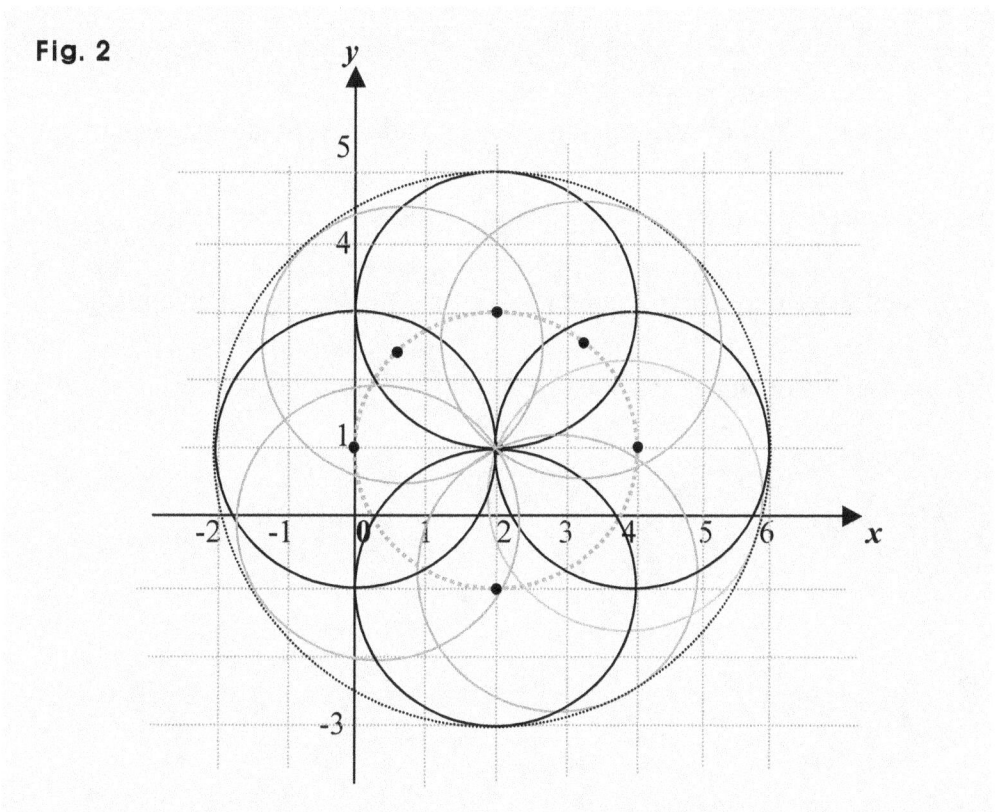

We can see all the circles above are within a circle of radius 4 centered at the point **(2, 1)**.

Also, $(a - 2)^2 + (b - 1)^2 = 2^2$ can be called the equation of the trace of the centers of all the circles in the group.

In short:

Since the radius is 2, we can put the circle in such an equation as follows:

$(x - a)^2 + (y - b)^2 = 2^2$ where **(a, b)** is the center.

Then, since the point **(2, 1)** is in the circle, we get:

$(2 - a)^2 + (1 - b)^2 = 2^2$, which is the same as $(a - 2)^2 + (b - 1)^2 = 2^2$.

Therefore, the solution to this problem is every circle that satisfies an equation below:

$(x - a)^2 + (y - b)^2 = 2^2$ where *a* and *b* satisfy $(a - 2)^2 + (b - 1)^2 = 2^2$.

Note:

A connective expression shows a relationship between variables. So does an equation.

However, some connective expressions are equations, and some others are not.

For instance, a connective expression can be $s^2 - 3s < t - 2$, which is not an equation.

Then, putting the expression in a graph, we can get the one shown below:

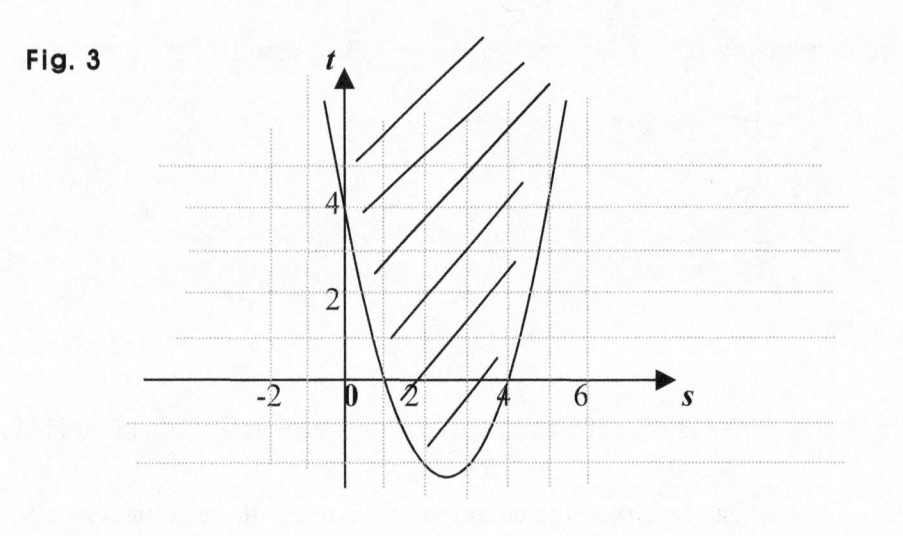

Fig. 3

So the expression, $s^2 - 3s < t - 2$ explains the relationship between the coordinates of each of all the points in the area above the parabola in the *s-t* plane above.

All the points in the parabola itself are not included in the area, of course.

Examples 4 in Circles

Find the circle in each case below assume the circle is in the *x-y* plane.

0. The circle has the center in a line $y = \frac{x}{2} + 1$, and is tangent to both coordinate axes.

1. The circle includes two points **(1, 2)** and **(2, -1)**, and the center is in a line $y = \frac{x}{2} + 1$.

2. Two points **(0, 3)** and **(2, -1)** are in the circle where the center is in a line $y = x$.

3. The circle has two points **(1, 4)** and **(3, 2)**, and the center is in a line $y = x + 1$.

Suggestions or Solutions
To the Problem in the Example 0

The circle has the center in a line $y = \frac{x}{2} + 1$, and is tangent to both coordinate axes.

If a circle is tangent to both of the axes, the center is the same distance away from the two axes. So the center is in either of the two lines $y = x$ and $y = -x$.

Suppose L is the line $y = \frac{x}{2} + 1$, and C is the circle to be found.

Then, since the center of C is in L, the center is either of two points where L meets either of the two lines above.

To begin with, finding the point where the line L, $y = \frac{x}{2} + 1$ meets the line $y = -x$, we get:
$\frac{x}{2} + 1 = -x \Rightarrow \frac{3x}{2} + 1 = 0 \Rightarrow x = -\frac{2}{3} \Rightarrow y = -x = -(-\frac{2}{3}) = \frac{2}{3}$.

So the center of C can be $(-\frac{2}{3}, \frac{2}{3})$, and thus, the radius can be $\frac{2}{3}$.

Therefore, the equation of C can be: $(x + \frac{2}{3})^2 + (y - \frac{2}{3})^2 = \frac{4}{9}$.

Next, finding the point where the line L meets the other line $y = x$, we get:
$\frac{x}{2} + 1 = x \Rightarrow \frac{x}{2} = 1 \Rightarrow x = 2 \Rightarrow y = x = 2$.

So the center of C can be **(2, 2)**, and thus, the radius can be 2.

Thus, the equation of C can be: $(x - 2)^2 + (y - 2)^2 = 2^2$.

Therefore, the circle C is: $(x + \frac{2}{3})^2 + (y - \frac{2}{3})^2 = \frac{4}{9}$ or $(x - 2)^2 + (y - 2)^2 = 2^2$.

If not quite sure of the idea behind the processes above, follow the steps below:

Let's begin with putting in a graph some circles tangent to both coordinate axes.
Then, we can see better where the solution is around.

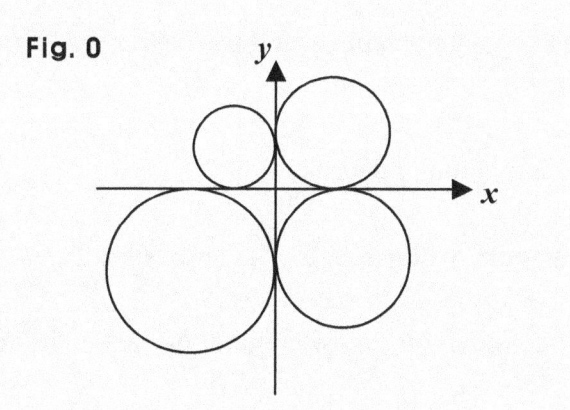

Fig. 0

If a circle is tangent to both of the axes, the center of the circle is the same distance away from the two axes respectively.   Then, where does the center have to be?

Since both of the axes are the same distance away from the center, the magnitudes of the coordinates of the center are the same as each other. So at the center, the magnitude of the x-coordinate equals that of the y-coordinate.

In short, $|y| = |x|$.   In other words, we get $y = x$ or $y = -x$.

Therefore, the center of the circle can be in either of the two lines $y = x$ and $y = -x$. So let's put the two lines in the graph, too.

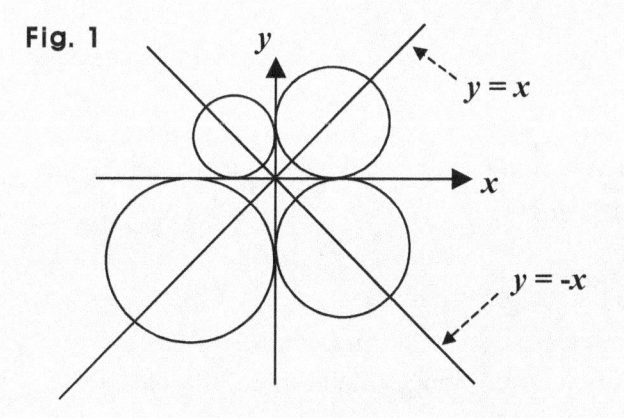

Fig. 1

Suppose now, C is the circle we are after in this problem, and L is the line $y = \frac{x}{2} + 1$.

Then, the circle **C** is either of the four circles in the graph above, and has the center in either of the two lines $y = x$ and $y = -x$.

And also, the problem says that the line **L** passes through the center of **C**.

Thus, we can see that the line **L** meets either of the two lines at the center of **C**.

So let's add the line **L** to the graph above, and see how **L** can meet either of the two lines.

Since the y-intercept of the line **L** is positive, **L** cannot pass through the centers of the two circles below the x-axis. So we want to consider the two circles above the x-axis.

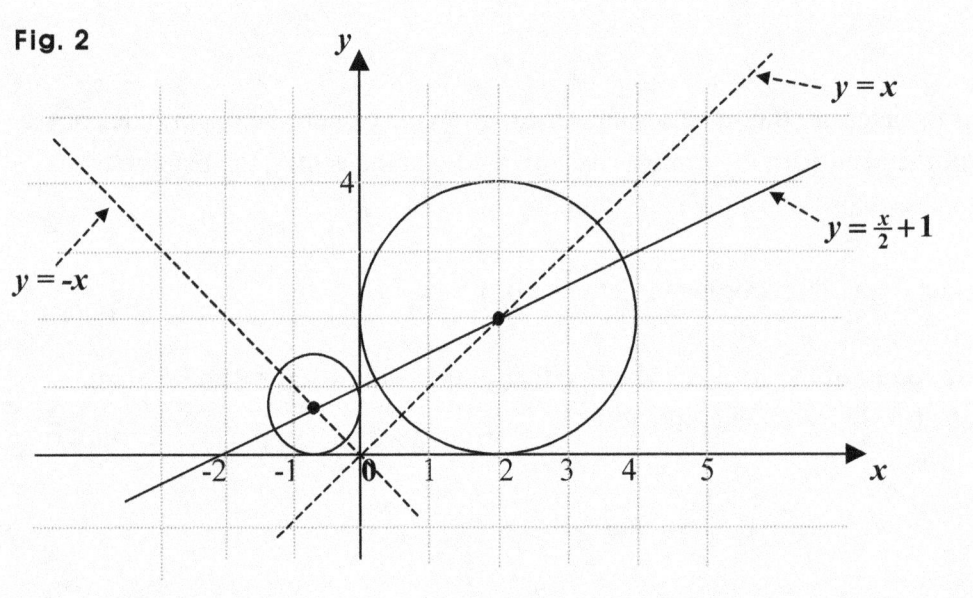

Fig. 2

Then, we can see that the line **L** meets the two lines $y = x$ and $y = -x$ at two points, and that the two points can be the center of the circle **C**.

That is to say that the two circles above can be the circle **C**.

In sum, **L** passes through two centers of two circles tangent to both coordinate axes.

Specifically:

In one circle, the center is at the point where the line L meets the line $y = -x$.

In the other, the center is at the point where the line L meets the other line $y = x$.

So let's now find the two points where the line L meets the two lines.

To begin with, let's find the point where the line L, $y = \frac{x}{2} + 1$ meets the line $y = -x$.

Then, the circle C will be the smaller of the two in the graph above, and we get:

$$\frac{x}{2} + 1 = -x \Rightarrow \frac{3x}{2} + 1 = 0 \Rightarrow x = -\frac{2}{3} \Rightarrow y = -x = -(-\frac{2}{3}) = \frac{2}{3}.$$

So the center of the circle C can be at $(-\frac{2}{3}, \frac{2}{3})$, and thus, its radius is $\frac{2}{3}$.
How come the radius is $\frac{2}{3}$, though?

Since the circle C is tangent to both axes, the radius of C is the same as the magnitude of either of the two coordinates at the center, which is the point where the line L meets the line $y = -x$.   How come either of the two?

The center is a point in the line $y = -x$, where every point has a pair of coordinates of the same magnitude. For instance, **(1, -1)**, **(-1, 1)**, **(3, -3)**, **(-5, 5)**, etc. are in the line $y = -x$.

Therefore, the equation of the circle C can be: $\{x - (-\frac{2}{3})\}^2 + (y - \frac{2}{3})^2 = (\frac{2}{3})^2$.

Next, let's find the point where the line L: $y = \frac{x}{2} + 1$ meets the other line $y = x$.

Then, the circle C will be the larger of the two circles in the graph above, and we get:

$$\frac{x}{2} + 1 = x \Rightarrow \frac{x}{2} = 1 \Rightarrow x = 2 \Rightarrow y = x = 2.$$

Thus, the center of C can be **(2, 2)**, and thus, its radius can be 2.

So the equation of the circle C can be: $(x-2)^2 + (y-2)^2 = 2^2$.

Therefore, the solution is: $(x+\frac{2}{3})^2 + (y-\frac{2}{3})^2 = \frac{4}{9}$ or $(x-2)^2 + (y-2)^2 = 2^2$.

In short:

I If a circle is tangent to both of the axes, the center is the same distance away from the two axes. So the center is in either of the two lines $y = x$ and $y = -x$.

Suppose L is the line $y = \frac{x}{2} + 1$, and C is the circle to be found.

Then, since the center of C is in L, the center is either of two points where L meets either of the two lines above.

To begin with, finding the point where the line L, $y = \frac{x}{2} + 1$ meets the line $y = -x$, we get:

$\frac{x}{2} + 1 = -x \Rightarrow \frac{3x}{2} + 1 = 0 \Rightarrow x = -\frac{2}{3} \Rightarrow y = -x = -(-\frac{2}{3}) = \frac{2}{3}$.

So the center of C can be $(-\frac{2}{3}, \frac{2}{3})$, and thus, the radius can be $\frac{2}{3}$.

Therefore, the equation of C can be: $(x+\frac{2}{3})^2 + (y-\frac{2}{3})^2 = \frac{4}{9}$.

Next, finding the point where the line L meets the other line $y = x$, we get:

$\frac{x}{2} + 1 = x \Rightarrow \frac{x}{2} = 1 \Rightarrow x = 2 \Rightarrow y = x = 2$.

So the center of C can be **(2, 2)**, and thus, the radius can be 2.

Thus, the equation of C can be: $(x-2)^2 + (y-2)^2 = 2^2$.

Therefore, the circle C is: $(x+\frac{2}{3})^2 + (y-\frac{2}{3})^2 = \frac{4}{9}$ or $(x-2)^2 + (y-2)^2 = 2^2$.

Suggestions or Solutions
To the Problem in the Example 1

The circle includes two points (1, 2) and (2, -1), and the center is in a line $y = \frac{x}{2} + 1$.

Suppose C is the circle to be found, (s, t) is the center, and D is the distance from the center of C to each of the two points given.

Then, we get: $D^2 = (s - 1)^2 + (t - 2)^2$, and $D^2 = (s - 2)^2 + \{t - (-1)\}^2 = (s - 2)^2 + (t + 1)^2$.
So we get: $(s - 1)^2 + (t - 2)^2 = (s - 2)^2 + (t + 1)^2$.
Thus, we get: $(s - 1)^2 - (s - 2)^2 = (t + 1)^2 - (t - 2)^2$
$\Rightarrow \{(s - 1) + (s - 2)\}\{(s - 1) - (s - 2)\} = \{(t + 1) + (t - 2)\}\{(t + 1) - (t - 2)\}$
$\Rightarrow (2s - 3)1 = 2s - 3 = (2t - 1)3 = 6t - 3 \Rightarrow 2s = 6t \Rightarrow s = 3t$.

We know that the center (s, t) is in the given line $y = \frac{x}{2} + 1$. So we get: $t = \frac{s}{2} + 1$.

Thus, we get: $s = 3t \Rightarrow t = \frac{s}{2} + 1 = \frac{3t}{2} + 1 \Rightarrow t = \frac{3t}{2} + 1 \Rightarrow \frac{t}{2} = -1 \Rightarrow t = -2 \Rightarrow s = -6$.

Therefore, the center of C is **(-6, -2)**.

Since a radius is the distance from the center to a point in a circle, the radius is D.
$D^2 = (s - 1)^2 + (t - 2)^2 = (-6 - 1)^2 + (-2 - 2)^2 = 49 + 16 = 65$.

Therefore, the circle C is: $(x + 6)^2 + (y + 2)^2 = 65$.

If not quite sure of the idea behind the processes above, follow the steps below:

Basically, finding a circle, we need the center and radius. Then, using the standard form, we can get the circle.   What about the general form, then?

Using the general form, we basically need three points in the circle. We are given only two points only though. Thus, we may want to try the standard form.

So let's get the center and radius.   Then to begin with, where is the radius?

Every point in a circle is the radius away from the center.
So the radius is between the center and either of the two points **(1, 2)** and **(2, -1)**.
We have only the whereabouts of the center, though, which is somewhere in the
line $y = \frac{x}{2} + 1$.   Where then, exactly is it?

Let's put in a graph the line and the two points, together with some candidates for the
circle we are after. Such a candidate doesn't have to be a complete circle. Just some part
of a circle can help, too. Then, we can see better where the solution can be around.

Fig. 0

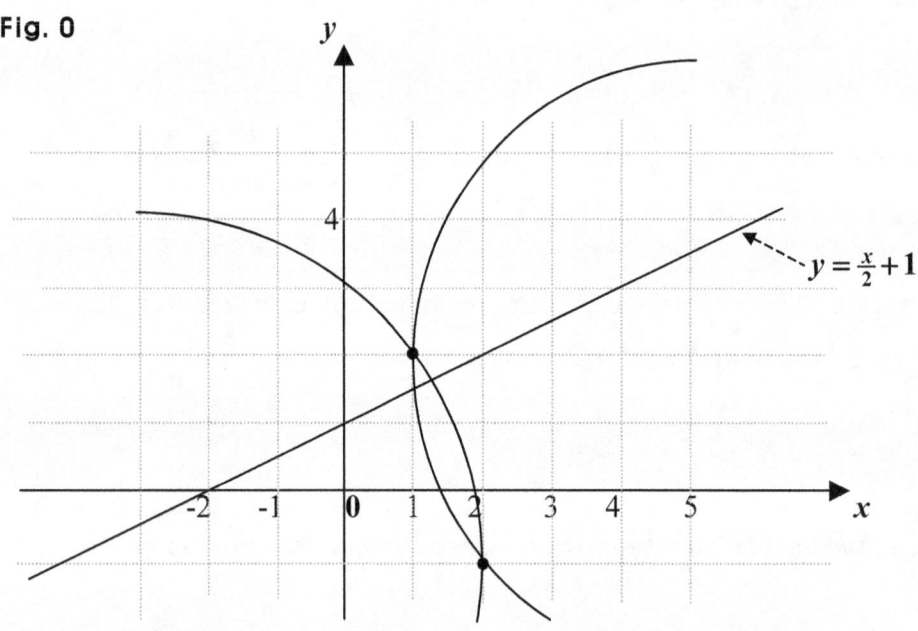

Now, whereabouts does the center of the circle have to be?

It seems to be somewhere in the third quadrant, and doesn't seem to be in the first.
Anyway, it's not quite easy to see it. Then, what should we do?

Getting stuck, we may want to get back to the basics, which is the definition for circles.

The definition says:
- Each of all the points in a circle is the same distance away from the center.
- The same distance is the radius, of course.

So each of the two points given in the problem needs to be the same distance away from the center, too. What then, do we mean by the same distance?

It means the radius, of course.   What else can we think of, though?

The distance from the center to one of the two points is the same as that from the center to the other point.   So what?

We can find the center by means of the famous distance formula.

Suppose now, C is the circle we are after, (s, t) is the center of C, and D is the distance from the center to either of the two points given.

Then, beginning with the distance between the center and the point $(1, 2)$, we get:

$$D^2 = (s - 1)^2 + (t - 2)^2.$$

Next is the distance between the center and the point $(2, -1)$.

$$D^2 = (s - 2)^2 + \{t - (-1)\}^2 = (s - 2)^2 + (t + 1)^2.$$

Now, we know the two distances above have to be the same.

So we get: $(s - 1)^2 + (t - 2)^2 = (s - 2)^2 + (t + 1)^2.$

Meanwhile, we have a factorization identity, $A^2 - B^2 = (A + B)(A - B).$

So taking advantage of it, we get:

$(s-1)^2 - (s-2)^2 = (t+1)^2 - (t-2)^2$

$\Rightarrow \{(s-1)+(s-2)\}\{(s-1)-(s-2)\} = \{(t+1)+(t-2)\}\{(t+1)-(t-2)\}$

$\Rightarrow (2s-3)1 = 2s-3 = (2t-1)3 = 6t-3 \Rightarrow 2s = 6t \Rightarrow s = 3t.$

Now, we have one equation, but we have two unknowns.   What then, about the other equation?

The problem says that the center is in the line $y = \frac{x}{2}+1$.

That is, the center **(s, t)** is in the line **L**, $y = \frac{x}{2}+1$.   So we get: $t = \frac{s}{2}+1.$

So we get: $s = 3t \Rightarrow t = \frac{s}{2}+1 = \frac{3t}{2}+1 \Rightarrow t = \frac{3t}{2}+1 \Rightarrow \frac{t}{2} = -1 \Rightarrow t = -2 \Rightarrow s = -6.$

Therefore, the center is **(-6, -2)**, which is in the third quadrant.

Now, what's left is the radius.

Since a radius is the distance from the center to a point in a circle, the radius is **D**. So we can make use of D^2, which has been set above, and is as follows:

$D^2 = (s-1)^2 + (t-2)^2 = (-6-1)^2 + (-2-2)^2 = 49 + 16 = 65.$

Now, we have the radius squared, and the center.
So we can use the standard form, and put the circle **C** in such a way as follows:

$(x+6)^2 + (y+2)^2 = 65.$

By the way, we can find the center a bit differently, too.

To begin with, the center **(s, t)** is in the line $y = \frac{x}{2}+1$, so we get: $t = \frac{s}{2}+1$.

Therefore, the center can be put this way: $(s, \frac{s}{2}+1)$.

Next, the distance **D** from the center to **(1, 2)** is the same as that from the center to **(2, -1)**.

So we get: $D^2 = (s-1)^2 + (\frac{s}{2}+1-2)^2 = (s-1)^2 + (\frac{s}{2}-1)^2$, and also:

$D^2 = (s-2)^2 + \{\frac{s}{2}+1-(-1)\}^2 = (s-2)^2 + (\frac{s}{2}+2)^2.$

Therefore, we get: $(s-1)^2 + (\frac{s}{2}-1)^2 = (s-2)^2 + (\frac{s}{2}+2)^2$.

Then again, we get to use the identity, $A^2 - B^2 = (A+B)(A-B)$.
(The identity above is used quite frequently, and often convenient, so we may want to put it in our memory. We should be able to know how to get it, too, though.)

So we get: $(s-1)^2 - (s-2)^2 + (\frac{s}{2}-1)^2 - (\frac{s}{2}+2)^2 = 0$

$\Rightarrow \{(s-1)+(s-2)\}\{(s-1)-(s-2)\} + \{(\frac{s}{2}-1)+(\frac{s}{2}+2)\}\{(\frac{s}{2}-1)-(\frac{s}{2}+2)\} = 0$

$\Rightarrow (2s-3)1 + (s+1)(-3) = 2s - 3 - 3s - 3 = -s - 6 = 0 \Rightarrow s = -6$.

Thus, we get: $s = -6 \Rightarrow t = \frac{s}{2} + 1 = -\frac{6}{2} + 1 = -2$.

Therefore, the center is **(-6, -2)**.

In short:

Suppose C is the circle to be found, (s, t) is the center, and D is the distance from the center of C to each of the two points given.

Then, we get: $D^2 = (s-1)^2 + (t-2)^2$, and $D^2 = (s-2)^2 + \{t-(-1)\}^2 = (s-2)^2 + (t+1)^2$.

So we get: $(s-1)^2 + (t-2)^2 = (s-2)^2 + (t+1)^2$.

Thus, we get: $(s-1)^2 - (s-2)^2 = (t+1)^2 - (t-2)^2$
$\Rightarrow \{(s-1)+(s-2)\}\{(s-1)-(s-2)\} = \{(t+1)+(t-2)\}\{(t+1)-(t-2)\}$
$\Rightarrow (2s-3)1 = 2s - 3 = (2t-1)3 = 6t - 3 \Rightarrow 2s = 6t \Rightarrow s = 3t$.

We know that the center (s, t) is in the given line $y = \frac{x}{2} + 1$.   So we get: $t = \frac{s}{2} + 1$.

Thus, we get: $s = 3t \Rightarrow t = \frac{s}{2} + 1 = \frac{3t}{2} + 1 \Rightarrow t = \frac{3t}{2} + 1 \Rightarrow \frac{t}{2} = -1 \Rightarrow t = -2 \Rightarrow s = -6$.

Therefore, the center of C is **(-6, -2)**.

Since a radius is the distance from the center to a point in a circle, the radius is D.
$D^2 = (s-1)^2 + (t-2)^2 = (-6-1)^2 + (-2-2)^2 = 49 + 16 = 65$.

Therefore, the circle C is: $(x+6)^2 + (y+2)^2 = 65$.

Suggestions or Solutions
To the Problem in the Example 2

Two points (0, 3) and (2, -1) are in the circle where the center is in a line $y = x$.

Suppose the radius is **R**, and since the center is in the line $y = x$, the center is **(a, a)**.

Then, $R^2 = (a - 0)^2 + (a - 3)^2$, and also, $R^2 = (a - 2)^2 + (a + 1)^2$.

So we get:

$a^2 + (a - 3)^2 = (a - 2)^2 + (a + 1)^2 \Rightarrow a^2 - (a - 2)^2 + (a - 3)^2 - (a + 1)^2 = 0$

$\Rightarrow \{a + (a - 2)\}\{a - (a - 2)\} + \{(a - 3) + (a + 1)\}\{(a - 3) - (a + 1)\}$

$= 2(2a - 2) + (-4)(2a - 2) = 4a - 4 - 8a + 8 = -4a + 4 = 0 \Rightarrow a = 1.$

Thus, $R^2 = (a - 2)^2 + (a + 1)^2 = (1 - 2)^2 + (1 + 1)^2 = 1 + 4 = 5.$

Therefore, the circle is: $(x - 1)^2 + (y - 1)^2 = 5.$

If not quite sure of the idea behind the processes above, follow the steps below:

Putting the problem in a graph, we can see better the solution's whereabouts, and reduce mistakes capturing the solution. So let's begin with putting in a graph the two points and the line given.

Fig. 0

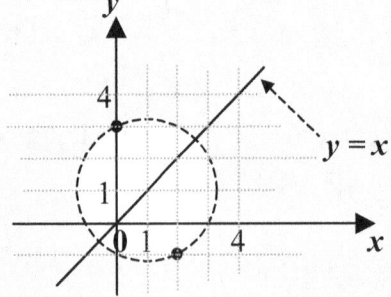

Then, even looking at the graph only, we can readily see that the center of the circle we are after is **(1, 1)**.

Besides, by the distance formula, we can quickly see that (**the radius**)$^2 = 1^2 + 2^2 = 5$.

Giving a name to an object in math, we can work with it conveniently.
Thus next, assuming C is the circle to be found, by means of the standard form, we can put the circle C in an equation where $(x - 1)^2 + (y - 1)^2 = 5$.

So if the problem is multiple-choice, we can get to the solution fast by simply putting the problem in a graph. What if the equation has to be in the general form, though?

Expanding the one in the standard form and simplifying the result, we can readily get it.

What if the problem is asking not only the circle but the procedure to get it, too, though?

Then. we need to show that the center is **(1, 1)**, and how to get the radius, too.

So let's see now, how to get the center and the radius analytically. What do we mean by 'analytically'?

We mean 'by calculations' if you will. That is, we get the solution by algebra.

We know all the points in a circle are respectively the same distance, that is, the radius away from the center, so the two points given have to be the radius away from the center, also.   What then, do we want to take advantage of?

The distance formula, called Pythagorean theorem, can help us find the center and radius. So we can find them using the formula, together with the two points given.

Suppose that the radius of the circle C is R, and that since the center is in the line $y = x$, the center is (a, a), where a is constant, of course.

Then, beginning with the distance from (a, a) to $(0, 3)$, we get: $R^2 = (a - 0)^2 + (a - 3)^2$.

Next, we have another point $(2, -1)$ in the circle C, so we get: $R^2 = (a - 2)^2 + (a + 1)^2$.

Thus, we get: $a^2 + (a - 3)^2 = (a - 2)^2 + (a + 1)^2$.

Meanwhile, we have a factorization identity, where $A^2 - B^2 = (A + B)(A - B)$.

So we get:

$a^2 + (a - 3)^2 = (a - 2)^2 + (a + 1)^2 \Rightarrow a^2 - (a - 2)^2 + (a - 3)^2 - (a + 1)^2 = 0$

$\Rightarrow \{a + (a - 2)\}\{a - (a - 2)\} + \{(a - 3) + (a + 1)\}\{(a - 3) - (a + 1)\}$

$= 2(2a - 2) + (-4)(2a - 2) = 4a - 4 - 8a + 8 = -4a + 4 = 0 \Rightarrow a = 1$.

Next, using one of the two expressions for R above, we get the radius as follows:

$R^2 = (a - 2)^2 + (a + 1)^2 = (1 - 2)^2 + (1 + 1)^2 = 1 + 4 = 5$.

So the center is $(1, 1)$, and the radius is 5.

Therefore, the circle C is: $(x - 1)^2 + (y - 1)^2 = 5$.

In short:

Suppose the radius is R, and since the center is in the line $y = x$, the center is (a, a).

Then, $R^2 = (a - 0)^2 + (a - 3)^2$, and also, $R^2 = (a - 2)^2 + (a + 1)^2$.

So we get:

$a^2 + (a - 3)^2 = (a - 2)^2 + (a + 1)^2 \Rightarrow a^2 - (a - 2)^2 + (a - 3)^2 - (a + 1)^2 = 0$

$\Rightarrow \{a + (a - 2)\}\{a - (a - 2)\} + \{(a - 3) + (a + 1)\}\{(a - 3) - (a + 1)\}$

$= 2(2a - 2) + (-4)(2a - 2) = 4a - 4 - 8a + 8 = -4a + 4 = 0 \Rightarrow a = 1$.

Thus, $R^2 = (a - 2)^2 + (a + 1)^2 = (1 - 2)^2 + (1 + 1)^2 = 1 + 4 = 5$.

Therefore, the circle is: $(x - 1)^2 + (y - 1)^2 = 5$.

Suggestions or Solutions
To the Problem in the Example 3

The center of the circle is in a line $y = x + 1$, and (1, 4) and (3, 2) are in the circle.

Suppose that the radius is **R**, and that the center is **(a, b)** where **a** and **b** are constant. Then, $R^2 = (a-1)^2 + (b-4)^2$, and also, $R^2 = (a-3)^2 + (b-2)^2$. So we get:

$$(a-1)^2 + (b-4)^2 = (a-3)^2 + (b-2)^2 \Rightarrow (a-1)^2 - (a-3)^2 + (b-4)^2 - (b-2)^2 = 0$$

$$\Rightarrow \{(a-1) + (a-3)\}\{(a-1) - (a-3)\} + \{(b-4) + (b-2)\}\{(b-4) - (b-2)\}$$

$$= 2(2a-4) - 2(2b-6) = 4a - 8 - 4b + 12 = 4a - 4b + 4 = 0 \Rightarrow a - b + 1 = 0.$$

Besides, the center is in the line $y = x + 1$, so we get: **b = a + 1**, which however, is the same as $a - b + 1 = 0$. Therefore, each and every point in the line given can be the center, and all the circles share the two given points. The center is **(a, b)**, and we have: **b = a + 1**.

So the center can be put in **(a, a + 1)** where **a** is constant. Thus, the radius is as follows.

$$R^2 = (a-3)^2 + (b-2)^2 = (a-3)^2 + (a+1-2)^2 = (a-3)^2 + (a-1)^2 = 2a^2 - 8a + 10$$

$$= 2(a^2 - 4a) + 10 = 2(a^2 - 4a + 4 - 4) + 10 = 2(a-2)^2 + 2.$$

Therefore, the circle is: $(x-a)^2 + (y-a-1)^2 = 2(a-2)^2 + 2$ where **a** is a constant.

If not quite sure of the idea behind the processes above, follow the steps below:

Let's put in a graph the given line and the two points given.
Then, we can see better where the solution has to be around.

Fig. 0

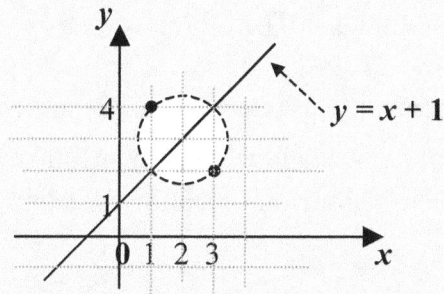

Now, in this example, too, even merely seeing the graph only, we can see that the center can be **(2, 3)**, and that the radius can be $\sqrt{2}$, so the circle can be: $(x - 2)^2 + (y - 3)^2 = 2$.

It can be the circle we want, of course. Is that it, though?

It is not the only circle that can be the solution. That is, another circle can be centered at a point in the line and passing through the two points. In fact, we can get infinitely many of such circles. So this time, some analytic approach to the center and radius is needed.

• Finding the center and radius, we can put the circle in the standard form.

• All the points in a circle are respectively the same distance, that is, the radius away from the center. So the two points given are respectively the radius away from the center.

• We can find the distance between two points by means of the distance formula.

Suppose now, **C** is the circle to be found, the radius is **R**, and the center is **(a, b)**.

Then, beginning with the distance from **(a, b)** to **(1, 4)**, we get: $R^2 = (a - 1)^2 + (b - 4)^2$.

Next, we have another point **(3, 2)** in the circle **C**, so we get: $R^2 = (a - 3)^2 + (b - 2)^2$.

Meanwhile, we have: $A^2 - B^2 = (A + B)(A - B)$. So we get:

$(a - 1)^2 + (b - 4)^2 = (a - 3)^2 + (b - 2)^2 \Rightarrow (a - 1)^2 - (a - 3)^2 + (b - 4)^2 - (b - 2)^2 = 0$

$\Rightarrow \{(a - 1) + (a - 3)\}\{(a - 1) - (a - 3)\} + \{(b - 4) + (b - 2)\}\{(b - 4) - (b - 2)\}$

$= 2(2a - 4) - 2(2b - 6) = 4a - 8 - 4b + 12 = 4a - 4b + 4 = 0 \Rightarrow a - b + 1 = 0$.

Also, the center **(a, b)** is in the line $y = x + 1$. Therefore, we get: $b = a + 1$.

So we now have $a - b + 1 = 0$ and $b = a + 1$, which however, are the same as each other. Finding **a** and **b**, we need a system of two equations for **a** and **b**, and the two equations have to be different, of course. We have only one equation for **a** and **b**, though. Where then, is the other?

It doesn't exist, which means, there can be many circles, each of which can be the circle *C*. There are in fact, infinitely many of such circles.   How come?

We have ended up with only one equation $b = a + 1$, so the solution to that equation is any pair of values can satisfy "$b = a + 1$."

For instance, $(a, b) = (1, 2), (1.1, 2.1), (-2, -1)$, etc.

So there can be infinitely many values can be assigned to *a* and *b*. In fact, each and every point in the line $y = x + 1$ is the center of the circle *C*.   How come?

The point (a, b) is the center of *C*, *a* is the *x*-coordinate, and *b* is the *y*-coordinate.

The solution to the equation $b = a + 1$ is the same as the one to the equation $y = x + 1$, which indicates a line, so the coordinates of every point in the line satisfy the equation where $y = x + 1$.

So the solution is not a particular circle but a group of infinitely many circles, and every point in the line $y = x + 1$ is the center of each circle in the group.

Of course, each of all the circles passes through the two given points, **(1, 4)** and **(3, 2)**.

So all the circles meet each other altogether at the two given points, after all.

That is to say that all the circles share altogether the two points given.

Now, how do we put in an equation the group of such infinitely many circles?

We first, get the radius, and then, use the standard form.   What then, about the center?

The center has already been chosen to be (a, b), so we have: $b = a + 1$ since the center is in the line $y = x + 1$.   Thus, the center can also, be put in $(a, a + 1)$ where *a* is constant.

Now, since the point **(3, 2)** is in the circle *C*, finding the distance from the center to the point **(3, 2)**, we get the radius.

To begin with, we get:

$$R^2 = (a - 3)^2 + (b - 2)^2 = (a - 3)^2 + (a + 1 - 2)^2 \text{ since } b = a + 1.$$

So we get:

$$R^2 = (a - 3)^2 + (a - 1)^2 = 2a^2 - 8a + 10 = 2(a^2 - 4a) + 10$$
$$= 2(a^2 - 4a + 4 - 4) + 10 = 2(a - 2)^2 + 2.$$

Next, using the standard form, we get: $(x - a)^2 + \{y - (a + 1)\}^2 = 2(a - 2)^2 + 2$ where a is constant. So the circle C is a circle indicated by the equation below for each value of a:

$$(x - a)^2 + (y - a - 1)^2 = 2(a - 2)^2 + 2 \text{ where } a \text{ is constant.}$$

For each value of a, we get one circle by means of the equation above, and the circle passes through the two points $(1, 4)$ and $(3, 2)$, and is centered at a point in the line.

So the equation above represents a group of circles passing through the two points.

We can notice in this equation, that when $a = 2$, the radius is the smallest, so the circle is the smallest when $a = 2$. Thus, the smallest circle is centered at $(2, 3)$, and has a radius of $\sqrt{2}$. Let's now, put in a graph some of the circles in the group.

Fig. 1

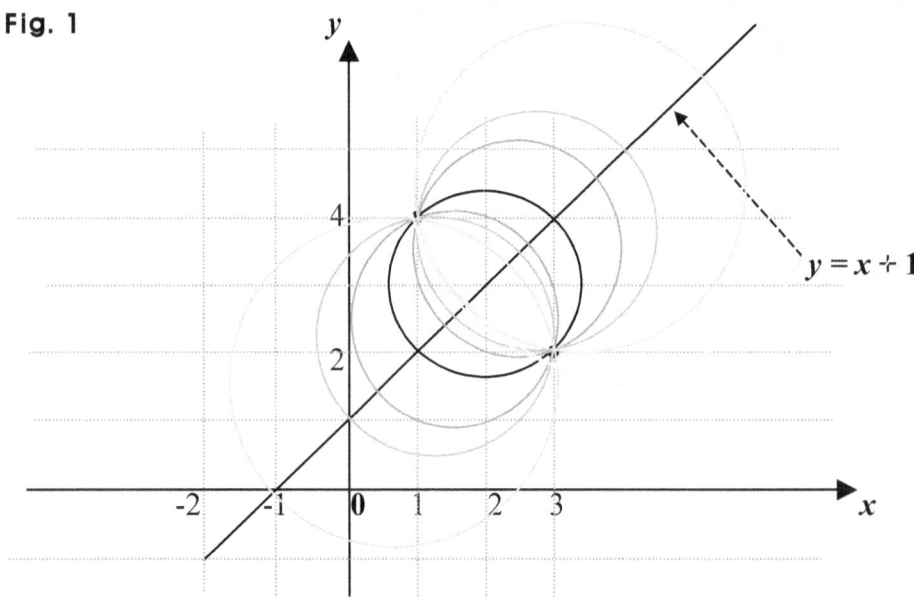

Examples 5 in Circles

Find the circle in each case below assume the circle is in the x-y plane.

0. The circle passes through **(1, 4)** and **(3, 2)**, and the center is in a line $x = 0$.

1. Two points **(1, 3)** and **(2, -1)** are in the circle where the center is in a line $y = 0$.

2. The center of the circle is in a line $y = 1 - x$, and a point **(1, 3)** is in the circle.

3. The circle has **(1, 3)**, and is tangent to the x-axis, and the center is in a line $y = 1 - x$.

4. The circle has **(1, 3)**, and is tangent to the y-axis, and the center is in a line $y = 1 - x$.

Suggestions or Solutions
To the Problem in the Example 0

The circle passes through (1, 4) and (3, 2), and the center is in a line $x = 0$.

Suppose that **R** is the radius, and that the center is **(0, b)**, where **b** is constant since the center is in the *y*-axis.

Then, $R^2 = (0-1)^2 + (b-3)^2$, and also, $R^2 = (0-2)^2 + (b+1)^2$. So we get:

$(0-1)^2 + (b-3)^2 = (0-2)^2 + (b+1)^2 \Rightarrow 1 + (b-3)^2 = 4 + (b+1)^2$

$\Rightarrow (b-3)^2 - (b+1)^2 - 3 = 0 \Rightarrow b^2 - 6b + 9 - b^2 - 2b - 1 - 3 = -8b + 5 = 0 \Rightarrow b = \frac{5}{8}$.

Thus, the center is $(0, \frac{5}{8})$, and we get:

$R^2 = 4 + (b+1)^2 = 4 + (\frac{5}{8}+1)^2 = 4 + (\frac{13}{8})^2 = 4 + \frac{169}{64} = \frac{256+169}{64} = \frac{425}{64}$.

Therefore, the circle is: $x^2 + (y - \frac{5}{8})^2 = (\frac{5\sqrt{17}}{8})^2$.

If not quite sure of the idea behind the processes above, follow the steps below:

Putting in a graph the line and two points given, together with a probable circle, we can see better where the solution can be around. The line $x = 0$ is the *y*-axis. It's quite easy to put a probable circle in the graph since the center of the circle to be found is in the *y*-axis.

Fig. 0

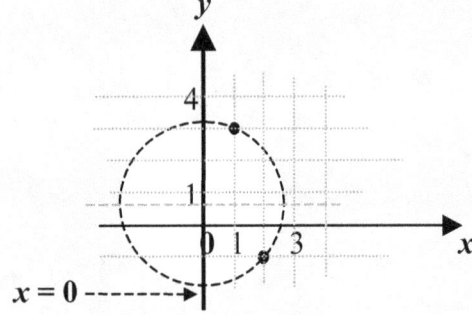

$x = 0$

Even looking at the graph only, we can see that the center should be somewhere between 0 and 1 in the *y*-axis. So the *x*-coordinate at the center is **0**, and **0** < the *y*-coordinate < **1**.

Let's now, begin with finding precisely where the center is.

The definition for circles says the two points **(1, 3)** and **(2, -1)** are respectively the same distance away from the center, and the same distance is the radius.

Suppose now, that *C* is the circle we are after, that *b* is constant, that the center of the circle *C* is **(0, *b*)** since it is in the *y*-axis, and that *R* is the radius.

Then, beginning with the distance from **(0, *b*)** to **(1, 3)**, we get: $R^2 = (0-1)^2 + (b-3)^2$.

Next, we have another point **(2, -1)** in the circle *C*, so we get: $R^2 = (0-2)^2 + (b+1)^2$.

Thus, we get: $(0-1)^2 + (b-3)^2 = (0-2)^2 + (b+1)^2 \Rightarrow 1 + (b-3)^2 = 4 + (b+1)^2$

$\Rightarrow (b-3)^2 - (b+1)^2 - 3 = 0 \Rightarrow b^2 - 6b + 9 - b^2 - 2b - 1 - 3 = -8b + 5 = 0 \Rightarrow b = \frac{5}{8}$.

Therefore, the center of the circle *C* is $(0, \frac{5}{8})$.

Next, since we now have the value of *b*, the radius *R* is as follows:

$R^2 = 4 + (b+1)^2 = 4 + (\frac{5}{8}+1)^2 = 4 + (\frac{13}{8})^2 = 4 + \frac{169}{64} = \frac{256+169}{64} = \frac{425}{64}$.

Therefore, using the standard form, we can see that the circle *C* is as follows:

$x^2 + (y-\frac{5}{8})^2 = \frac{425}{64} = (\frac{5\sqrt{17}}{8})^2$.

Suggestions or Solutions
To the Problem in the Example 1

Two points (1, 3) and (2, -1) are in the circle where the center is in a line $y = 0$.

Suppose R is the radius, and the center is $(a, 0)$ where a is constant.

Then, $R^2 = (a-1)^2 + (0-3)^2$, and also, $R^2 = (a-2)^2 + (0+1)^2$.

So we get: $(a-1)^2 + (0-3)^2 = (a-2)^2 + (0+1)^2 \Rightarrow 9 + (a-1)^2 = 1 + (a-2)^2$

$\Rightarrow (a-1)^2 - (a-2)^2 + 8 = 0 \Rightarrow a^2 - 2a + 1 - a^2 + 4a - 4 + 8 = 2a + 5 = 0 \Rightarrow a = -\frac{5}{2}$.

Thus, the center is $(-\frac{5}{2}, 0)$, so we get:

$R^2 = (a-2)^2 + (0+1)^2 = (-\frac{5}{2} - 2)^2 + 1 = (-\frac{9}{2})^2 + 1 = \frac{81}{4} + 1 = \frac{85}{4}$.

Therefore, the circle is: $(x + \frac{5}{2})^2 + y^2 = \frac{85}{4}$.

If not quite sure of the idea behind the processes above, follow the steps below:

Making the problem visible, we can see better the solution's whereabouts. So we may want to begin with putting in a graph the line and two points given, together with some probable circles. The line is: $y = 0$, which is the x-axis itself, so the center is in the x-axis.

Fig. 0

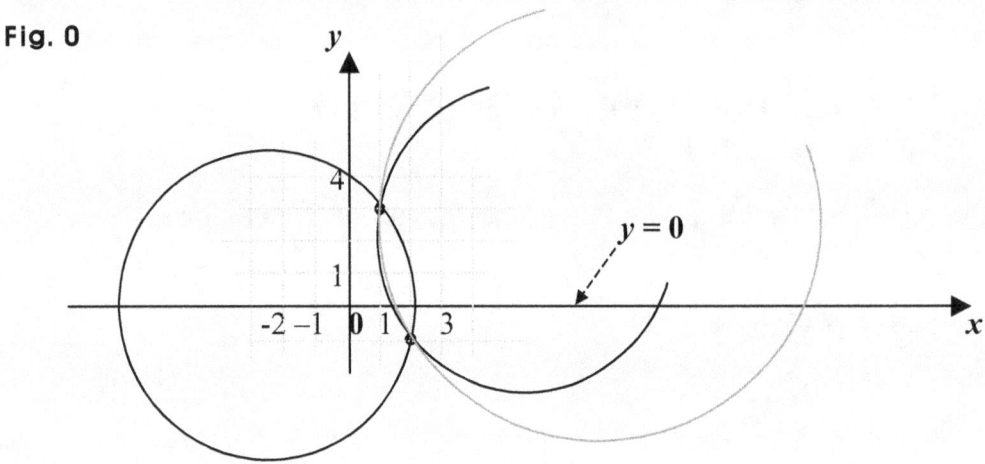

Looking at the graph, we can see that the center should be on the left of the origin in the *x*-axis, and quite clearly see that the center cannot be on the right of the origin.

It looks like the center should be somewhere between -3 and -2 in the *x*-axis.

So let's now find exactly where the center is positioned.

Suppose that *C* is the circle we want, that *R* is the radius of the circle *C*, and that since the center is in the *x*-axis, the center is *(a, 0)* where *a* is constant.

By the definition for circles, the two points **(1, 3)** and **(2, -1)** are respectively the radius away from the center.

So beginning with the distance from the center to **(1, 3)**, we get: $R^2 = (a-1)^2 + (0-3)^2$, and thus, finding *a*, we can get the radius *R*, too.

Next, we have another point **(2, -1)** in the circle *C*, so we get: $R^2 = (a-2)^2 + (0+1)^2$.

Thus, we get: $(a-1)^2 + (0-3)^2 = (a-2)^2 + (0+1)^2 \Rightarrow 9 + (a-1)^2 = 1 + (a-2)^2$

$\Rightarrow (a-1)^2 - (a-2)^2 + 8 = 0 \Rightarrow a^2 - 2a + 1 - a^2 + 4a - 4 + 8 = 2a + 5 = 0 \Rightarrow a = -\frac{5}{2}$.

Therefore, the center of the circle *C* is $(-\frac{5}{2}, 0)$.

Next, now that we have found *a*, the radius *R* is as follows:

$R^2 = (a-2)^2 + (0+1)^2 = (-\frac{5}{2}-2)^2 + 1 = (-\frac{9}{2})^2 + 1 = \frac{81}{4} + 1 = \frac{85}{4} = (\frac{\sqrt{85}}{2})^2 \approx 4.61^2$.

Therefore, using the standard form, we can see that the circle *C* is: $(x + \frac{5}{2})^2 + y^2 = \frac{85}{4}$.

Suggestions or Solutions
To the Problem in the Example 2

The center of the circle is in a line $y = 1 - x$, and a point (1, 3) is in the circle.

Suppose now, *C* is the circle to be found, the center of the circle *C* is (*a*, *b*), where *a* and *b* are constant, and the radius is *R*.

Then, first, (*a*, *b*) is in the line $y = 1 - x$, so we get: *b* = 1 − *a*, and (*a*, *b*) = (*a*, 1 − *a*).

Next, the point (1, 3) is the circle *C*, so it is *R* away from the center.

So we get:

$$R^2 = (a-1)^2 + (b-3)^2 = (a-1)^2 + (1-a-3)^2 = (a-1)^2 + (-a-2)^2$$

$$= (a-1)^2 + (a+2)^2 = 2a^2 + 2a + 5.$$

Therefore, the circle *C* is: $(x-a)^2 + (y-1+a)^2 = 2a^2 + 2a + 5$, where *a* is constant.

If not quite sure of the idea behind the processes above, follow the steps below:

To begin with, let's put in a graph the line and point given, along with some probable circles. Then, we can see better where the solution can be around, and quite frequently, we can even see the solution right away, or in just a few steps

Fig. 0

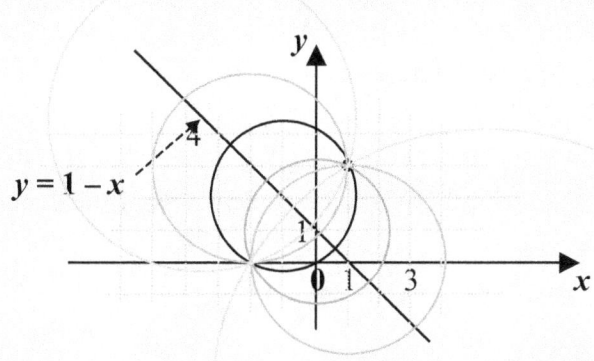

$y = 1 - x$

Now, we can see not just one but many circles, too, can be the solution. In fact, it can be infinitely many. So we are looking for not one particular circle but a group of circles.

Every circle in the group passes through the point **(1, 3)**, but is centered at a different point, which however, is always in the line $y = 1 - x$.   So we can put it the way below:

• As a circle moves along the line $y = 1 - x$ keeping its center in the line, the circle shrinks or expands, but always passes through the point **(1, 3)**.

Besides, every circle in the group seems to pass through another particular point, which is on the other sides of the line $y = 1 - x$, and the particular point looks like **(-2, 0)**.

Suppose now, *C* is the circle to be found, the center of the circle *C* is **(*a*, *b*)**, where *a* and *b* are constant, and the radius is *R*. Then, we want to find two things. What are the two?

One is the radius, and the other is the center, of course. However, we don't really want to find the values of *a* and *b*, do we? That's because, there can be infinitely many circles that can be the solution.   What then, do we need to do with *a* and *b*?

In fact, we should indicate the center **(*a*, *b*)** more specifically rather than find it.

We know **(*a*, *b*)** is a point in a plane, of course, but it is not just any point in the plane.

We know **(*a*, *b*)** is the center of the circle *C*, and the center is in the line $y = 1 - x$.

However, (a, b) alone doesn't show the fact that the center is in the line $y = 1 - x$.

So we want to put the center (a, b) in a different manner. How?

We can do so explaining the relationship between the coordinates of the center (a, b).

How do we explain it?

Coming up with an expression in terms of a and b, we explain the relationship.

Due to the relationship, (a, b) is in the line $y = 1 - x$.

A connective expression can explain such a relationship, and the connective expression connects a and b, of course.   So we need the connective expression between a and b.

We can readily find such an expression by means of the equation of the line given.

That's because (a, b) is the center of the circle C, and is in the line given.

The line is: $y = 1 - x$, and (a, b) is in that line, so we get: $b = 1 - a$, which is the very connective expression between a and b.

So we can put the center (a, b) in this way: $(a, 1 - a)$.

Now, $(a, 1 - a)$ is the center, and $(a, 1 - a)$ alone can show the fact that the center is in the line $y = 1 - x$.

Next, let's move on to the radius R.

All the points in a circle are respectively the radius away from the center.

So the point $(1, 3)$ in the circle C is R away from the center (a, b), which is $(a, 1 - a)$.

We can find the radius R by means of the distance formula.

Taking the distance from the point **(1, 3)** to the center **(a, b) = (a, 1 − a)**, we get:

$$R^2 = (a-1)^2 + (b-3)^2 = (a-1)^2 + (1-a-3)^2 = (a-1)^2 + (-a-2)^2$$

$$= (a-1)^2 + (a+2)^2 = 2a^2 + 2a + 5.$$

Now, the center is **(a, 1 − a)**, and the radius is $\sqrt{2a^2 + 2a + 5}$, where **a** is constant.

Thus, using the standard form, we can get the circle **C**, which is as follows:

$$(x-a)^2 + \{y-(1-a)\}^2 = (x-a)^2 + (y-1+a)^2 = 2a^2 + 2a + 5.$$

Therefore, the circle **C** can be any of all the circles in the group represented by such an equation as follows: $(x-a)^2 + (y-1+a)^2 = 2a^2 + 2a + 5$, where **a** is constant.

By the way, we have:

$$2a^2 + 2a + 5 = 2(a^2 + a) + 5 = 2(a^2 + a + \tfrac{1}{4} - \tfrac{1}{4}) + 5 = 2(a+\tfrac{1}{2})^2 - \tfrac{1}{2} + 5 = 2(a+\tfrac{1}{2})^2 + \tfrac{9}{2}$$

$$= 2(a+\tfrac{1}{2})^2 + (\tfrac{3}{\sqrt{2}})^2 = 2(a+\tfrac{1}{2})^2 + (\tfrac{3\sqrt{2}}{2})^2.$$

So when $a = -\tfrac{1}{2}$, such a circle has the minimum radius, which is $\tfrac{3\sqrt{2}}{2} \approx 2.12$.

In fact, all the circles in the group share the two points **(1, 3)** and **(-2, 0)**, and the smallest of all the circles has a diameter where the two points above are the two endpoints. That is, the distance from **(1, 3)** to **(-2, 0)** is the diameter of the smallest circle in the group.

Suggestions or Solutions
To the Problem in the Example 3

The circle has (1, 3), and is tangent to the *x*-axis, and the center is in a line $y = 1 - x$.

Suppose that the center is (a, b), where a and b are constant, that $(a, 0)$ is the point where the circle is tangent to the x-axis, and that the radius is R.

Then, we get: $R^2 = (a - 1)^2 + (b - 3)^2$, and also, $R^2 = (a - a)^2 + (b - 0)^2 = b^2 \Rightarrow R^2 = b^2$.

So we get: $(a - 1)^2 + (b - 3)^2 = b^2$.

Since (a, b) is in the given line $y = 1 - x$, we get: $b = 1 - a$.

So $(a - 1)^2 + (b - 3)^2 = b^2 \Rightarrow b^2 + (b - 3)^2 = b^2 \Rightarrow b = 3 \Rightarrow R^2 = 9$ since $R^2 = b^2$.

Thus, $b = 1 - a \Rightarrow a = 1 - b = 1 - 3 = -2$. So the center is (-2, 3).

Therefore, the circle is: $(x + 2)^2 + (y - 3)^2 = 3^2$.

If not quite sure of the idea behind the processes above, follow the steps below:

Putting the problem in a graph, we can see better where the solution can be around. Not only that, but we can reduce mistakes approaching the solution, too. Besides, it is often the case where we can even see the solution right away. So let's first, put in a graph the line, the point given, and some probable circles, together with a circle tangent to the *x*-axis, of course.

Fig. 0

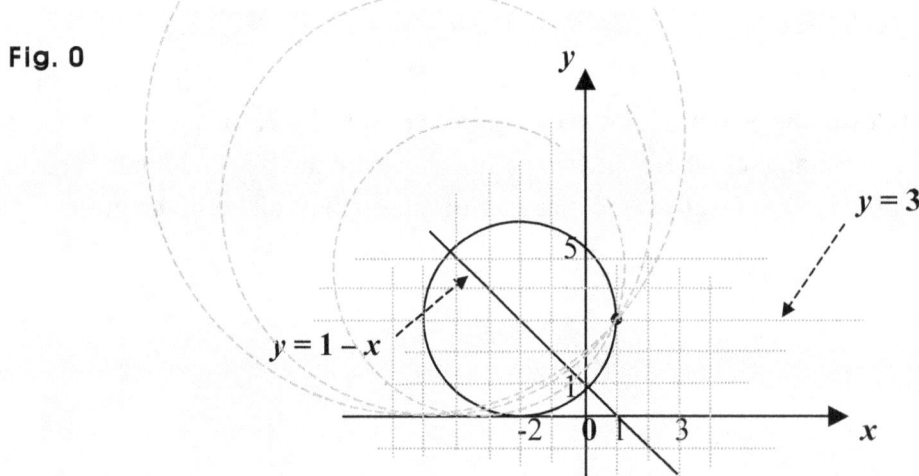

Even without examining the graph closely, it's not hard to see that the center is **(-2, 3)**.

So a circle passing through the point **(1, 3)** and tangent to the *x*-axis is centered at **(-2, 3)**.

What about the radius, then?

The *y*-coordinate of the center is 3. So is the radius. That's because the circle is tangent to the *x*-axis.

That is, if a circle is tangent to the *x*-axis, the radius is the magnitude of the *y*-coordinate of the center.

And if a circle is tangent to the *y*-axis, the radius is the magnitude of the *x*-coordinate of the center. So for instance, if a circle is centered at **(-1, 2)**, and is tangent to the *y*-axis, the radius is |**-1**| = **1**. What if a circle is tangent to both axes, then?

Both coordinates at the center are the same in magnitude since the center is in either of two lines $y = x$, and $y = -x$. So the radius is the magnitude of either coordinate.

How come the circle is centered at **(-2, 3)**, though? In other words, how do we know if it is the only circle centered at a point in the line given, passing through **(1, 3)**, and tangent to the *x*-axis? It's not quite clear that some other circle cannot be the solution, is it?

Suppose *A* is a circle tangent to the *x*-axis and centered at a point in the line $y = 1 - x$, and the radius is bigger than 3.

Then, it seems that the bigger the radius gets, the more the center gets away from the line $y = 1 - x$. So it seems that only the circle centered at **(-2, 3)** is the solution.

However, we still need to show the fact analytically.

Suppose now, *C* is the circle we are after, the center is (*a*, *b*), where *a* and *b* are constant, (*a*, **0**) is the point where the circle is tangent to the *x*-axis, and the radius is *R*.

Then, $(1, 3)$ and $(a, 0)$ are in the circle C, so each of the two points is R away from (a, b). How come?

Every point in a circle is the radius away from the center.

So beginning with the distance from the center to $(1, 3)$, we get: $R^2 = (a - 1)^2 + (b - 3)^2$.

And next, we have another point $(a, 0)$ in the circle C, too, so we get:

$R^2 = (a - a)^2 + (b - 0)^2 = b^2 \Rightarrow R^2 = b^2$.

So we can see that $(a - 1)^2 + (b - 3)^2 = b^2$.

Next, since (a, b) is in the given line $y = 1 - x$, we get: $b = 1 - a$.

Therefore, we get: $(a - 1)^2 + (b - 3)^2 = b^2 \Rightarrow b^2 + (b - 3)^2 = b^2 \Rightarrow b = 3$.

Thus, we get: $b = 1 - a \Rightarrow a = 1 - b = 1 - 3 = -2$. So the center is $(-2, 3)$.

Besides, we have: $R^2 = b^2$, too, so we get: $R^2 = 9$.

Therefore, the circle C is: $(x + 2)^2 + (y - 3)^2 = 3^2$.

In short:

Suppose that the center is (a, b), where a and b are constant, that $(a, 0)$ is the point where the circle is tangent to the x-axis, and that the radius is R.

Then, we get: $R^2 = (a - 1)^2 + (b - 3)^2$, and also, $R^2 = (a - a)^2 + (b - 0)^2 = b^2 \Rightarrow R^2 = b^2$.
So we get: $(a - 1)^2 + (b - 3)^2 = b^2$.
Since (a, b) is in the given line $y = 1 - x$, we get: $b = 1 - a$.
So $(a - 1)^2 + (b - 3)^2 = b^2 \Rightarrow b^2 + (b - 3)^2 = b^2 \Rightarrow b = 3 \Rightarrow R^2 = 9$ since $R^2 = b^2$.
Thus, $b = 1 - a \Rightarrow a = 1 - b = 1 - 3 = -2$. So the center is $(-2, 3)$.
Therefore, the circle is: $(x + 2)^2 + (y - 3)^2 = 3^2$.

Suggestions or Solutions
To the Problem in the Example 4

The circle has (1, 3), and is tangent to the *y*-axis, and the center is in a line $y = 1 - x$.

Suppose the radius is **R**, and the center is **(a, b)**, where **a** and **b** are constant. Then, first, since the circle is tangent to the *y*-axis, the tangent point is **(0, b)**.   Next, we get:

The distance from the center to **(1, 3)** $\Rightarrow R^2 = (a-1)^2 + (b-3)^2$
The distance from the center to **(0, b)** $\Rightarrow R^2 = (a-0)^2 + (b-b)^2 = a^2$.
So we get: $(a-1)^2 + (b-3)^2 = a^2$.

Also, since **(a, b)** is in the given line $y = 1 - x$, we get: $b = 1 - a$.
So $(a-1)^2 + (b-3)^2 = a^2 \Rightarrow (a-1)^2 + (1-a-3)^2 = a^2 \Rightarrow (a-1)^2 + (a+2)^2 = a^2$.
However, $a^2 - 2a + 1 + a^2 + 4a + 4 = a^2 \Rightarrow a^2 + 2a + 5 = (a+1)^2 + 5$, which cannot be 0.

Therefore, there is no circle that satisfies the problem.

If not quite sure of the idea behind the processes above, follow the steps below:

Let's first, put in a graph the line and point given, along with some probable circles. It is often the case though where we can't put in a graph a whole circle as a probable one. We may want to try however, at least a part of such a circle if a whole circle is not possible.

Fig. 0

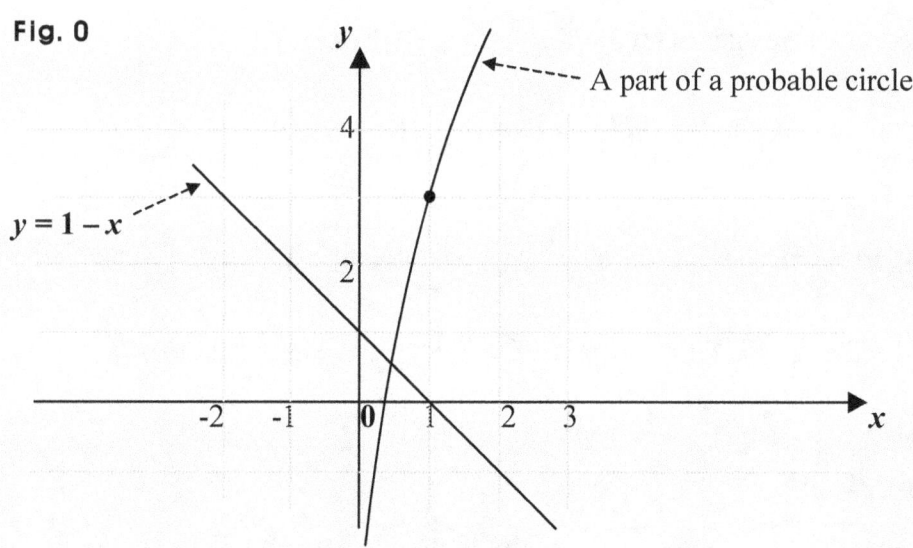

A part of a probable circle

$y = 1 - x$

Now, examining the graph, we can see that the center might be in the fourth quadrant if the circle exists. However, it also looks quite clear that the circle does not exist.

Running math though, we can just assume it is the case no matter how it may look clear since it only looks clear. Math is exact science.

So we want to be sure if the circle exists, and if it is the case, we want to get the precise location of the center. So let's now, pinpoint the center.

Suppose now, C is the circle we are after, the radius is R, and the center is (a, b), where a and b are constant.

Then, the circle C is tangent to the y-axis, so at the tangent point, the x-coordinate is 0, and the y-coordinate is the same as the y-coordinate at the center.

Thus, the tangent point is $(0, b)$.   How come?

The center is assumed to be (a, b).

By the definition for circles, the tangent point $(0, b)$ and the given point $(1, 3)$ are the same distance away from the center, and the same distance is: R.

So putting the ideas above in equations, we get:

First, the distance from the center to $(1, 3) \Rightarrow R^2 = (a-1)^2 + (b-3)^2$

Next, the distance from the center to $(0, b) \Rightarrow R^2 = (a-0)^2 + (b-b)^2 = a^2$.

So we get $(a-1)^2 + (b-3)^2 = a^2$.

Besides, since the center (a, b) is in the given line $y = 1 - x$, we get: $b = 1 - a$.

So we get: $(a-1)^2 + (b-3)^2 = a^2 \Rightarrow (a-1)^2 + (1-a-3)^2 = a^2$
$\Rightarrow (a-1)^2 + (a+2)^2 = a^2$.

Thus, we get: $a^2 - 2a + 1 + a^2 + 4a + 4 = a^2 \Rightarrow 2a^2 + 2a + 5 = a^2$.

So we get: $a^2 + 2a + 5 = 0$.

Then, the equation $a^2 + 2a + 5 = 0$ has to get at least one root if any value of a exists.

Then, the discriminant has to be ≥ 0.

And the discriminant is: $2^2 - 4 \cdot 1 \cdot 5 = -16$, which is not even 0 but negative.

So a doesn't exist, and in turn, neither does b, since we have: $b = 1 - a$.

Consequently, there is no circle that can satisfy this problem.

If we try putting it in a complete square, we get: $(a + 1)^2 + 5$, which cannot be 0 since no real number squared can be -5.

And of course, no real number squared can be any number negative.

If a number is real, the number squared is positive or 0 only.

In short:

Suppose the radius is R, and the center is (a, b), where a and b are constant. Then, first, since the circle is tangent to the y-axis, the tangent point is $(0, b)$.   Next, we get:

The distance from the center to $(1, 3) \Rightarrow R^2 = (a - 1)^2 + (b - 3)^2$
The distance from the center to $(0, b) \Rightarrow R^2 = (a - 0)^2 + (b - b)^2 = a^2$.

So we get: $(a - 1)^2 + (b - 3)^2 = a^2$.
Also, since (a, b) is in the given line $y = 1 - x$, we get: $b = 1 - a$.
So $(a - 1)^2 + (b - 3)^2 = a^2 \Rightarrow (a - 1)^2 + (1 - a - 3)^2 = a^2 \Rightarrow (a - 1)^2 + (a + 2)^2 = a^2$.
However, $a^2 - 2a + 1 + a^2 + 4a + 4 = a^2 \Rightarrow a^2 + 2a + 5 = (a + 1)^2 + 5$, which cannot be 0.
Therefore, there is no circle that satisfies the problem.

Examples 6 in Circles

Find the circle in each case below.

0. The circle has $(3, 1)$, and is tangent to the y-axis, and the center is in a line $y = 1 - x$.

1. Two points $(1, 0)$ and $(3, 0)$ are in the circle tangent to the y-axis.

2. Two points $(0, 2)$ and $(0, 4)$ are in the circle tangent to the x-axis.

3. Two points $(1, 2)$ and $(3, 5)$ are in the circle tangent to the x-axis.

4. Two points $(1, 2)$ and $(3, 5)$ are in the circle tangent to the y-axis.

Suggestions or Solutions
To the Problem in the Example 0

The circle has (3, 1), and is tangent to the y-axis, and the center is in a line $y = 1 - x$.

Suppose the radius is R, and the center is (a, b), where a and b are constant.

Then, first, since the circle is tangent to the y-axis, the tangent point is $(0, b)$.

Next, we get:

First, the distance from the center to $(3, 1) \Rightarrow R^2 = (a - 3)^2 + (b - 1)^2$.

Next, the distance from the center to $(0, b) \Rightarrow R^2 = (a - 0)^2 + (b - b)^2 = a^2 \Rightarrow R^2 = a^2$.

So we get $(a - 3)^2 + (b - 1)^2 = a^2$.

Besides, since (a, b) is in the given line $y = 1 - x$, we get $b = 1 - a$.

So we get: $(a - 3)^2 + (b - 1)^2 = a^2 \Rightarrow (a - 3)^2 + (1 - a - 1)^2 = a^2 \Rightarrow (a - 3)^2 + a^2 = a^2$
$\Rightarrow (a - 3)^2 = 0 \Rightarrow a = 3$.

Thus, we get: $b = 1 - a = 1 - 3 = -2$, and $R^2 = a^2 = 3^2$.

Therefore, the center is $(3, -2)$, and the circle is: $(x - 3)^2 + (y + 2)^2 = 9$.

If not quite sure of the idea behind the processes above, follow the steps below:

Putting the problem in a graph, we can get the solution safely as well as quickly.

So to begin with, let's put in a graph the line, the point given, and some probable circles, tangent to the y-axis.

Fig. 0

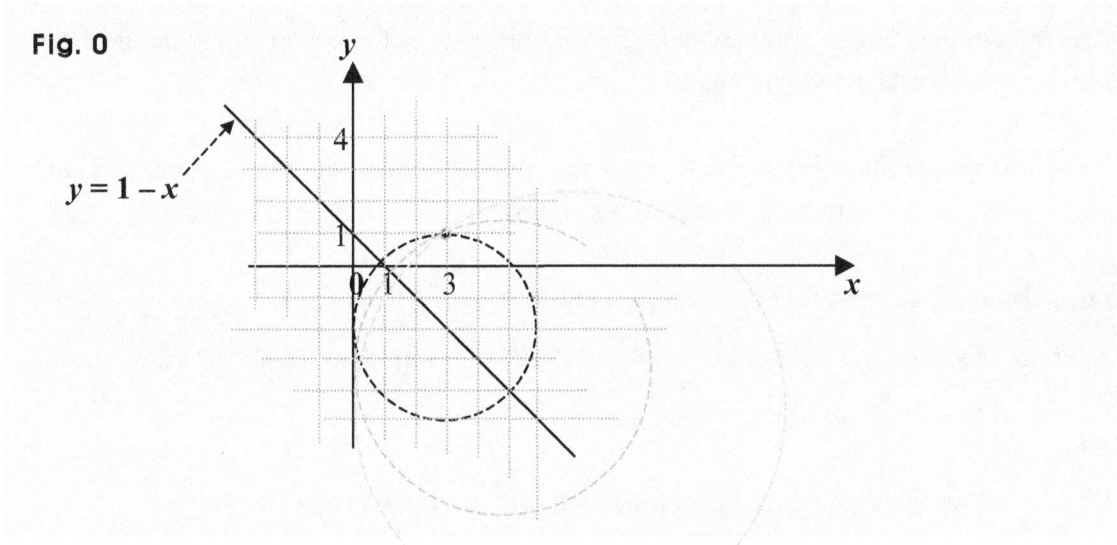

$y = 1 - x$

Now, examining the graph, we can readily see that the center should be at **(3, -2)**.
It is quite clear that the circle centered at **(3, -2)** is tangent to the *y*-axis, and is passing
through the point **(3, 1)**.

So the radius should be 3, and thus, the circle seems to be $(x - 3)^2 + (y + 2)^2 = 9$.

We want to show however, that it is the case, since it is not completely clear, because
math is exact science.

It looks like in fact, there can be some other circle that can be the solution, too.

Examining the graph more closely, we can see that if a circle tangent to the *y*-axis passes
through the point **(3, 1)**, and the radius is bigger than 3, then the bigger the radius gets,
the more the center gets away from the line $y = 1 - x$.
We want to make sure though, that only the circle above can be the solution.

Suppose now, *C* is the circle we are after, the radius is *R*, and the center is **(a, b)**, where
a and *b* are constant.

Then, since the circle *C* is tangent to the *y*-axis, the tangent point is **(0, b)**. How come?

If a circle is tangent to the *y*-axis, the *y*-coordinate at the center has to be the same as the
y-coordinate at the tangent point.

And if a circle is tangent to the x-axis, the x-coordinate at the center has to be the same as the x-coordinate at the tangent point.

Now, by the definition for circles, the tangent point $(0, b)$ and the given point $(3, 1)$ are respectively R away from the center.   So putting the ideas above in equations, we get:

First, the distance from the center to $(3, 1) \Rightarrow R^2 = (a - 3)^2 + (b - 1)^2$.

Next, the distance from the center to $(0, b) \Rightarrow R^2 = (a - 0)^2 + (b - b)^2 = a^2 \Rightarrow R^2 = a^2$.

So we get $(a - 3)^2 + (b - 1)^2 = a^2$.

Besides, since the center (a, b) is in the given line $y = 1 - x$, we get $b = 1 - a$.

So we get $(a - 3)^2 + (b - 1)^2 = a^2 \Rightarrow (a - 3)^2 + (1 - a - 1)^2 = a^2 \Rightarrow (a - 3)^2 + a^2 = a^2$

$\Rightarrow (a - 3)^2 = 0 \Rightarrow a = 3 \Rightarrow b = 1 - a = 1 - 3 = -2$.

Therefore, the center of the circle C is $(3, -2)$.

Meanwhile, we get: $R^2 = a^2 = 3^2$.   Therefore, the circle C is: $(x - 3)^2 + (y + 2)^2 = 9$.

In short:

Suppose the radius is R, and the center is (a, b), where a and b are constant.

Then, first, since the circle is tangent to the y-axis, the tangent point is $(0, b)$.
Next, we get:

First, the distance from the center to $(3, 1) \Rightarrow R^2 = (a - 3)^2 + (b - 1)^2$.
Next, the distance from the center to $(0, b) \Rightarrow R^2 = (a - 0)^2 + (b - b)^2 = a^2 \Rightarrow R^2 = a^2$.

So we get $(a - 3)^2 + (b - 1)^2 = a^2$.
Besides, since (a, b) is in the given line $y = 1 - x$, we get $b = 1 - a$.

So we get: $(a - 3)^2 + (b - 1)^2 = a^2 \Rightarrow (a - 3)^2 + (1 - a - 1)^2 = a^2 \Rightarrow (a - 3)^2 + a^2 = a^2$
$\Rightarrow (a - 3)^2 = 0 \Rightarrow a = 3$.

Thus, we get: $b = 1 - a = 1 - 3 = -2$, and $R^2 = a^2 = 3^2$.

Therefore, the center is $(3, -2)$, and the circle is: $(x - 3)^2 + (y + 2)^2 = 9$.

Suggestions or Solutions
To the Problem in the Example 1

Two points (1, 0) and (3, 0) are in the circle tangent to the *y*-axis.

Suppose *C* is the circle we want, and is tangent to the *y*-axis at **(0, b)**, where **b** is constant. Then, the center of *C* is in the line *y = b*.

The center is also, in the line passing through the midpoint between the two points given since the two points are in the circle *C*.

The midpoint is $(\frac{3+1}{2}, \frac{0+0}{2}) =$ **(2, 0)**, so the center is in a line *x = 2*, also.

Therefore, the center is at **(2, b)**, where the two lines *x = 2* and *y = b* meet each other.

Thus, the tangent point is 2 away from the center, so the radius of the circle *C* is 2. The two points given are in *C*, so the two are respectively 2 away from the center **(2, b)**. Thus, 2 is the distance from the point **(1, 0)** to the center of *C*.

So we get: $2^2 = (2-1)^2 + (b-0)^2 \Rightarrow 4 = 1 + b^2 \Rightarrow b = \pm\sqrt{3}.$

Thus, the center of *C* is $(2, \sqrt{3})$ or $(2, -\sqrt{3})$.

Of the two centers above, $(2, \sqrt{3})$ is the center of a circle above the *x*-axis, and the other, $(2, -\sqrt{3})$ is the center of a circle below the *x*-axis.

Therefore, the circle *C* is: $(x-2)^2 + (y \pm \sqrt{3})^2 = 4.$

If not quite sure of the idea behind the processes above, follow the steps below:

Let's first, put in a graph the two points given. Then, we can see better how a circle has to be placed so that it can be tangent to the *y*-axis passing through the two points.

Fig. 0

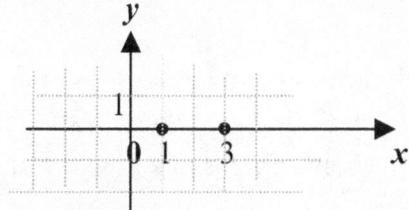

Then, adding some circles to the graph, we will see a circle likely to satisfy the problem.

And we will quickly see that not one but two circles can satisfy the problem. Putting a circle in a graph, it's always a good idea to indicate (or show, mark, etc.) the point where the center is. For instance, we can put two dashed lines perpendicular to each other and meeting at the center.

Fig. 1

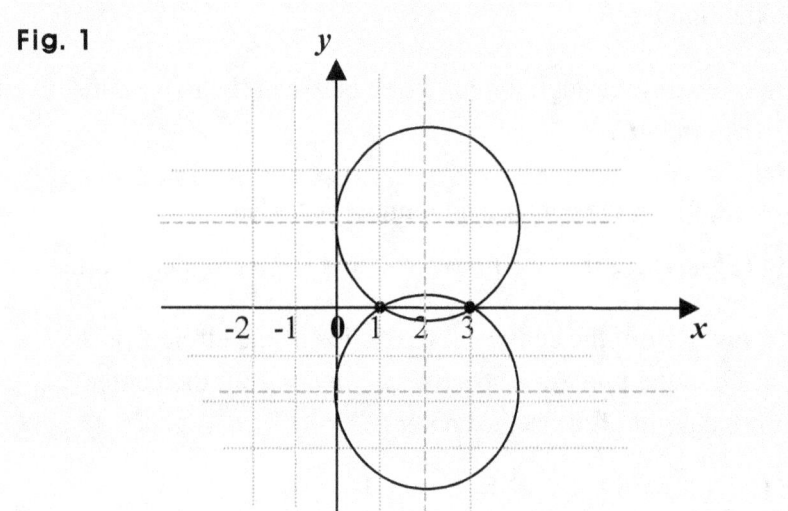

Now, we have two circles tangent to the y-axis, so we may want to find the points where the circles are tangent to the y-axis.

Finding each tangent point, we will have three points in total to work with since we are already given two points **(1, 0)** and **(3, 0)**. So we can find each of the two circles at a time, for three points not in a line can determine a circle. Of course, finding the center and radius, we can get the same, too, by the standard form.

Let's now, begin with the upper circle in the graph above.
So assuming **C** is the circle, and examining the graph, we can clearly see that the center is in the line $x = 2$.   How come the center is in the line?

• Suppose **L** is a line that passes through the midpoint between two points in a circle, and is perpendicular to the line segment connecting the two points. Then, the line **L** passes through the center of the circle. Put some circles in a graph, and try it.

Now, the midpoint between $(1, 0)$ and $(3, 0)$ is $(2, 0)$, so the center of the circle C has to be in the line $x = 2$, which is parallel to the y-axis.

Therefore, the tangent point is 2 away from the center, so the radius of the circle C is 2.

Suppose now, that the circle C is tangent to the y-axis at $(0, b)$, where b is constant.

Then, the center of C is in the line $y = b$, too. How come?

• A line tangent to a circle is perpendicular to a line passing through the tangent point and the center of the circle. Put some circles in a graph again, and try it.

Now, the y-axis is a line tangent to the circle C at the point $(0, b)$.

The line $y = b$ is perpendicular to the y-axis and passes through the tangent point $(0, b)$.

So the line $y = b$ passes through the center, which therefore, is in the line $y = b$.

Thus, the center of C is not only in the line $x = 2$ but in the line $y = b$, too.
Where is the center, then?

The center is at a point $(2, b)$, where the two lines $x = 2$ and $y = b$ meet each other.

Now, finding the center (that is, finding b), we can find the circle C by means of the standard form since we know the radius, which is 2. How do we find the center, then?

We can find it by means of the definition for circles:

• All the points in a circle are respectively the radius away from the center.

So the two points given are respectively the radius away from the center $(2, b)$.

The radius is 2, and is the distance from the point $(1, 0)$ to the center $(2, b)$, so we get:
$2^2 = (2 - 1)^2 + (b - 0)^2 \Rightarrow 4 = 1 + b^2 \Rightarrow b = \pm\sqrt{3}$.

Thus, we can see that the center of the circle **C** is **(2, $\sqrt{3}$)** or **(2, -$\sqrt{3}$)**.

Looking at the **y**-coordinates at the two centers above, we can readily see that **(2, $\sqrt{3}$)** is the center of the upper circle, and **(2, -$\sqrt{3}$)** is the center of the lower circle.

Therefore, the circle **C** is as follows: **(x – 2)² + (y – $\sqrt{3}$)² = 4** or **(x – 2)² + (y + $\sqrt{3}$)² = 4**.

The circle **C** can be put this way, too: **(x – 2)² + (y ± $\sqrt{3}$)² = 4**.

Fig. 2

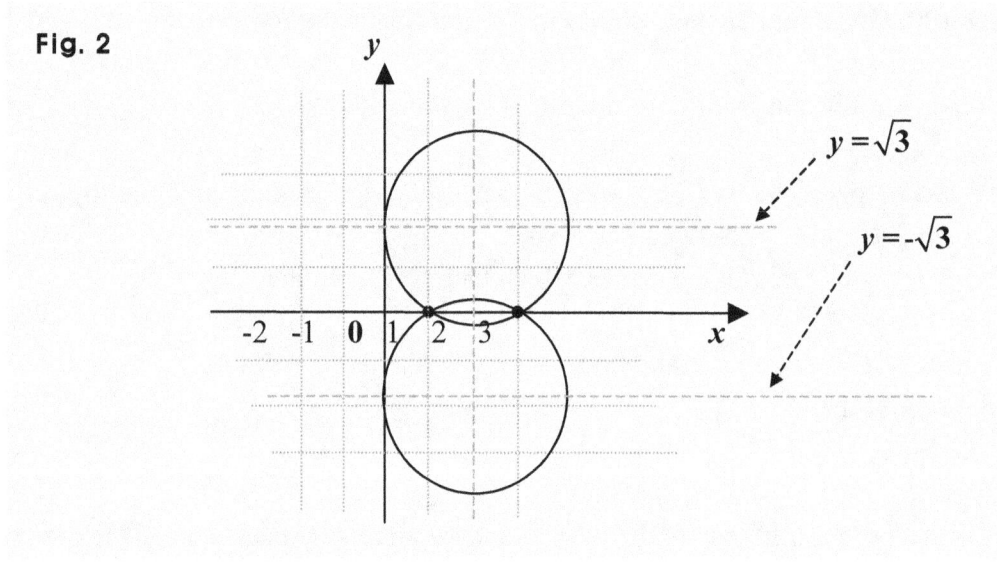

Suggestions or Solutions
To the Problem in the Example 2

Two points (0, 2) and (0, 4) are in the circle tangent to the *x*-axis.

Suppose *C* is the circle we want, and is tangent to the *x*-axis at **(*a*, 0)**, where *a* is constant.
Then, the center of *C* is in a line *x* = *a*, and also, is in a line passing through the midpoint
between the two points given, and perpendicular to the line segment connecting the two.

The midpoint is $(\frac{0+0}{2}, \frac{4+2}{2})$ = **(0, 3)**, and the line perpendicular to the line segment is
parallel to the *x*-axis, so the center is in a line *y* = 3, too. Thus, the center is **(*a*, 3)**.

Thus, the tangent point is 3 away from the center, so the radius is 3.
The point **(0, 2)** is in the circle *C*, so the point is 3 away from the center **(*a*, 3)**.
Thus, we get: $3^2 = (a-0)^2 + (3-2)^2 \Rightarrow 9 = a^2 + 1 \Rightarrow a = \pm\sqrt{8} = \pm 2\sqrt{2}$. So the center is
($2\sqrt{2}$, 3), which is for the circle on the right of the *y*-axis, or is **(-$2\sqrt{2}$, 3)**, which is for
the one on the left. Therefore, the circle *C* is: $(x \pm 2\sqrt{2})^2 + (y-3)^2 = 3^2$.

If not quite sure of the idea behind the processes above, follow the steps below:

The solution to this problem is just about the same as the one to the previous problem.
Comparing the two solutions, we will see another way of handling problems with circles.

Now, let's put in a graph the two points given, together with probable circles, of course.

Fig. 0

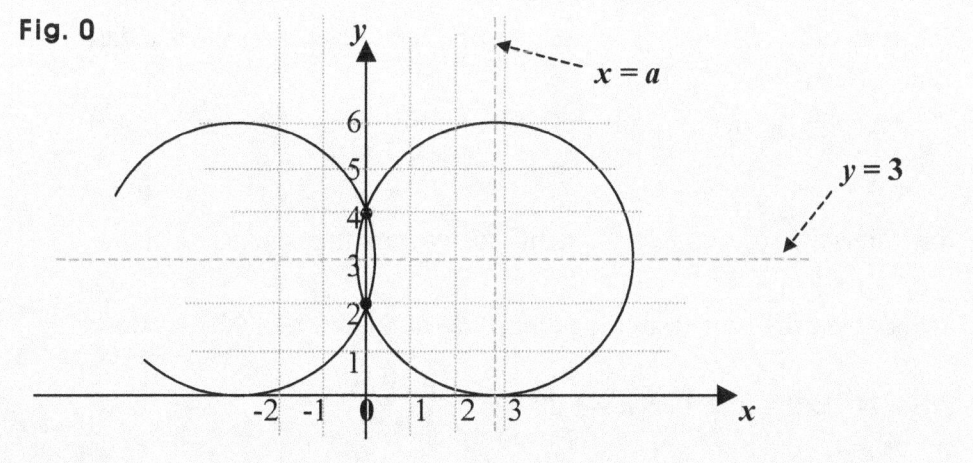

Then, we can see two circles can be the solution. Each of the two is tangent to the *x*-axis.

Considering the tangent point, along with the two points given, we can work with three points in total which are not in a line, so we can find the circle. And of course, finding the center and radius, we can get the circle, too, by the template standard.

Let's now, begin with the circle on the right in the graph above.

Assuming *C* is the circle, and looking at the graph, we can clearly see that the center is in a line *y* = 3. How come the center is in the line, though?

• A line passing through the midpoint between two points in a circle and perpendicular to the line segment connecting the two points passes through the center of the circle.

Now, the midpoint between (0, 2) and (0, 4) is (0, 3), so the center of the circle *C* has to be in the line *y* = 3, which is parallel to the *x*-axis.   Thus, the tangent point is 3 away from the center, so the radius is 3.   Suppose now, *C* is tangent to the *x*-axis at (*a*, 0), where *a* is constant. Then, the center is in a line *x* = *a*, too.   How come?

• A line tangent to a circle is perpendicular to a line passing through the tangent point and the center of the circle.

The line perpendicular to the *x*-axis and passing through the tangent point (*a*, 0) is the line *x* = *a*.
So the center of *C* is in the line *x* = *a*, and is a point (*a*, 3) where the two lines *x* = *a* and *y* = 3 meet each other.
Thus, finding the center (that is, finding *a*), we can find the circle using the standard form since we know the radius, which is 3.
And we can get the center by means of the definition: every point is the radius away from the center.

So the two given points are respectively the radius away from the center (*a*, 3).

The radius is 3, and is the distance from the point (0, 2) to the center (*a*, 3), so we get:

$$3^2 = (a-0)^2 + (3-2)^2 \Rightarrow 9 = a^2 + 1 \Rightarrow a = \pm\sqrt{8} = \pm 2\sqrt{2}\,.$$

Therefore, we can see that the center is $(2\sqrt{2}, 3)$, which is for the circle on the right, or $(-2\sqrt{2}, 3)$, which is for the one on the left.

So the circle C is: $(x - 2\sqrt{2})^2 + (y - 3)^2 = 9$ or $(x + 2\sqrt{2})^2 + (y - 3)^2 = 9$.

We can put it this way, too: $(x \pm 2\sqrt{2})^2 + (y - 3)^2 = 3^2$.

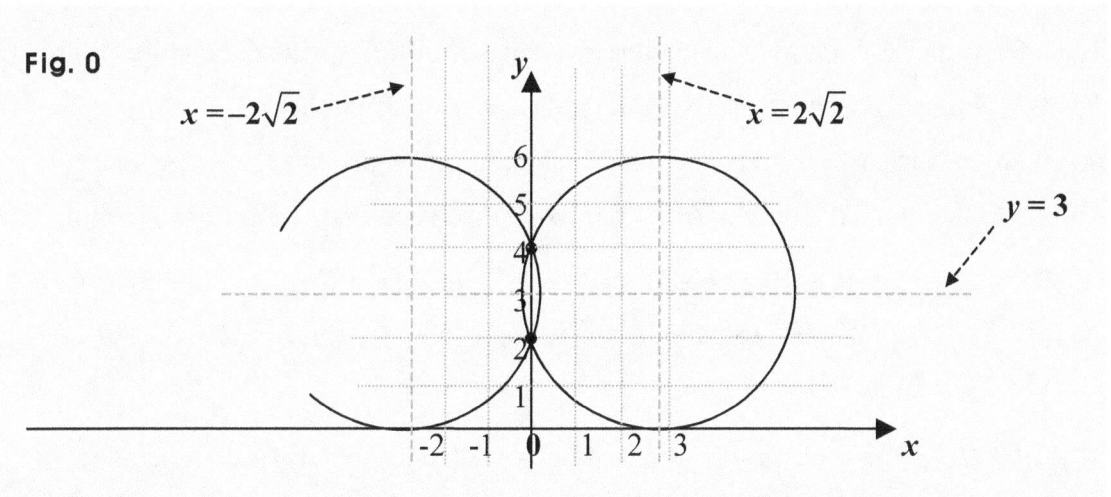

In short:

Suppose C is the circle we want, and is tangent to the x-axis at $(a, 0)$, where a is constant. Then, the center of C is in a line $x = a$, and also, is in a line passing through the midpoint between the two points given, and perpendicular to the line segment connecting the two.

The midpoint is $(\frac{0+0}{2}, \frac{4+2}{2}) = (0, 3)$, and the line perpendicular to the line segment is parallel to the x-axis, so the center is in a line $y = 3$, too. Thus, the center is $(a, 3)$.

Thus, the tangent point is 3 away from the center, so the radius is 3.

The point $(0, 2)$ is in the circle C, so the point is 3 away from the center $(a, 3)$.

Thus, we get: $3^2 = (a - 0)^2 + (3 - 2)^2 \Rightarrow 9 = a^2 + 1 \Rightarrow a = \pm\sqrt{8} = \pm2\sqrt{2}$. So the center is $(2\sqrt{2}, 3)$, which is for the circle on the right of the y-axis, or is $(-2\sqrt{2}, 3)$, which is for the one on the left. Therefore, the circle C is: $(x \pm 2\sqrt{2})^2 + (y - 3)^2 = 3^2$.

Suggestions or Solutions
To the Problem in the Example 3

Two points (1, 2) and (3, 5) are in the circle tangent to the *x*-axis.

Suppose the center is in a line $y = b$, where b is constant, and the point where the circle is tangent to the *x*-axis is $(a, 0)$, where a is constant.

Then, the center is at (a, b), where a line $x = a$ meets a line $y = b$, and the radius is $|b|$.

The two points given are respectively the radius $|b|$ away from the center (a, b).

So taking the distances from (a, b) to $(1, 2)$ and $(3, 5)$, we get:

$b^2 = (a-1)^2 + (b-2)^2$, and $b^2 = (a-3)^2 + (b-5)^2$. Then first, we get:

$(a-1)^2 + (b-2)^2 = (a-3)^2 + (b-5)^2 \Rightarrow (a-1)^2 - (a-3)^2 + (b-2)^2 - (b-5)^2 = 0$

$\Rightarrow \{(a-1) - (a-3)\}\{(a-1) + (a-3)\} + \{(b-2) - (b-5)\}\{(b-2) + (b-5)\} = 0$

$\Rightarrow 2(2a-4) + 3(2b-7) = 4a + 6b - 29 = 0$. Next, we get:

$(a-1)^2 + (b-2)^2 = b^2 \Rightarrow (a-1)^2 + b^2 - 4b + 4 = b^2 \Rightarrow (a-1)^2 + 4 - 4b = 0$.

So we need to solve a system where $4a + 6b - 29 = 0$ and $(a-1)^2 + 4 - 4b = 0$.

Then, we get first: $4a + 6b - 29 = 0 \Rightarrow b = \frac{29-4a}{6}$. Next, we get:

$(a-1)^2 + 4 - 4b = a^2 - 2a + 1 + 4 - 4\frac{(29-4a)}{6} = a^2 - 2a + \frac{8a}{3} + 5 - \frac{58}{3} = 0 \Rightarrow a^2 + \frac{2a}{3} - \frac{43}{3} = 0$.

Then, $a^2 + \frac{2a}{3} - \frac{43}{3} = a^2 + \frac{2a}{3} + \frac{1}{9} - \frac{1}{9} - \frac{43}{3} = (a + \frac{1}{3})^2 - \frac{1}{9} - \frac{43}{3} = (a + \frac{1}{3})^2 - \frac{130}{9} = 0$

$\Rightarrow (a + \frac{1}{3})^2 = \frac{130}{9} \Rightarrow a + \frac{1}{3} = \pm\sqrt{\frac{130}{9}} \Rightarrow a = -\frac{1}{3} \pm \frac{\sqrt{130}}{3}$.

So we get: $a = \frac{-1+\sqrt{130}}{3}$ or $a = \frac{-1-\sqrt{130}}{3}$. Also, we have: $b = \frac{29-4a}{6}$. So we get:

$a = \frac{-1+\sqrt{130}}{3} \Rightarrow b = \frac{1}{6}(29 - 4 \cdot \frac{-1+\sqrt{130}}{3}) = \frac{1}{6} \cdot \frac{87 + 4 - 4\sqrt{130}}{3} = \frac{91 - 4\sqrt{130}}{18}$.

$a = \frac{-1-\sqrt{130}}{3} \Rightarrow b = \frac{1}{6}(29 - 4 \cdot \frac{-1-\sqrt{130}}{3}) = \frac{1}{6} \cdot \frac{87 + 4 + 4\sqrt{130}}{3} = \frac{91 + 4\sqrt{130}}{18}$.

And we know that the radius is $|b|$, and that the center is (a, b).

Therefore, the circle is: $(x - a)^2 + (y - b)^2 = b^2$ where:

$a = \frac{-1+\sqrt{130}}{3}$ and $b = \frac{91-4\sqrt{130}}{18}$ or $a = \frac{-1-\sqrt{130}}{3}$ and $b = \frac{91+4\sqrt{130}}{18}$.

If not quite sure of the idea behind the processes above, follow the steps below:

This example is quite similar to the Example E. So putting in a graph the two points first, we can see better what circle can be tangent to the *x*-axis passing through the two points.

Fig. 0

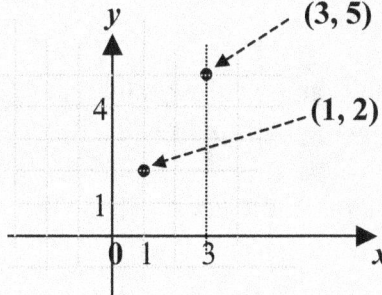

We don't really have to use compasses, and a part of a circle can serve the purpose, too. Sketching some probable circles, we can see quickly that two circles can be the solution.

Fig. 1

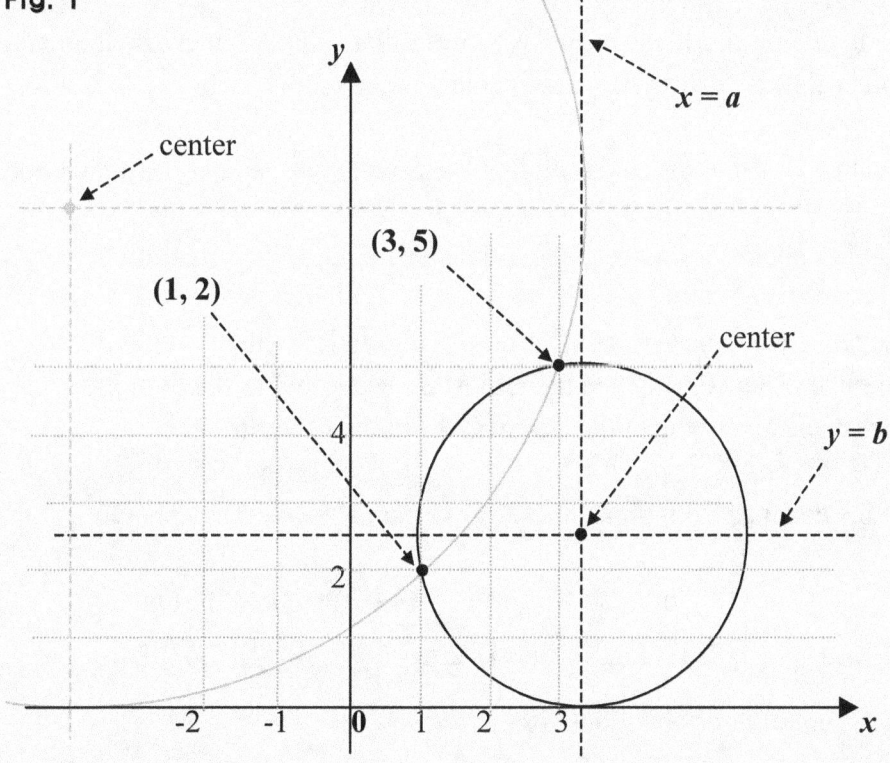

Let's for now, begin with the circle on the right, which is the smaller of the two.

And since the circle is tangent to the x-axis, we may want to begin with the tangent point.

Suppose C is the circle we want, and is tangent to the x-axis at a point $(a, 0)$, where a is constant. Then, the center of C is in a line $x = a$. How come?

• A line tangent to a circle is perpendicular to the line passing through the tangent point and the center of the circle.

The x-axis is the line tangent to the circle C at the point $(a, 0)$, and the line perpendicular to the x-axis and passing through the tangent point $(a, 0)$ is the line $x = a$.

So the center of C is in the line $x = a$. That is, the x-coordinate at the center is a. What then, about the x-coordinate at the center?

We want to find it, too, of course, so suppose now, the center is in the line $y = b$, too. That is, we assume that the y-coordinate at the center is b.

Then, the center is at a point (a, b), where the two lines $x = a$ and $y = b$ meet each other, and the tangent point $(a, 0)$ is $|b|$ away from the center, so the radius of the circle C is $|b|$.

Thus, finding the values of a and b (that is, finding the center), we can get the radius, too.

Then, by means of the standard form, we can get the circle C.

We can find the center of C by means of the definition for circles, which keeps repeating the story about the same distance, which is the radius. So according to the definition, the two points given are respectively the radius, $|b|$ away from the center (a, b).

So beginning with the distance from (a, b) to $(1, 2)$, we get: $b^2 = (a - 1)^2 + (b - 2)^2$.

Next, we have another point $(3, 5)$ in C, so we get: $b^2 = (a - 3)^2 + (b - 5)^2$, too.

Therefore, we get: $(a - 1)^2 + (b - 2)^2 = (a - 3)^2 + (b - 5)^2$.

Then, do we get a system of three equations for a an b?

No, we don't.   The three systems shown below are actually equivalent to each other:

- $b^2 = (a-1)^2 + (b-2)^2$, and $b^2 = (a-3)^2 + (b-5)^2$.
- $b^2 = (a-1)^2 + (b-2)^2$, and $(a-1)^2 + (b-2)^2 = (a-3)^2 + (b-5)^2$.
- $b^2 = (a-3)^2 + (b-5)^2$, and $(a-1)^2 + (b-2)^2 = (a-3)^2 + (b-5)^2$.

So solving either of the three above, we get the same solution. Let's solve the second one of the three. Then, beginning with the first of the two equations, we get:

$$(a-1)^2 + (b-2)^2 = b^2 \Rightarrow (a-1)^2 + b^2 - 4b + 4 = b^2 \Rightarrow (a-1)^2 + 4 - 4b = 0.$$

Next, moving on to the other equation, we get:

$$(a-1)^2 + (b-2)^2 = (a-3)^2 + (b-5)^2 \Rightarrow (a-1)^2 - (a-3)^2 + (b-2)^2 - (b-5)^2 = 0$$

$$\Rightarrow \{(a-1)-(a-3)\}\{(a-1)+(a-3)\} + \{(b-2)-(b-5)\}\{(b-2)+(b-5)\} = 0$$

$$\Rightarrow 2(2a-4) + 3(2b-7) = 4a + 6b - 29 = 0.$$

Now, the system has been reduced to another system for *a* and *b* as follows:
$$4a + 6b - 29 = 0 \text{ and } (a-1)^2 + 4 - 4b = 0.$$

The system above is equivalent to any of the three systems stated above, of course. The system above is now smaller, but is still quadratic. So we get to do some algebra. To begin with, $4a + 6b - 29 = 0 \Rightarrow b = \frac{29-4a}{6}$.   Then, we get:

$$(a-1)^2 + 4 - 4b = a^2 - 2a + 1 + 4 - 4\frac{(29-4a)}{6} = a^2 - 2a + \frac{8a}{3} + 5 - \frac{58}{3} = 0 \Rightarrow a^2 + \frac{2a}{3} - \frac{43}{3} = 0.$$

Then, we get: $a^2 + \frac{2a}{3} - \frac{43}{3} = a^2 + \frac{2a}{3} + \frac{1}{9} - \frac{1}{9} - \frac{43}{3} = (a+\frac{1}{3})^2 - \frac{1}{9} - \frac{43}{3} = (a+\frac{1}{3})^2 - \frac{130}{9} = 0$

$$\Rightarrow (a+\frac{1}{3})^2 = \frac{130}{9} \Rightarrow a + \frac{1}{3} = \pm\sqrt{\frac{130}{9}} \Rightarrow a = -\frac{1}{3} \pm \frac{\sqrt{130}}{3}.$$

So we get: $a = \frac{-1+\sqrt{130}}{3}$ or $a = \frac{-1-\sqrt{130}}{3}$. In short, we have: $a = \frac{-1\pm\sqrt{130}}{3}$.

And looking at the beginning above, we can see we have: $b = \frac{29-4a}{6}$, too.   So we get:

$$a = \frac{-1+\sqrt{130}}{3} \Rightarrow b = \frac{1}{6}(29 - 4\cdot\frac{-1+\sqrt{130}}{3}) = \frac{1}{6}\cdot\frac{87+4-4\sqrt{130}}{3} = \frac{91-4\sqrt{130}}{18}.$$

$$a = \frac{-1-\sqrt{130}}{3} \Rightarrow b = \frac{1}{6}(29 - 4\cdot\frac{-1-\sqrt{130}}{3}) = \frac{1}{6}\cdot\frac{87+4+4\sqrt{130}}{3} = \frac{91+4\sqrt{130}}{18}.$$

Now, we know that the radius is $|b|$, and that the center is (a, b).

Therefore, the circle C is: $(x - a)^2 + (y - b)^2 = b^2$ where:

$a = \frac{-1+\sqrt{130}}{3} \approx 3.467$ and $b = \frac{91-4\sqrt{130}}{18} \approx 2.522$

or $a = \frac{-1-\sqrt{130}}{3} \approx -4.134$ and $b = \frac{91+4\sqrt{130}}{18} \approx 7.589$.

And putting the circles in a graph, we can put them the way below:

Fig. 2

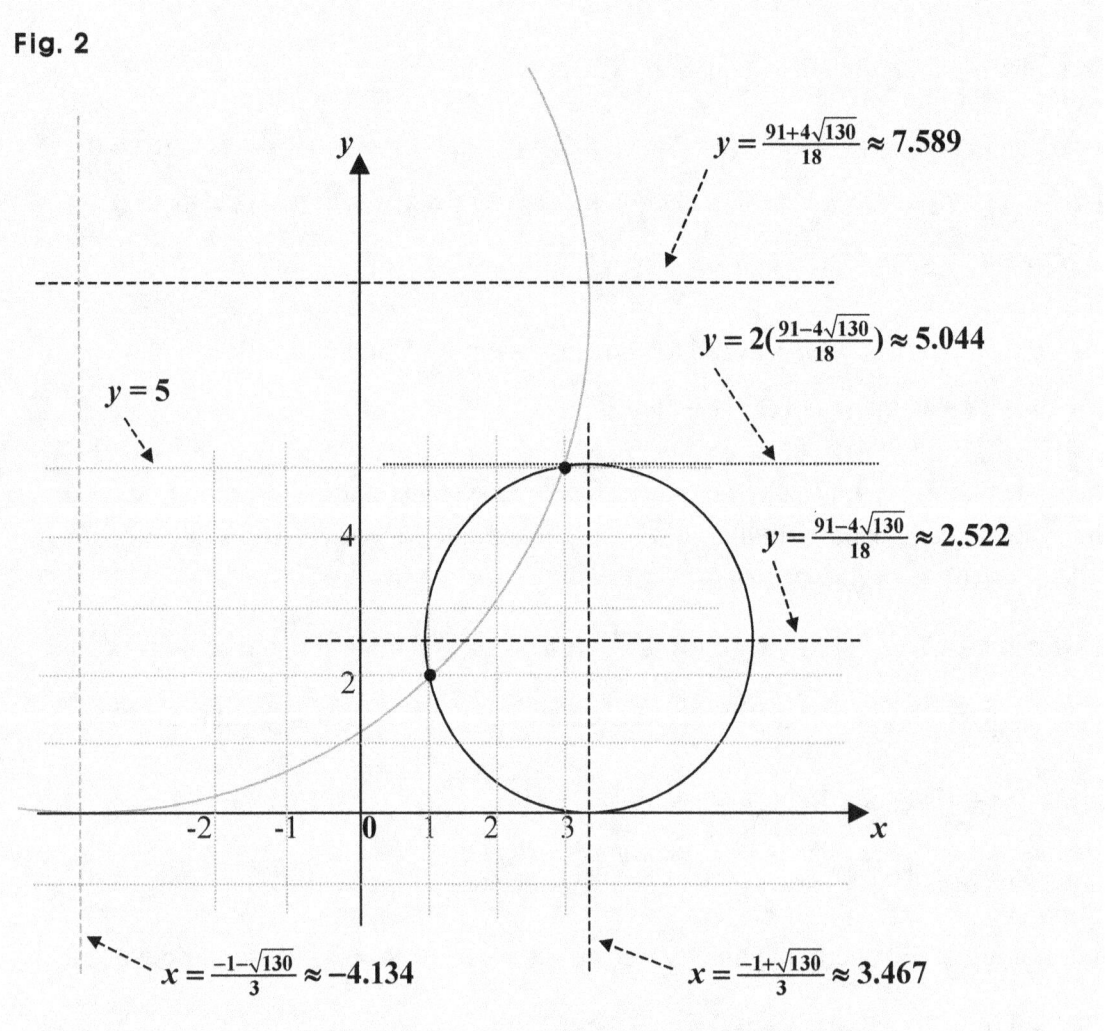

Suggestions or Solutions
To the Problem in the Example 4

Two points (1, 2) and (3, 5) are in the circle tangent to the y-axis

Suppose that the center is in a line $x = a$, and that the point where the circle is tangent to the y-axis is $(0, b)$. Then, the center is at a point (a, b), where the line $x = a$ meets the line $y = b$, and the radius is $|a|$.

Each of the two points given is the radius $|a|$ away from (a, b).
So taking the distances from (a, b) to $(1, 2)$ and $(3, 5)$, we get:
$b^2 = (a-1)^2 + (b-2)^2$, and $b^2 = (a-3)^2 + (b-5)^2$.

Then first, we get:
$(a-1)^2 + (b-2)^2 = a^2 \Rightarrow a^2 - 2a + 1 + b^2 - 4b + 4 = a^2 \Rightarrow b^2 - 4b - 2a + 5 = 0$.

Next, we get:
$(a-1)^2 + (b-2)^2 = (a-3)^2 + (b-5)^2 \Rightarrow (a-1)^2 - (a-3)^2 + (b-2)^2 - (b-5)^2 = 0$
$\Rightarrow \{(a-1) - (a-3)\}\{(a-1) + (a-3)\} + \{(b-2) - (b-5)\}\{(b-2) + (b-5)\} = 0$
$\Rightarrow 2(2a - 4) + 3(2b - 7) = 4a + 6b - 29 = 0$.

Thus, we need to solve a system where $4a + 6b - 29 = 0$ and $b^2 - 4b - 2a + 5 = 0$.

Then first, we get: $4a + 6b - 29 = 0 \Rightarrow a = \frac{29 - 6b}{4}$. Next, we get:

$b^2 - 4b - 2a + 5 = b^2 - 4b - 2 \cdot \frac{(29 - 6b)}{4} + 5 = b^2 - 4b + 3b + 5 - \frac{29}{2} = 0 \Rightarrow b^2 - b - \frac{19}{2} = 0$.

Then, $b^2 - b - \frac{19}{2} = b^2 - b + \frac{1}{4} - \frac{1}{4} - \frac{19}{2} = (b - \frac{1}{2})^2 - \frac{1}{4} - \frac{19}{2} = (b - \frac{1}{2})^2 - \frac{39}{4} = 0$

$\Rightarrow (b - \frac{1}{2})^2 = \frac{39}{4} \Rightarrow b - \frac{1}{2} = \pm\sqrt{\frac{39}{4}} \Rightarrow b = \frac{1}{2} \pm \frac{\sqrt{39}}{2}$.

So we get: $b = \frac{1 + \sqrt{39}}{2}$ or $b = \frac{1 - \sqrt{39}}{2}$. And we have: $a = \frac{29 - 6b}{4}$, too. Thus, we get:

$b = \frac{1 + \sqrt{39}}{2} \Rightarrow a = \frac{1}{4}(29 - 6 \cdot \frac{1 + \sqrt{39}}{2}) = \frac{1}{4} \cdot \frac{58 - 6 - 6\sqrt{39}}{2} = \frac{52 - 6\sqrt{39}}{8} = \frac{26 - 3\sqrt{39}}{4}$.

$b = \frac{1 - \sqrt{39}}{2} \Rightarrow a = \frac{1}{4}(29 - 6 \cdot \frac{1 - \sqrt{39}}{2}) = \frac{1}{4} \cdot \frac{58 - 6 + 6\sqrt{39}}{2} = \frac{52 + 6\sqrt{39}}{8} = \frac{26 + 3\sqrt{39}}{4}$.

We know that the radius is $|a|$, and that the center is (a, b).

Therefore, the circle C is: $(x - a)^2 + (y - b)^2 = a^2$, where $a = \frac{26 \mp 3\sqrt{39}}{4}$ and $b = \frac{1 \pm \sqrt{39}}{2}$.

If not quite sure of the idea behind the processes above, follow the steps below:

First, putting in a graph the two points given, we can see better what circle can be tangent to the **x**-axis passing through the two points.

Fig. 0

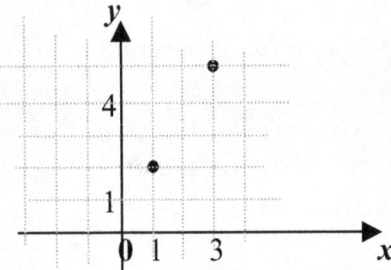

This example is in fact, almost the same as the previous one. Sketching some probable circles in the graph, we can see that two circles can be the solution.

Fig. 1

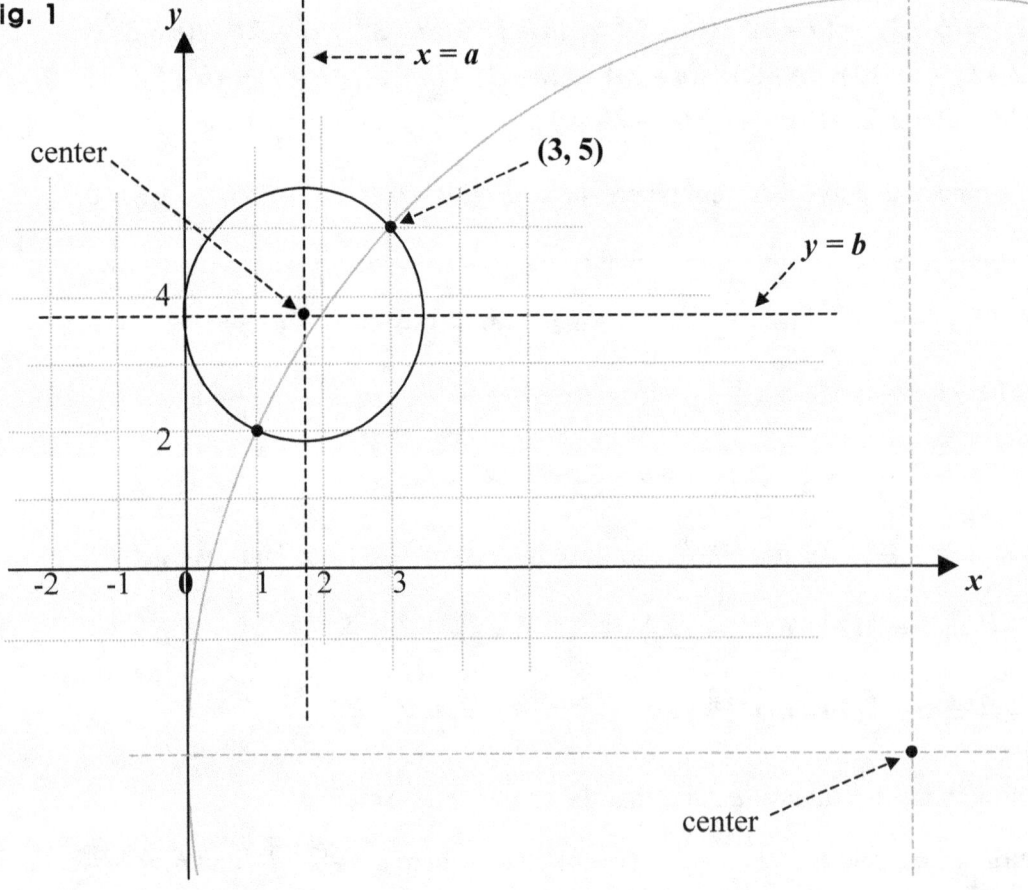

Let's begin with the smaller of the two circles above.

Suppose C is the circle, and is tangent to the y-axis at $(0, b)$, where b is constant.

Then, the center of the circle C is in a line $y = b$.   How come?

We have a fact that a line tangent to a circle is perpendicular to a line passing through the tangent point and the center.

Now, the line $y = b$ is perpendicular to the y-axis, which is tangent to C, and passes through the tangent point $(0, b)$. So the center of C is in the line $y = b$.

Suppose now, that the center of C is in a line $x = a$, too.

Then, the center is a point (a, b), where the line $x = a$ meets the line $y = b$.

So the tangent point is $|a|$ away from the center, and therefore, the radius of C is $|a|$.

Thus, finding the values of a and b (that is, finding the center), we can get the radius, too.

Then, by means of the standard form, we can get the circle C.

We can find the center of C using the definition for circles. By the definition, each of the two points given is the radius $|a|$ away from the center (a, b).

So beginning with the distance from (a, b) to $(1, 2)$, we get: $a^2 = (a - 1)^2 + (b - 2)^2$.

Next, we have another point $(3, 5)$ in C, so we get: $a^2 = (a - 3)^2 + (b - 5)^2$, too.

Therefore, we get: $(a - 1)^2 + (b - 2)^2 = (a - 3)^2 + (b - 5)^2$.

Then first, we get:

$(a - 1)^2 + (b - 2)^2 = a^2 \Rightarrow a^2 - 2a + 1 + b^2 - 4b + 4 = a^2 \Rightarrow b^2 - 4b - 2a + 5 = 0$.

Next, we get:

$$(a-1)^2 + (b-2)^2 = (a-3)^2 + (b-5)^2 \Rightarrow (a-1)^2 - (a-3)^2 + (b-2)^2 - (b-5)^2 = 0$$

$$\Rightarrow \{(a-1)-(a-3)\}\{(a-1)+(a-3)\} + \{(b-2)-(b-5)\}\{(b-2)+(b-5)\} = 0$$

$$\Rightarrow 2(2a-4) + 3(2b-7) = 4a + 6b - 29 = 0.$$

Thus, we get to solve a system of two equations for a and b as follows:

$$4a + 6b - 29 = 0 \text{ and } b^2 - 4b - 2a + 5 = 0.$$

To begin with, $4a + 6b - 29 = 0 \Rightarrow a = \frac{(29-6b)}{4}$. Then, we get:

$$b^2 - 4b - 2a + 5 = b^2 - 4b - 2 \cdot \frac{(29-6b)}{4} + 5 = b^2 - 4b + 3b + 5 - \frac{29}{2} = 0 \Rightarrow b^2 - b - \frac{19}{2} = 0.$$

Then, we get: $b^2 - b - \frac{19}{2} = b^2 - b + \frac{1}{4} - \frac{1}{4} - \frac{19}{2} = (b-\frac{1}{2})^2 - \frac{1}{4} - \frac{19}{2} = (b-\frac{1}{2})^2 - \frac{39}{4} = 0$

$$\Rightarrow (b-\tfrac{1}{2})^2 = \tfrac{39}{4} \Rightarrow b - \tfrac{1}{2} = \pm\sqrt{\tfrac{39}{4}} \Rightarrow b = \tfrac{1}{2} \pm \tfrac{\sqrt{39}}{2}.$$

So we get: $b = \frac{1+\sqrt{39}}{2}$ or $b = \frac{1-\sqrt{39}}{2}$.   Also, we have: $a = \frac{(29-6b)}{4}$.   Thus, we get:

$$b = \tfrac{1+\sqrt{39}}{2} \Rightarrow a = \tfrac{1}{4}(29 - 6 \cdot \tfrac{1+\sqrt{39}}{2}) = \tfrac{1}{4} \cdot \tfrac{58-6-6\sqrt{39}}{2} = \tfrac{52-6\sqrt{39}}{8} = \tfrac{26-3\sqrt{39}}{4}.$$

$$b = \tfrac{1-\sqrt{39}}{2} \Rightarrow a = \tfrac{1}{4}(29 - 6 \cdot \tfrac{1-\sqrt{39}}{2}) = \tfrac{1}{4} \cdot \tfrac{58-6+6\sqrt{39}}{2} = \tfrac{52+6\sqrt{39}}{8} = \tfrac{26+3\sqrt{39}}{4}.$$

We know that the radius is $|a|$, and that the center is (a, b).

Therefore, the circle C is: $(x-a)^2 + (y-b)^2 = a^2$ where:

$$a = \tfrac{26-3\sqrt{39}}{4} \approx 1.816 \text{ and } b = \tfrac{1+\sqrt{39}}{2} \approx 3.622$$

or $a = \tfrac{26+3\sqrt{39}}{4} \approx 11.184 \text{ and } b = \tfrac{1-\sqrt{39}}{2} \approx -2.622.$

Let's now, put the circles in a graph.

Fig. 2

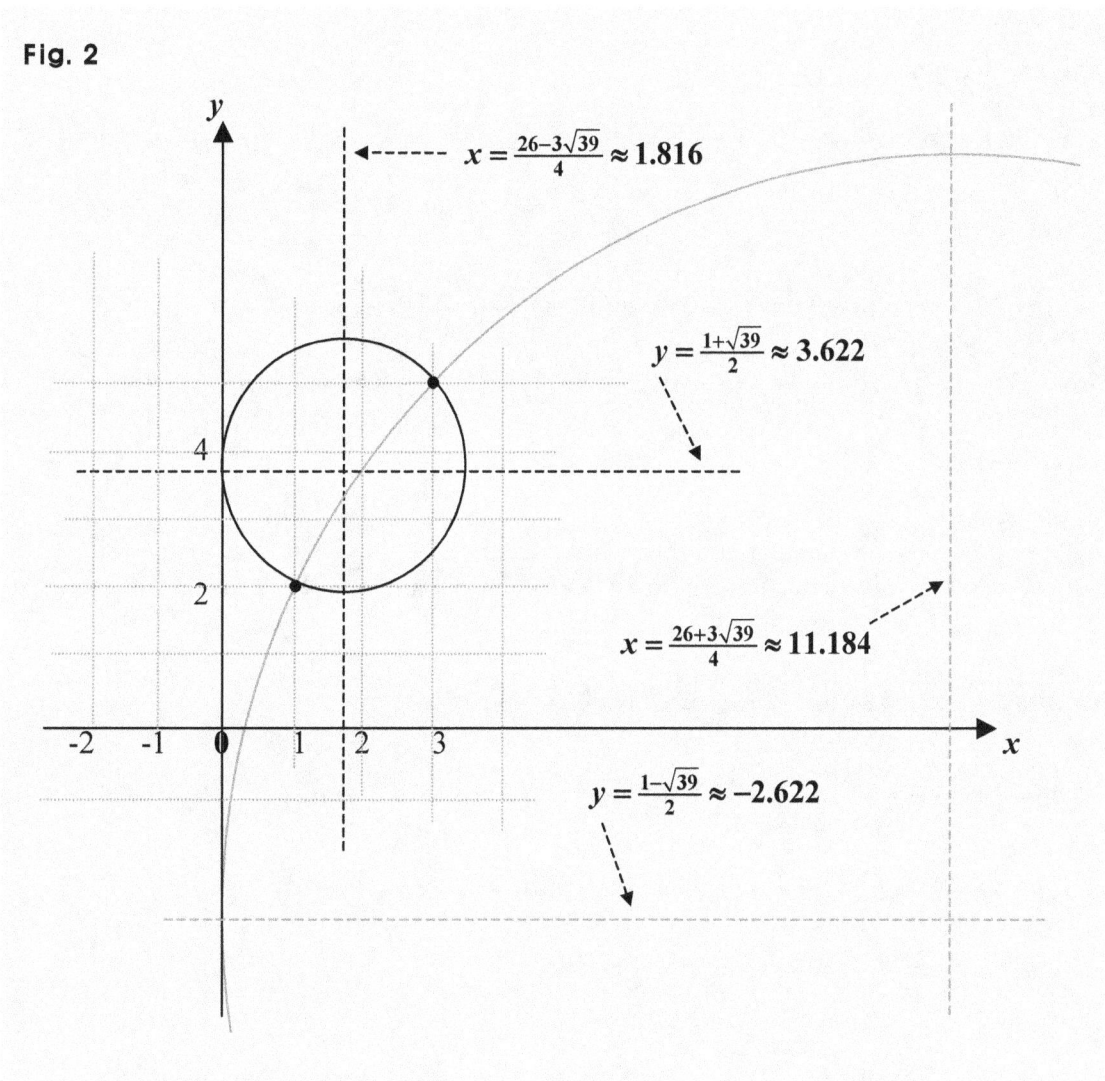

In short:

Suppose that the center is in a line $x = a$, and that the point where the circle is tangent to the y-axis is $(0, b)$.   Then, the center is at a point (a, b), where the line $x = a$ meets the line $y = b$, and the radius is $|a|$.

Each of the two points given is the radius $|a|$ away from (a, b).

So taking the distances from (a, b) to $(1, 2)$ and $(3, 5)$, we get:

$b^2 = (a - 1)^2 + (b - 2)^2$, and $b^2 = (a - 3)^2 + (b - 5)^2$.

Then first, we get:

$(a - 1)^2 + (b - 2)^2 = a^2 \Rightarrow a^2 - 2a + 1 + b^2 - 4b + 4 = a^2 \Rightarrow b^2 - 4b - 2a + 5 = 0$.

Next, we get:

$(a - 1)^2 + (b - 2)^2 = (a - 3)^2 + (b - 5)^2 \Rightarrow (a - 1)^2 - (a - 3)^2 + (b - 2)^2 - (b - 5)^2 = 0$

$\Rightarrow \{(a - 1) - (a - 3)\}\{(a - 1) + (a - 3)\} + \{(b - 2) - (b - 5)\}\{(b - 2) + (b - 5)\} = 0$

$\Rightarrow 2(2a - 4) + 3(2b - 7) = 4a + 6b - 29 = 0$.

Thus, we need to solve a system where $4a + 6b - 29 = 0$ and $b^2 - 4b - 2a + 5 = 0$.

Then first, we get: $4a + 6b - 29 = 0 \Rightarrow a = \frac{29 - 6b}{4}$. Next, we get:

$b^2 - 4b - 2a + 5 = b^2 - 4b - 2 \cdot \frac{(29 - 6b)}{4} + 5 = b^2 - 4b + 3b + 5 - \frac{29}{2} = 0 \Rightarrow b^2 - b - \frac{19}{2} = 0$.

Then, $b^2 - b - \frac{19}{2} = b^2 - b + \frac{1}{4} - \frac{1}{4} - \frac{19}{2} = (b - \frac{1}{2})^2 - \frac{1}{4} - \frac{19}{2} = (b - \frac{1}{2})^2 - \frac{39}{4} = 0$

$\Rightarrow (b - \frac{1}{2})^2 = \frac{39}{4} \Rightarrow b - \frac{1}{2} = \pm\sqrt{\frac{39}{4}} \Rightarrow b = \frac{1}{2} \pm \frac{\sqrt{39}}{2}$.

So we get: $b = \frac{1 + \sqrt{39}}{2}$ or $b = \frac{1 - \sqrt{39}}{2}$.   And we have: $a = \frac{29 - 6b}{4}$, too.   Thus, we get:

$b = \frac{1 + \sqrt{39}}{2} \Rightarrow a = \frac{1}{4}(29 - 6 \cdot \frac{1 + \sqrt{39}}{2}) = \frac{1}{4} \cdot \frac{58 - 6 - 6\sqrt{39}}{2} = \frac{52 - 6\sqrt{39}}{8} = \frac{26 - 3\sqrt{39}}{4}$.

$b = \frac{1 - \sqrt{39}}{2} \Rightarrow a = \frac{1}{4}(29 - 6 \cdot \frac{1 - \sqrt{39}}{2}) = \frac{1}{4} \cdot \frac{58 - 6 + 6\sqrt{39}}{2} = \frac{52 + 6\sqrt{39}}{8} = \frac{26 + 3\sqrt{39}}{4}$.

We know that the radius is $|a|$, and that the center is (a, b).

Therefore, the circle C is: $(x - a)^2 + (y - b)^2 = a^2$, where $a = \frac{26 \mp 3\sqrt{39}}{4}$ and $b = \frac{1 \pm \sqrt{39}}{2}$.

Examples 7 in Circles

0.   Suppose that:

A is a circle $(x - a_1)^2 + (y - b_1)^2 = c_1{}^2$, where a_1, b_1, and c_1 are constant, and $c_1 > 0$.

B is a circle $(x - a_2)^2 + (y - b_2)^2 = c_2{}^2$, where a_2, b_2, and c_2 are constant, and $c_2 > 0$.

d is the distance between the centers of the two circles A and B.

Then, find the relationship among c_1, c_2, and d in each of the cases below:

0.0.   The two circles meet at two points.

0.1.   The two circle meet at one point (i.e., they are tangent to each other).

0.2.   The two circles do not meet each other.

1.   Suppose that:

A is a circle $x^2 + y^2 + ax + by + c = 0$, where a, b, and c are constant.

B is a circle $x^2 + y^2 + dx + ey + f = 0$, where d, e, and f are constant.

Suppose also, that the two circles meet each other at two particular points.

Then, find another circle passing through the two particular points.

Suggestions or Solutions
To the Problem 0 in the Example 0

Suppose that:

A is a circle $(x - a_1)^2 + (y - b_1)^2 = c_1{}^2$, where a_1, b_1, and c_1 are constant, and $c_1 > 0$.

B is a circle $(x - a_2)^2 + (y - b_2)^2 = c_2{}^2$, where a_2, b_2, and c_2 are constant, and $c_2 > 0$.

d is the distance between the centers of the two circles A and B.

Then, find the relationship among c_1, c_2, and d if the two circles meet at two points.

Let's first, put some circles in a graph so that two circles can meet at two points. Then, we can see better what we can do about the problem. Now, what matters in a circle?

Fig. 0

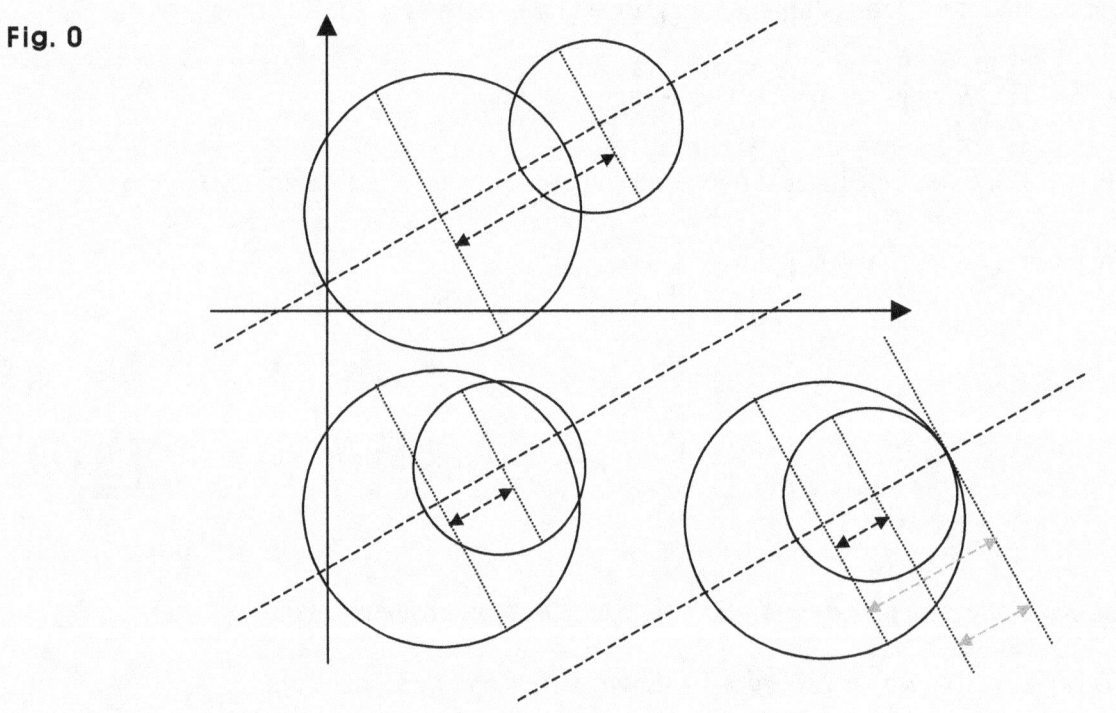

The center as well as the radius matters. So let's take a closer look at those two. Examining the radii and the distance between the centers, we can see how the radii and the distance should be related if two circles have to meet each other at two points. First, the distance between the centers has to be less than the sum of the two radii. However, the distance has to be greater than the difference between the two radii. So if the two circle meet each other at two points, we need to have:

$|c_1 - c_2| < d < c_1 + c_2$.

Suggestions or Solutions
To the Problem 1 in the Example 0

Suppose that:

A is a circle $(x - a_1)^2 + (y - b_1)^2 = c_1^2$, where a_1, b_1, and c_1 are constant, and $c_1 > 0$.

B is a circle $(x - a_2)^2 + (y - b_2)^2 = c_2^2$, where a_2, b_2, and c_2 are constant, and $c_2 > 0$.

d is the distance between the centers of the two circles A and B.

Then, find the relationship among c_1, c_2, and d if the two circles meet at one point.

To begin with, let's put some circles in a graph so that two circles meet each other at one point. Then, we can see better where the solution can be around.

Fig. 0

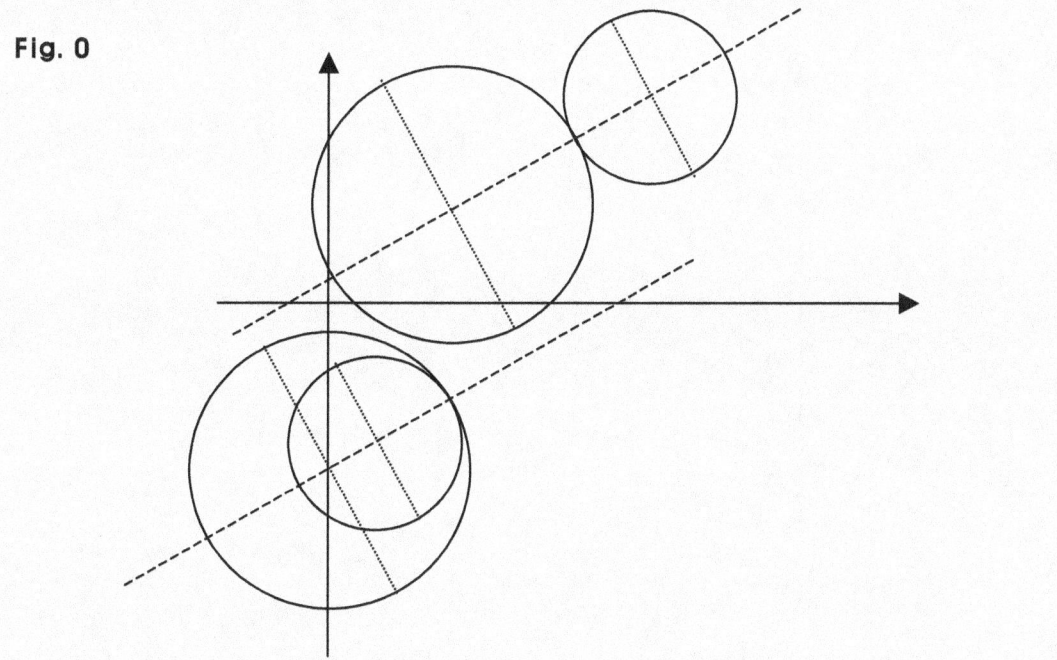

Now, two circles meet each other at one point in either of two cases below:

• The distance between the centers is the same as the sum of the two radii

• The distance between the centers is the same as the difference between the two radii.

So if the two circles meet at one point, we need to have: $d = |c_1 - c_2|$, or $d = c_1 + c_2$.

More specifically:

• If $d = |c_1 - c_2|$, then the two circles are tangent to each other internally.

That is, the smaller circle is tangent to the larger circle inside the larger.

• If $d = c_1 + c_2$, then the two circles are tangent to each other externally.

Suggestions or Solutions
To the Problem 2 in the Example 0

Suppose that:
A **is a circle** $(x - a_1)^2 + (y - b_1)^2 = c_1^2$, **where** a_1, b_1, **and** c_1 **are constant, and** $c_1 > 0$.
B **is a circle** $(x - a_2)^2 + (y - b_2)^2 = c_2^2$, **where** a_2, b_2, **and** c_2 **are constant, and** $c_2 > 0$.
d **is the distance between the centers of the two circles** A **and** B.

Then, find the relationship among c_1, c_2, **and** d **if the two circles do not meet each other.**

Let's this time, put some circles in a graph so that two circles do not meet each other.

Fig. 0

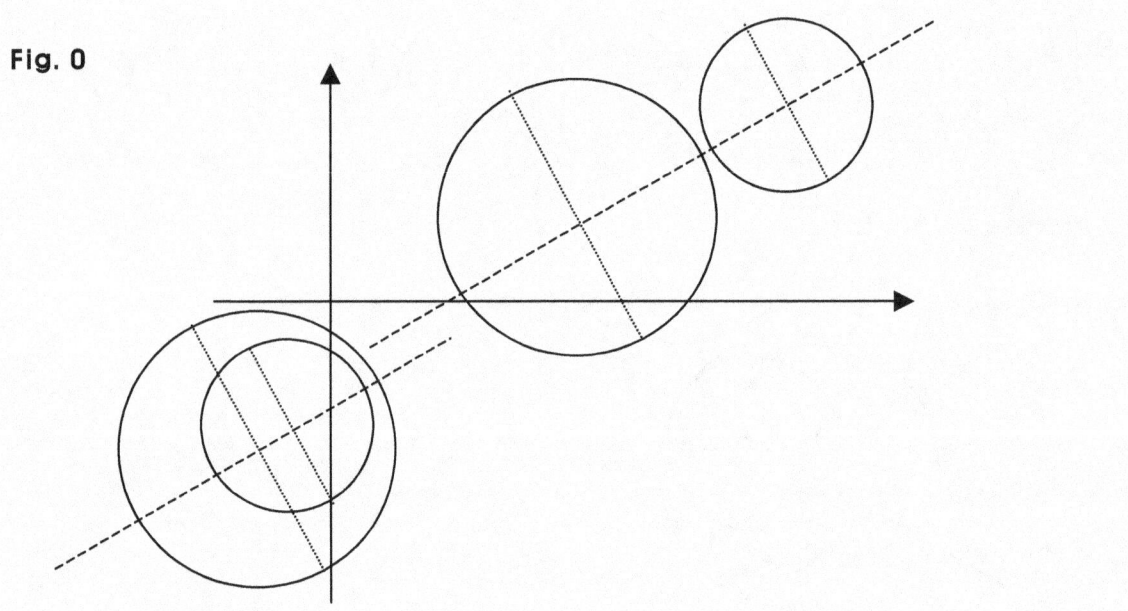

Now, we can see that two circles do not meet each other in either of two cases below.
• The distance between the centers is greater than the sum of the two radii.
• The distance between the centers is less than the difference between the two radii.

So if the two circles do not meet each other, we need to have: $d < |c_1 - c_2|$, or $d > c_1 + c_2$. More specifically:

• If $d < |c_1 - c_2|$, then the two circles are away from each other internally.
That is, the smaller circle is inside the larger circle, but is not tangent to the larger.

• If $d > c_1 + c_2$, then the two circles are away from each other externally. That is, the two circles are completely outside each other.

Suggestions or Solutions
To the Problem in the Example 1

Suppose that:

A **is a circle** $x^2 + y^2 + ax + by + c = 0$**, where** a**,** b**, and** c **are constant.**

B **is a circle** $x^2 + y^2 + dx + ey + f = 0$**, where** d**,** e**, and** f **are constant.**

Suppose also, the two circles meet each other at two particular points.

Then, find another circle passing through the two particular points.

Suppose first, m and n are constant, and $m + n \neq 0$.

Then, $m(x^2 + y^2 + ax + by + c) + n(x^2 + y^2 + dx + ey + f) = 0$, too, indicates a circle.

That's because:

First, $m(x^2 + y^2 + ax + by + c) + n(x^2 + y^2 + dx + ey + f)$

$= (m + n)x^2 + (m + n)y^2 + (am + dn)x + (bm + en)y + cm + fn$

$= (m + n)\{x^2 + y^2 + \frac{(am+dn)}{(m+n)}x + \frac{(bm+en)}{(m+n)}y + \frac{(cm+fn)}{(m+n)}\} = 0$. And we have: $m + n \neq 0$.

So next, we get: $x^2 + y^2 + \frac{(am+dn)}{(m+n)}x + \frac{(bm+en)}{(m+n)}y + \frac{(cm+fn)}{(m+n)} = 0$, which is a circle since a, b, c, d, e, f, m, and n are constants.

Suppose next, C is the circle $m(x^2 + y^2 + ax + by + c) + n(x^2 + y^2 + dx + ey + f) = 0$ where $m + n \neq 0$. Suppose also, (s, t) and (u, v) are the two points where two circles A and B meet each other.   Then:

First, we get: $s^2 + t^2 + as + bt + c = 0$ and $s^2 + t^2 + ds + et + f = 0$ since each of the two circles A and B passes through the point (s, t).

So we get: $m(s^2 + t^2 + as + bt + c) + n(s^2 + t^2 + ds + et + f) = 0$, which means C passes through the point (s, t), too.

Next, we get: $u^2 + v^2 + au + bv + c = 0$ and $u^2 + v^2 + du + ev + f = 0$ since each of the two circles A and B passes through the other point (u, v), too.

162

So we get: $m(u^2 + v^2 + au + bv + c) + n(u^2 + v^2 + du + ev + f) = 0$, which means C passes through the point (u, v), too.

Thus, the circle C passes through the two points (s, t) and (u, v), too.

So the three circles A, B, and C share the two points (s, t) and (u, v).

Therefore, the solution to this problem is as follows:

$m(x^2 + y^2 + ax + by + c) + n(x^2 + y^2 + dx + ey + f) = 0$ where m & n are constants, and $m + n \neq 0$.

If not quite sure of the idea behind the processes above, follow the steps below:

Knowing three points not in a line, we can find a particular circle, which passes through the three points, of course.

Three points not in a line can determine one circle only. That is, one circle only can pass through such three points, and no two different circles can share such three points.

What then, about two points?

Infinitely many circles can share the two, and of course, one point, too, can be shared by that many circles. Thus, infinitely many circles can be the solution to this problem.

It doesn't mean that any circle can be the solution, though.
So let's find out about what circles can be the solution.

Suppose now, that m and n are constant, that $m + n \neq 0$, and that curves of two equations $f(x, y) = 0$ and $g(x, y) = 0$ meet each other at particular points (that is, they share the particular points).
The two equations f and g are in terms of the two variables x and y, and for instance, the two equations can be: $3xy + 2 = 0$ and $x^2 - xy + 2y^3 - 1 = 0$.

Then, the curve of another equation $mf(x, y) + ng(x, y) = 0$ meets at the particular points the curves of $f(x, y) = 0$ and $g(x, y) = 0$.

That is, the curve of $mf(x, y) + ng(x, y) = 0$, too, includes the very particular points, so all the three curves share the particular points.

Now, the same is true for circles, too. In this example, the two circles A and B meet each other at two particular points, and their equations are as follows:

$x^2 + y^2 + ax + by + c = 0$ and $x^2 + y^2 + dx + ey + f = 0$.

Thus, if $m + n \neq 0$, $m(x^2 + y^2 + ax + by + c) + n(x^2 + y^2 + dx + ey + f) = 0$ is a circle, and meets two circles A and B at the two particular points.   How come?

Rearranging the equation above, we get:

$m(x^2 + y^2 + ax + by + c) + n(x^2 + y^2 + dx + ey + f)$

$= (m + n)x^2 + (m + n)y^2 + (am + dn)x + (bm + en)y + cm + fn$

$= (m + n)\{x^2 + y^2 + \frac{(am+dn)}{(m+n)} x + \frac{(bm+en)}{(m+n)} y + \frac{(cm+fn)}{(m+n)}\} = 0$. And we have: $m + n \neq 0$.

So we get: $x^2 + y^2 + \frac{(am+dn)}{(m+n)} x + \frac{(bm+en)}{(m+n)} y + \frac{(cm+fn)}{(m+n)} = 0$, which is a circle since $a, b, c, d, e,$ $f, m,$ and n are constant, and in turn, the fractional expressions are constant, too.

Suppose next, C is the circle $m(x^2 + y^2 + ax + by + c) + n(x^2 + y^2 + dx + ey + f) = 0$, where $m + n \neq 0$, and the circle A meets the circle B at two points (s, t) and (u, v).

We know A is: $x^2 + y^2 + ax + by + c = 0$, and B is: $x^2 + y^2 + dx + ey + f = 0$.

Then, first, we get: $s^2 + t^2 + as + bt + c = 0$ and $s^2 + t^2 + ds + et + f = 0$ since each of the two circles A and B passes through the point (s, t).

Thus, we get: $m(s^2 + t^2 + as + bt + c) + n(s^2 + t^2 + ds + et + f) = 0$, which means the circle C passes through the point (s, t), too.

Also, we get: $u^2 + v^2 + au + bv + c = 0$ and $u^2 + v^2 + du + ev + f = 0$ since each of the two circles A and B passes through the other point (u, v), too.

So we get: $m(u^2 + v^2 + au + bv + c) + n(u^2 + v^2 + du + ev + f) = 0$, which means the circle C passes through the point (u, v), too.

Therefore, the curve of $m(x^2 + y^2 + ax + by + c) + n(x^2 + y^2 + dx + ey + f) = 0$, that is, the circle C passes through the two points (s, t) and (u, v).

So the three circles A, B and C meet altogether at the two points (s, t) and (u, v). In other words, the three circles A, B, and C share the two points (s, t) and (u, v).

Therefore, the solution to this problem is as follows:

$m(x^2 + y^2 + ax + by + c) + n(x^2 + y^2 + dx + ey + f) = 0$ where m & n are constants, and $m + n \neq 0$.

What if $m + n = 0$, though?

Then, $m(x^2 + y^2 + ax + by + c) + n(x^2 + y^2 + dx + ey + f) = 0$ is an equation of a line passing through the two points (s, t) and (u, v). How come?

To begin with, $m(x^2 + y^2 + ax + by + c) + n(x^2 + y^2 + dx + ey + f)$
$= (m + n)x^2 + (m + n)y^2 + (am + dn)x + (bm + en)y + cm + fn$
$= (am + dn)x + (bm + en)y + cm + fn = 0$, since $m + n = 0$.

Next, we know $am + dn$, $bm + en$, and $cm + fn$ are constant since a, b, c, d, e, f, m, and n are constants. So $(am + dn)x + (bm + en)y + cm + fn = 0$ is an equation of a line.

Now, since the two circles $x^2 + y^2 + ax + by + c = 0$ and $x^2 + y^2 + dx + ey + f = 0$ include respectively the point (s, t), we get: $s^2 + t^2 + as + bt + c = 0$ and $s^2 + t^2 + ds + et + f = 0$.

So we get: $m(s^2 + t^2 + as + bt + c) + n(s^2 + t^2 + ds + et + f) = 0$.
Thus, since $m + n = 0$, we get: $(am + dn)s + (bm + en)t + cm + fn = 0$.

Also, since the two circles $x^2 + y^2 + ax + by + c = 0$ and $x^2 + y^2 + dx + ey + f = 0$ share the point (u, v), we get: $u^2 + v^2 + au + bv + c = 0$ and $u^2 + v^2 + du + ev + f = 0$.

So we get: $m(u^2 + v^2 + au + bv + c) + n(u^2 + v^2 + du + ev + f) = 0$.

Therefore, since $m + n = 0$, we get: $(am + dn)u + (bm + en)v + cm + fn = 0$.

Thus, if $m + n = 0$, then $m(x^2 + y^2 + ax + by + c) + n(x^2 + y^2 + dx + ey + f) = 0$ indicates a line, which is $(am + dn)x + (bm + en)y + cm + fn = 0$, and also, the line passes through the two points (s, t) and (u, v) at which the two circles $x^2 + y^2 + ax + by + c = 0$ and $x^2 + y^2 + dx + ey + f = 0$ meet each other.

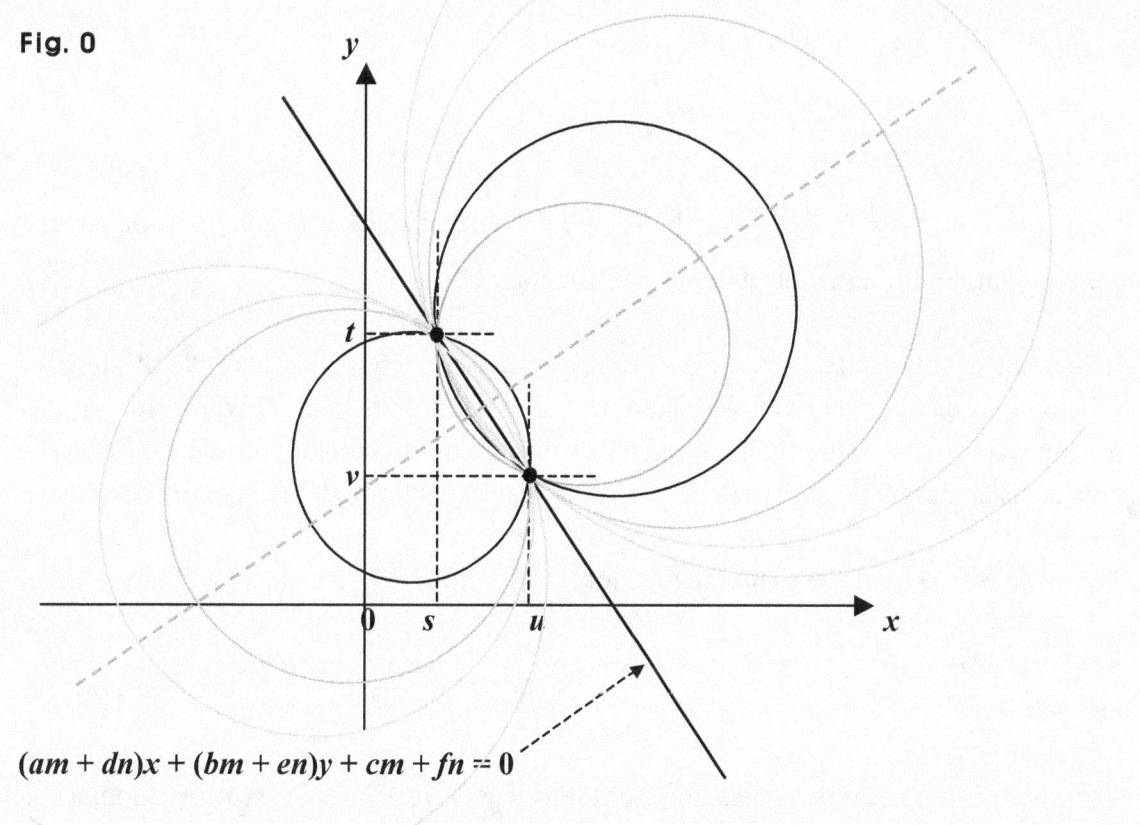

Fig. 0

$(am + dn)x + (bm + en)y + cm + fn = 0$

The centers of all the circles in the graph above are in a line, which is dashed in gray, passes through the midpoint between the two points (s, t) and (u, v), and is perpendicular to the line $(am + dn)x + (bm + en)y + cm + fn = 0$ passing through (s, t) and (u, v).

What about a circle and a line? That is, what if a circle meets a line at particular points?

The same is true for a circle and a line, too.

Suppose now, a circle $x^2 + y^2 + ax + by + c = 0$ meets a line $ux + vy + w = 0$ at particular points.

Then, an equation $m(x^2 + y^2 + ax + by + c) + n(ux + vy + w) = 0$ where $m \neq 0$ indicates a circle, and the circle passes through the particular points, too. That is to say that the two circles and the line share the particular points.

How come the equation above indicates a circle, though?

To begin with, $m(x^2 + y^2 + ax + by + c) + n(ux + vy + w) = 0$

$\Rightarrow m(x^2 + y^2) + (ma + nu)x + (mb + nv)y + mc + nw = 0$

$\Rightarrow x^2 + y^2 + \frac{(am+nu)}{m}x + \frac{(bm+nv)}{m}y + \frac{(cm+nw)}{m} = 0.$

Next, we know $\frac{(am+nu)}{m}$, $\frac{(bm+nv)}{m}$, and $\frac{(cm+nw)}{m}$ are constant since a, b, c, u, v, w, m, and n are constant. So the equation above is an equation of circle.

In fact, the equation $m(x^2 + y^2 + ax + by + c) + n(ux + vy + w) = 0$ where $m \neq 0$ can be said to represent a group of circles share the particular points. In other words, for a pair of values for m and n, the equation indicates one circle, for another pair, the equation indicates another circle, and each of all the circles passes through the particular points.

The same is true for the equation below, too:

$m(x^2 + y^2 + ax + by + c) + n(x^2 + y^2 + dx + ey + f) = 0$ where m & n are constants, and $m + n \neq 0$.

Suppose C is the equation above, and a circle $x^2 + y^2 + ax + by + c = 0$ meets another circle $x^2 + y^2 + dx + ey + f = 0$ at particular points.

Then, the equation C can be said to represent a group of circles that share the particular points. That is, for one pair of values of m and n, the equation C indicates one circle, for another pair, C indicates another circle, and so forth, and each of all the circles passes through the particular points.

Now, we can simplify the whole story above into such a fact as follows:

Suppose two circles $x^2 + y^2 + ax + by + c = 0$ and $x^2 + y^2 + dx + ey + f = 0$ meet each other at particular points, and m and n are constants.

Then, if $m + n \neq 0$, for each pair of the values of m and n, $m(x^2 + y^2 + ax + by + c) + n(x^2 + y^2 + dx + ey + f) = 0$ indicates another circle passing through the particular points, and the center is in the line passing through the centers of the two circles. And the line is perpendicular to the line passing through the particular points.

In short:

Suppose first, m and n are constant, and $m + n \neq 0$.

Then, $m(x^2 + y^2 + ax + by + c) + n(x^2 + y^2 + dx + ey + f) = 0$, too, indicates a circle.

That's because:

First, $m(x^2 + y^2 + ax + by + c) + n(x^2 + y^2 + dx + ey + f)$
$= (m + n)x^2 + (m + n)y^2 + (am + dn)x + (bm + en)y + cm + fn$
$= (m + n)\{x^2 + y^2 + \frac{(am+dn)}{(m+n)}x + \frac{(bm+en)}{(m+n)}y + \frac{(cm+fn)}{(m+n)}\} = 0$. And we have: $m + n \neq 0$.

So next, we get: $x^2 + y^2 + \frac{(am+dn)}{(m+n)}x + \frac{(bm+en)}{(m+n)}y + \frac{(cm+fn)}{(m+n)} = 0$, which is a circle since a, b, c, d, e, f, m, and n are constants.

Suppose next, C is the circle $m(x^2 + y^2 + ax + by + c) + n(x^2 + y^2 + dx + ey + f) = 0$ where $m + n \neq 0$. Suppose also, (s, t) and (u, v) are the two points where two circles A and B meet each other.   Then:

First, we get: $s^2 + t^2 + as + bt + c = 0$ and $s^2 + t^2 + ds + et + f = 0$ since each of the two circles A and B passes through the point (s, t).

So we get: $m(s^2 + t^2 + as + bt + c) + n(s^2 + t^2 + ds + et + f) = 0$, which means C passes through the point (s, t), too.

Next, we get: $u^2 + v^2 + au + bv + c = 0$ and $u^2 + v^2 + du + ev + f = 0$ since each of the two circles A and B passes through the other point (u, v), too.

So we get: $m(u^2 + v^2 + au + bv + c) + n(u^2 + v^2 + du + ev + f) = 0$, which means C passes through the point (u, v), too.

Thus, the circle C passes through the two points (s, t) and (u, v), too.

So the three circles A, B, and C share the two points (s, t) and (u, v).

Therefore, the solution to this problem is as follows:

$m(x^2 + y^2 + ax + by + c) + n(x^2 + y^2 + dx + ey + f) = 0$ where m & n are constants, and $m + n \neq 0$.

Examples 8 in Circles

See if there is any particular point the curve of each equation below passes through for any value of **k**, which is constant, and specify the particular point if any.

0. $(k + 1)(x^2 + y^2) + 2(2k + 1)x + 2(k - 1)y + 4k - 2 = 0.$

1. $9(k + 1)(x^2 + y^2) - 36(k + 2)x - 18(k + 3)y + 37k + 193 = 0.$

2. $(k + 1)(x^2 + y^2) - 2(3k + 4)x - 2(k + 2)y + 2k + 18 = 0.$

3. $(k + 1)(x^2 + y^2) - 2(k + 4)x - 2(k + 3)y + k + 21 = 0.$

4. $(k + 1)(x^2 + y^2) - 8(k + 1)x - 6(k + 1)y + 24k + 21 = 0.$

5. $(k + 1)(x^2 + y^2) - 2(2k + 3)x - 2(2k + 3)y - k + 17 = 0.$

Suggestions or Solutions
To the Problem in the Example 0

See if there is any particular point the curve of the equation below passes through for any value of k, which is constant, and specify the particular point if any.

$(k + 1)(x^2 + y^2) + 2(2k + 1)x + 2(k - 1)y + 4k - 2 = 0.$

$(k + 1)(x^2 + y^2) + 2(2k + 1)x + 2(k - 1)y + 4k - 2$

$= k(x^2 + y^2) + 4kx + 2ky + 4k + x^2 + y^2 + 2x - 2y - 2$

$= k(x^2 + y^2 + 4x + 2y + 4) + (x^2 + y^2 + 2x - 2y - 2) = 0$, which can be an equation of a circle for each value of k.

Suppose a circle $x^2 + y^2 + 4x + 2y + 4 = 0$ meets another circle $x^2 + y^2 + 2x - 2y - 2 = 0$ at (s, t).

Then, we get: $s^2 + t^2 + 4s + 2t + 4 = 0$, and $s^2 + t^2 + 2s - 2t - 2 = 0$.

So we get: $k(s^2 + t^2 + 4s + 2t + 4) + s^2 + t^2 + 2s - 2t - 2 = 0$.

So if the two circles above meet at a particular point, for any value of k, the equation given indicates a circle that includes the particular point. And the same is true for another particular point, too, if the two circles share the other particular point. So if the two circles meet at particular points, for any value of k, the equation given indicates a circle that includes the particular points.   So now, finding the points, we get:

First, $x^2 + y^2 + 4x + 2y + 4 = x^2 + y^2 + 2x - 2y - 2 \Rightarrow 2x + 4y + 6 = 0 \Rightarrow x + 2y + 3 = 0$

$\Rightarrow x = -2y - 3 \Rightarrow x^2 + y^2 + 2x - 2y - 2 = (-2y - 3)^2 + y^2 + 2(-2y - 3) - 2y - 2 = 0$

$\Rightarrow 4y^2 + 12y + 9 + y^2 - 4y - 6 - 2y - 2 = 5y^2 + 6y + 1 = (5y + 1)(y + 1) = 0$

$\Rightarrow y = -\frac{1}{5}$ or $y = -1$.

Next, $y = -\frac{1}{5} \Rightarrow x = -2y - 3 = \frac{2}{5} - 3 = -\frac{13}{5} \Rightarrow (-\frac{13}{5}, -\frac{1}{5})$, and

$y = -1 \Rightarrow x = -2y - 3 = -1 \Rightarrow (-1, -1)$.

Thus, the two circles meet at two points $(-1, -1)$ and $(-\frac{13}{5}, -\frac{1}{5})$, which are therefore, the particular points a circle indicated by the equation given includes for any value of k.

If not quite sure of the idea behind the processes above, follow the steps below:

The problems in this example 2 are no other than the one in the example 1 in the **Examples 7**. Doing those problems, we will see how a group of circles can share particular points, and see better how equations for circles work.

The number of circles in such a group can be infinite or finite depending on the way we set the constant *k* in the equation.

For instance, we can set: $3 \leq k < 7$ where *k* is integer, or can set $1 < k < 2$ where *k* is real.

Then in the first case, *k* can be 3, 4, 5, and 6 only, so the number of circles in the group can be up to 4, and in the second, since *k* can be any real number between 1 and 2, we can make infinite choices for *k*, so the group can have infinite circles.

Let's now, begin with a fact as follows:

• Suppose two circles $x^2 + y^2 + ax + by + c = 0$ and $x^2 + y^2 + dx + ey + f = 0$ meet at particular points, *m* and *n* are constants, and $m + n \neq 0$. Then, for each pair of values of *m* and *n*, a new equation $m(x^2 + y^2 + ax + by + c) + n(x^2 + y^2 + dx + ey + f) = 0$ indicates a circle passing through the particular points.

So every circle indicated by the new equation above passes through the particular points. The same is true for a circle and a line, too.

• Suppose a circle $x^2 + y^2 + ax + by + c = 0$ meets a line $ux + vy + w = 0$ at particular points. Then, for each pair of values of *m* and *n*, a new equation $m(x^2 + y^2 + ax + by + c) + n(ux + vy + w) = 0$ where $m \neq 0$ indicates a circle passes through the particular points.

So every circle indicated by the new equation above passes through the particular points.

• In particular, such a new equation can be called a linear combination of two equations.

Such two equations indicate respectively either 'two circles' or 'a circle and a line'.

So such a circle indicated by such a new equation can be said to be 'a linear combination of two circles' or 'a linear combination of a circle and a line'.

Besides, every center of every circle mentioned above is in the same line.

So for instance, if the center of every circle that is a linear combination of two circles A and B is in a line L, the line L passes through the centers of the two circles A and B, too.

The same is true for a linear combination of a circle and a line, also.

So for instance, if the center of every circle that is a linear combination of a circle C and a line N is in a line D, the line D passes through the center of the circle C, too.

In addition, the line D is perpendicular to the line N.

And the similar idea applies to the case of the linear combination of two circles.

Suppose for instance, two circles A and B meet each other at two points P and Q, a line R passes through P and Q, and a line L passes through the centers of A and B.

Then, the line L is perpendicular to the line R, and has all the centers of all the circles that are linear combinations of A and B.

Fig. 0

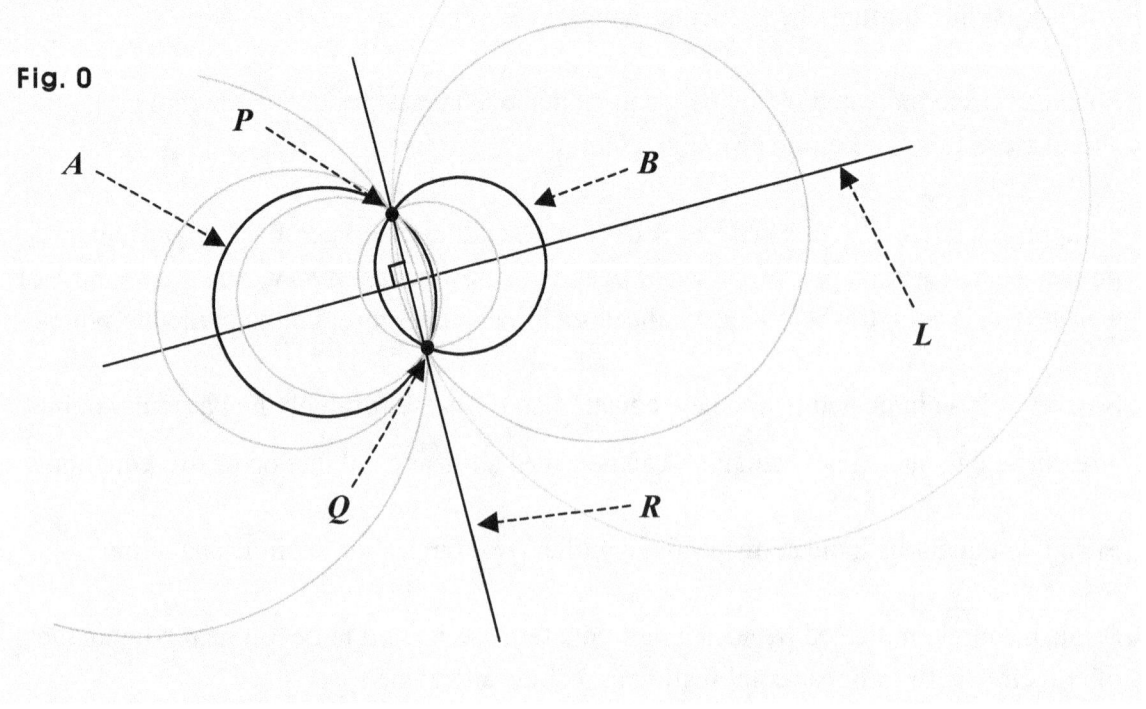

Fig. 1

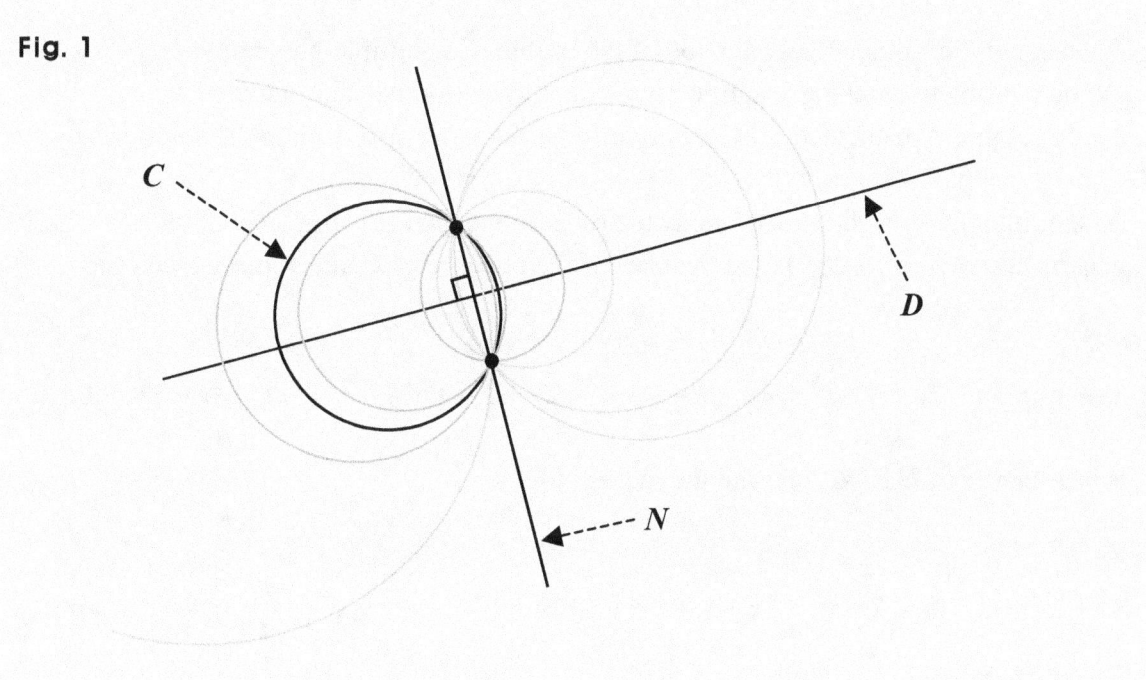

Now, in this problem, we have: $(k + 1)(x^2 + y^2) + 2(2k + 1)x + 2(k - 1)y + 4k - 2 = 0$ for k constant. For each value of k, the equation above can indicate a circle, which can be a linear combination of two curves, which can be two circles, or can be a circle and a line.

So let's extract the two curves from the equation given, and see what the curves are.

$$(k + 1)(x^2 + y^2) + 2(2k + 1)x + 2(k - 1)y + 4k - 2 = 0$$

$$\Rightarrow k(x^2 + y^2) + x^2 + y^2 + 4kx + 2x + 2ky - 2y + 4k - 2$$

$$= k(x^2 + y^2) + 4kx + 2ky + 4k + x^2 + y^2 + 2x - 2y - 2$$

$$= k(x^2 + y^2 + 4x + 2y + 4) + (x^2 + y^2 + 2x - 2y - 2) = 0.$$

Therefore, we can see that $m = k$, and $n = 1$ in the form of a linear combination of two equations, and that the curves are two circles as follows:

$$x^2 + y^2 + 4x + 2y + 4 = 0 \text{ and } x^2 + y^2 + 2x - 2y - 2 = 0.$$

What if however, we just put a couple of values into k in turn, get two circles, and then, find the two points?

We can get the solution that way, too, if the problem is multiple-choice.
Doing a problem in math though, we may want to justify the solution, too.
That is, doing it mathematically, we not only provide but also explain the solution.

Assuming now, U is the first of the two circles above, and V is the second, let's find the centers and radii of U and V and then, see if the two circles U and V meet each other.

$$x^2 + y^2 + 4x + 2y + 4 = x^2 + 4x + 4 - 4 + y^2 + 2y + 1 - 1 + 4 = (x + 2)^2 + (y + 1)^2 - 1 = 0$$

So the center of U is **(-2, -1)**, and the radius is 1.

$$x^2 + 2x + y^2 - 2y - 2 = x^2 + 2x + 1 - 1 + y^2 - 2y + 1 - 1 - 2 = (x + 1)^2 + (y - 1)^2 - 4 = 0$$

So the center of V is **(-1, 1)**, and the radius is 2.

Suppose now, D is the distance between the two centers **(-2, -1)** and **(-1, 1)**, S is the sum of the two radii, T is the difference between the radii, and Δ is the difference between two coordinates. We can find D by the distance formula, of course. Then, we get:

$$D^2 = (\Delta x)^2 + (\Delta y)^2 = \{-1 - (-2)\}^2 + \{1 - (-1)\}^2 = 1 + 4 = 5, \ S = 1 + 2 = 3, \ \& \ T = 2 - 1 = 1.$$

So we get: $S > D$ for S and D are nonnegative and $S^2 = 9 > D^2 = 5$. So we get: $T < D < S$.

Thus, we can see that the circle U meets the circle V at two particular points.

(If not sure, refer to the example 0 in **Examples 7**.)

Therefore, the equation $(k + 1)(x^2 + y^2) + 2(2k + 1)x + 2(k - 1)y + 4k - 2 = 0$ can be said to represent a group of circles sharing altogether the two particular points since for each value of k, the equation indicates a circle passing through the two particular points.

Let's now, get the two particular points. Then, we want to solve the system where:

$(x + 2)^2 + (y + 1)^2 - 1 = 0$ and $(x + 1)^2 + (y - 1)^2 - 4 = 0$. Then, we can get first:

$$(x + 1)^2 + (y - 1)^2 - 4 = (x + 2)^2 + (y + 1)^2 - 1$$

$$\Rightarrow (x + 1)^2 - (x + 2)^2 + (y - 1)^2 - (y + 1)^2 - 4 - (-1) = 0$$

$$\Rightarrow \{(x + 1) - (x + 2)\}\{(x + 1) + (x + 2)\} + \{(y - 1) - (y + 1)\}\{(y - 1) + (y + 1)\} - 3$$

$$= (-1)(2x + 3) - 2(2y) - 3 = -2x - 4y - 6 = 0 \Rightarrow x + 2y + 3 = 0 \Rightarrow x = -2y - 3.$$

Thus, we get: $x = -2y - 3 \Rightarrow (x + 2)^2 + (y + 1)^2 - 1 = (-2y - 3 + 2)^2 + (y + 1)^2 - 1 = 0$

$$\Rightarrow (-2y - 1)^2 + (y + 1)^2 - 1 = (2y + 1)^2 + (y + 1)^2 - 1 = 4y^2 + 4y + 1 + y^2 + 2y + 1 - 1$$

$$= 5y^2 + 6y + 1 = (5y + 1)(y + 1) = 0 \Rightarrow y = -\frac{1}{5} \text{ or } y = -1.$$

Therefore, $y = -\frac{1}{5} \Rightarrow x = -2y - 3 = \frac{2}{5} - 3 = -\frac{13}{5}$, and $y = -1 \Rightarrow x = -2y - 3 = -1$.

Now, all the circles in the group share altogether two points, **(-1, -1)** and $(-\frac{13}{5}, -\frac{1}{5})$.

Let's now, put in a graph some of the circles in the group, together with *U* and *V*.

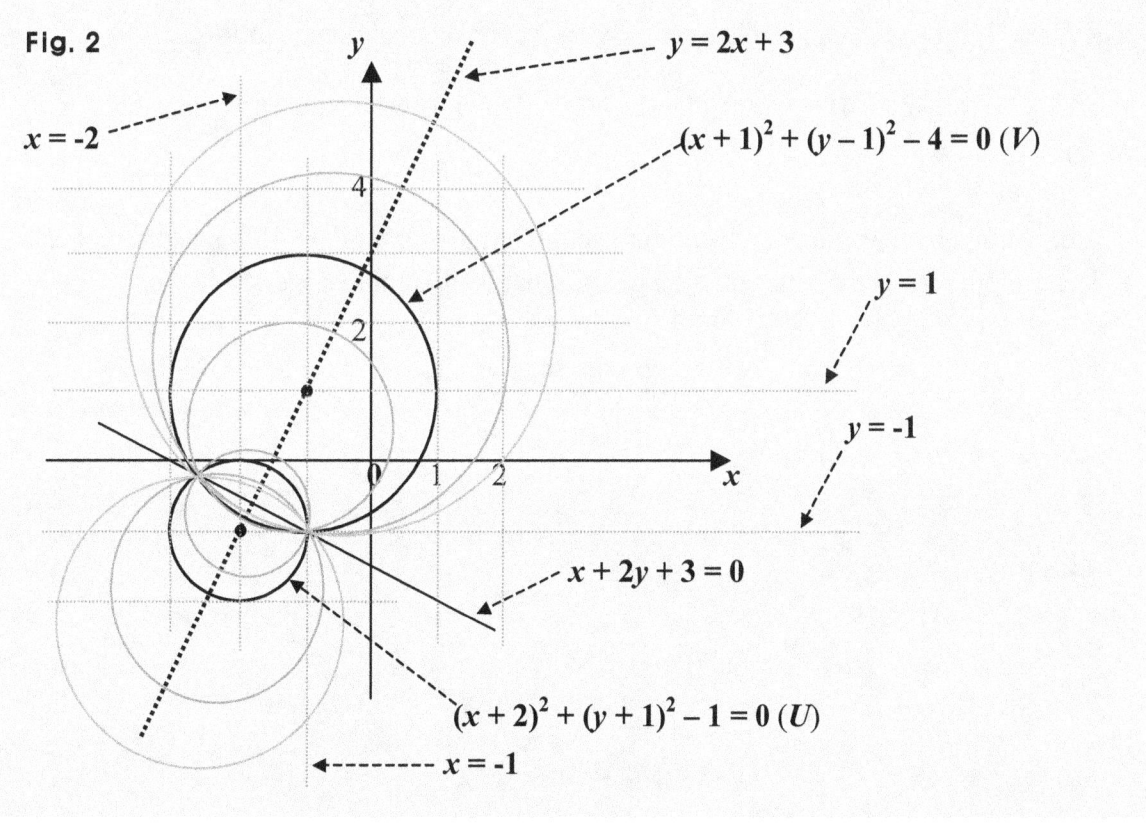

Fig. 2

In the graph above, the two circles U and V are in black. And the line $x + 2y + 3 = 0$ is passing through the two particular points where all the circles meet altogether. Besides, the centers of all the circles are in a line perpendicular to the line $x + 2y + 3 = 0$ and passing through the midpoint between the two particular points.

The perpendicular line is: $y = 2x + 3$, which passes through, of course, the centers of two circles U and V, of which each circle in the graph is a linear combination.

Note however, of the two circles U and V, the smaller one U, $x^2 + y^2 + 4x + 2y + 4 = 0$, which is the general equivalent to $(x + 2)^2 + (y + 1)^2 - 1 = 0$ cannot be made from the given equation $(k + 1)(x^2 + y^2) + 2(2k + 1)x + 2(k - 1)y + 4k - 2 = 0$ for any value of k.

Why not?

The equation given is the same as $k(x^2 + y^2 + 4x + 2y + 4) + (x^2 + y^2 + 2x - 2y - 2) = 0$, and for no value of k, we can get: $x^2 + y^2 + 4x + 2y + 4 = 0$, which is the smaller circle U.

The circle U however, can be obtained if we use such an equation as follows:

$m(x^2 + y^2 + 4x + 2y + 4) + n(x^2 + y^2 + 2x - 2y - 2) = 0$ where m and n are constants, and $m + n \neq 0$.

The equation above is not equivalent to the equation given, though. That's because it can indicate more circles than the equation given. That is, it is more general.

And in it, setting $m = 1$ and $n = 0$, we get the smaller circle U, $x^2 + y^2 + 4x + 2y + 4 = 0$.

Suggestions or Solutions
To the Problem in the Example 1

See if there is any particular point the curve of the equation below passes through for any value of k, which is constant, and specify the particular point if any.
$$9(k + 1)(x^2 + y^2) - 36(k + 2)x - 18(k + 3)y + 37k + 193 = 0.$$

To begin with, we can put the equation the way below:

$9(k + 1)(x^2 + y^2) - 36(k + 2)x - 18(k + 3)y + 37k + 193$

$= 9k(x^2 + y^2) - 36kx - 18ky + 37k + 9x^2 + 9y^2 - 72x - 54y + 193$

$= k(9x^2 + 9y^2 - 36x - 18y + 37) + (9x^2 + 9y^2 - 72x - 54y + 193) = 0$

$\Rightarrow k(x^2 + y^2 - 4x - 2y + \frac{37}{9}) + (x^2 + y^2 - 8x - 6y + \frac{193}{9}) = 0$, which can be an equation of a circle for each value of k.

Next, suppose two circles: $x^2 + y^2 - 4x - 2y + \frac{37}{9} = 0$ and $x^2 + y^2 - 8x - 6y + \frac{193}{9} = 0$ meet at (s, t). Then, we get: $s^2 + t^2 - 4s - 2t + \frac{37}{9} = 0$, and $s^2 + t^2 - 8s - 6t + \frac{193}{9} = 0$, so we get: $k(s^2 + t^2 - 4s - 2t + \frac{37}{9}) + s^2 + t^2 - 8s - 6t + \frac{193}{9} = 0$.

So if the two circles above meet at a particular point, for any value of k, the equation given indicates a circle that includes the particular point. And the same is true for another particular point, too, if the two circles share the other particular point. So if the two circles meet at particular points, for any value of k, the equation given indicates a circle that includes the particular points. So now, finding the points, we get:

$x^2 + y^2 - 4x - 2y + \frac{37}{9} = x^2 + y^2 - 8x - 6y + \frac{193}{9} \Rightarrow 4x + 4y - \frac{156}{9} = 0$

$\Rightarrow x + y - \frac{39}{9} = 0 \Rightarrow x = -y + \frac{13}{3}$

$\Rightarrow x^2 + y^2 - 4x - 2y + \frac{37}{9} = (-y + \frac{13}{3})^2 + y^2 - 4(-y + \frac{13}{3}) - 2y + \frac{37}{9}$

$= y^2 - \frac{26}{3}y + \frac{169}{9} + y^2 + 4y - \frac{52}{3} - 2y + \frac{37}{9} = 2y^2 - \frac{20}{3}y + \frac{50}{9}$

$= 2(y^2 - \frac{10}{3}y + \frac{25}{9}) = 2(y - \frac{5}{3})^2 = 0 \Rightarrow y = \frac{5}{3} \Rightarrow x = -y + \frac{13}{3} = -\frac{5}{3} + \frac{13}{3} = \frac{8}{3} \Rightarrow (\frac{8}{3}, \frac{5}{3})$.

So the two circles meet at one point $(\frac{8}{3}, \frac{5}{3})$, which is therefore, the only particular point a circle indicated by the equation given includes for any value of k.

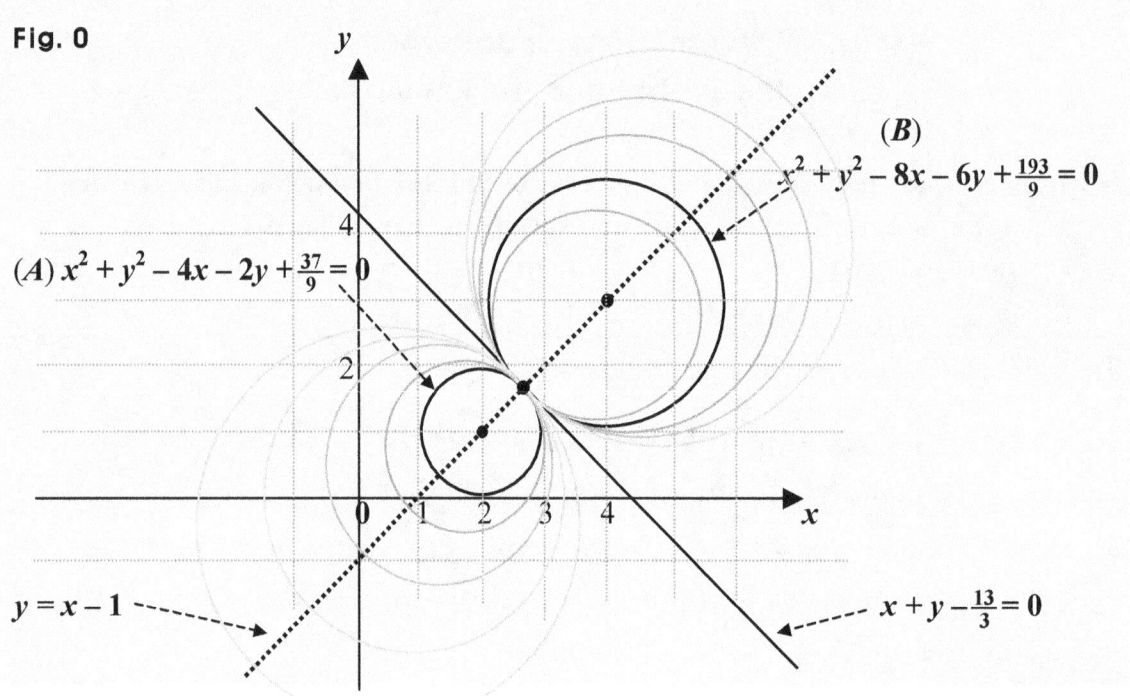

Fig. 0

$(A)\ x^2 + y^2 - 4x - 2y + \frac{37}{9} = 0$

(B)
$x^2 + y^2 - 8x - 6y + \frac{193}{9} = 0$

$y = x - 1$

$x + y - \frac{13}{3} = 0$

In the graph above, the two circles A and B are in black, and each of the other circles is a linear combination of A and B.

The centers of all the circles are in a line $y = x - 1$, which passes through the particular point, and is perpendicular to a line $x + y - \frac{13}{3} = 0$, which passes through the particular point, too, where all the circles in the group meet each other altogether.

Note however, of the two circles A and B, the smaller one A, $x^2 + y^2 - 4x - 2y + \frac{37}{9} = 0$, which is equivalent to $(x - 2)^2 + (y - 1)^2 - \frac{8}{9} = 0$ cannot be obtained from the given equation $9(k + 1)(x^2 + y^2) - 36(k + 2)x - 18(k + 3)y + 37k + 193 = 0$ for any value of k.

The given equation can be put in $k(x^2 + y^2 - 4x - 2y + \frac{37}{9}) + (x^2 + y^2 - 8x - 6y + \frac{193}{9}) = 0$, so for no value of k, we can get: $x^2 + y^2 - 4x - 2y + \frac{37}{9} = 0$, which is the smaller circle A.

The circle A however, can be obtained if we use the equation below:
$m(x^2 + y^2 - 4x - 2y + \frac{37}{9}) + n(x^2 + y^2 - 8x - 6y + \frac{193}{9}) = 0$, where $m + n \neq 0$.

The equation above is not equivalent to the equation given, though. That's because it can indicate more circles than the given equation can. So it is more general. In the equation, setting $m = 1$ and $n = 0$, we get the smaller circle A, $x^2 + y^2 - 4x - 2y + \frac{37}{9} = 0$.

Suggestions or Solutions
To the Problem in the Example 2

See if there is any particular point the curve of the equation below passes through for any value of k, which is constant, and specify the particular point if any.
$$(k+1)(x^2+y^2) - 2(3k+4)x - 2(k+2)y + 2k + 18 = 0.$$

To begin with, we can put the equation the way below:

$(k+1)(x^2+y^2) - 2(3k+4)x - 2(k+2)y + 2k + 18 = 0$

$= k(x^2+y^2) - 6kx - 2ky + 2k + x^2 + y^2 - 8x - 4y + 18$

$= k(x^2+y^2 - 6x - 2y + 2) + (x^2+y^2 - 8x - 4y + 18) = 0$, which can be an equation of a circle for each value of k.

Next, suppose two circles: $x^2+y^2 - 6x - 2y + 2 = 0$ and $x^2+y^2 - 8x - 4y + 18 = 0$ meet at (s, t).

Then, we get: $s^2+t^2 - 6s - 2t + 2 = 0$, and $s^2+t^2 - 8s - 4t + 18 = 0$.

So we get: $k(s^2+t^2 - 6s - 2t + 2) + s^2+t^2 - 8s - 4t + 18 = 0$.

So if the two circles above meet at a particular point, for any value of k, the equation given indicates a circle that includes the particular point. And the same is true for another particular point, too, if the two circles share the point. So if the two circles meet at particular points, for any value of k, the equation given indicates a circle that includes the particular points. Finding thus, the points now, we get:

$x^2+y^2 - 6x - 2y + 2 - (x^2+y^2 - 8x - 4y + 18) = -6x - 2y + 2 + 8x + 4y - 18$

$= 2x + 2y - 16 = 0 \Rightarrow x + y - 8 = 0 \Rightarrow x = 8 - y$

$\Rightarrow x^2+y^2 - 6x - 2y + 2 = (8-y)^2 + y^2 - 6(8-y) - 2y + 2 = 0$

$\Rightarrow y^2 - 16y + 64 + y^2 - 48 + 6y - 2y + 2 = 2y^2 - 12y + 18 = 2(y^2 - 6y + 9) = 2(y-3)^2 = 0$

$\Rightarrow y = 3 \Rightarrow x = 8 - y = 8 - 3 = 5 \Rightarrow (5, 3)$.

Thus, there is only one particular point, and it is **(5, 3)**.   And let's now, put in a graph, together with A and B, some of those circles that can be indicated by the equation given.

Fig. 0

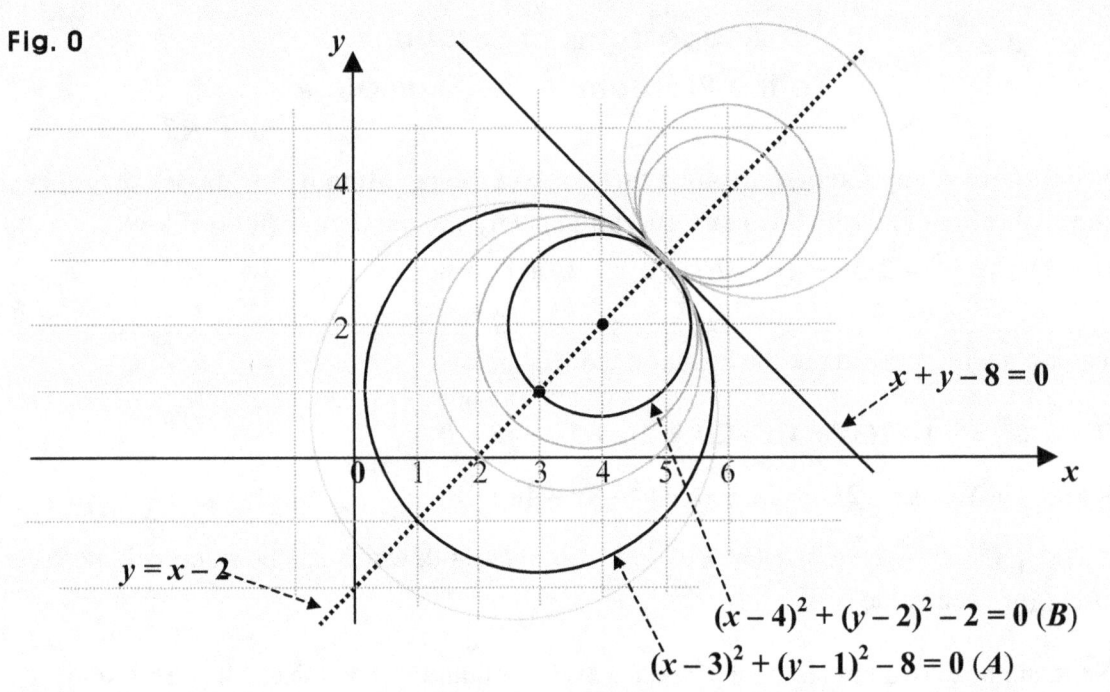

$x + y - 8 = 0$

$y = x - 2$

$(x - 4)^2 + (y - 2)^2 - 2 = 0 \ (B)$

$(x - 3)^2 + (y - 1)^2 - 8 = 0 \ (A)$

Suggestions or Solutions
To the Problem in the Example 3

See if there is any particular point the curve of the equation below passes through for any value of k, which is constant, and specify the particular point if any.
$$(k + 1)(x^2 + y^2) - 2(k + 4)x - 2(k + 3)y + k + 21 = 0.$$

First, we can put the equation the way as follows:

$(k + 1)(x^2 + y^2) - 2(k + 4)x - 2(k + 3)y + k + 21$
$= k(x^2 + y^2) - 2kx - 2ky + k + x^2 + y^2 - 8x - 6y + 21$
$= k(x^2 + y^2 - 2x - 2y + 1) + (x^2 + y^2 - 8x - 6y + 21) = 0$, which can be an equation of a circle for a value of k.

Next, suppose two circles $x^2 + y^2 - 2x - 2y + 1 = 0$ and $x^2 + y^2 - 8x - 6y + 21 = 0$ meet at (s, t).   Then, we get: $s^2 + t^2 - 2s - 2t + 1 = 0$, and $s^2 + t^2 - 8s - 6t + 21 = 0$,
So we get: $k(s^2 + t^2 - 2s - 2t + 1) + s^2 + t^2 - 8s - 6t + 21 = 0$.

So if the two circles above meet at a particular point, for any value of k, the equation given indicates a circle that includes the particular point. The same is true for another particular point, too, if the two circles share the point. So if the two circles meet at particular points, for any value of k, the equation given indicates a circle that includes the particular points.

$x^2 + y^2 - 2x - 2y + 1 = x^2 - 2x + 1 - 1 + y^2 - 2y + 1 - 1 + 1 = (x - 1)^2 + (y - 1)^2 - 1 = 0$
$\Rightarrow$ The center is at $(1, 1)$, and the radius is 1.

$x^2 + y^2 - 8x - 6y + 21 = x^2 - 8x + 16 - 16 + y^2 - 6y + 9 - 9 + 21$
$= (x - 4)^2 + (y - 3)^2 - 4 = 0 \Rightarrow$ The center is at $(4, 3)$, and the radius is 2.

Suppose D is the distance between the two centers, and S is the sum of the two radii.

Then, we get $D^2 = (\Delta x)^2 + (\Delta y)^2 = (4 - 1)^2 + (3 - 1)^2 = 9 + 4 = 13$, and $S = 1 + 2 = 3$.

Then, we get $S < D$ since S and D are nonnegative and $S^2 = 9 < D^2 = 13$.

So the two circles do not meet each other, and thus, there is no particular point shared by all the circles indicated by the given equation for all values of k.

182

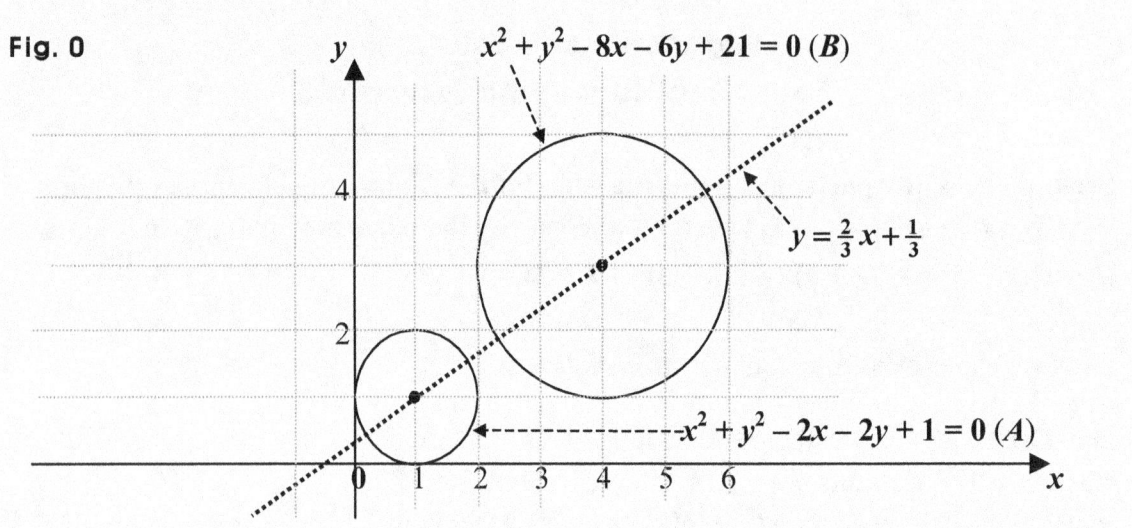

Fig. 0

$x^2 + y^2 - 8x - 6y + 21 = 0 \ (B)$

$y = \frac{2}{3}x + \frac{1}{3}$

$x^2 + y^2 - 2x - 2y + 1 = 0 \ (A)$

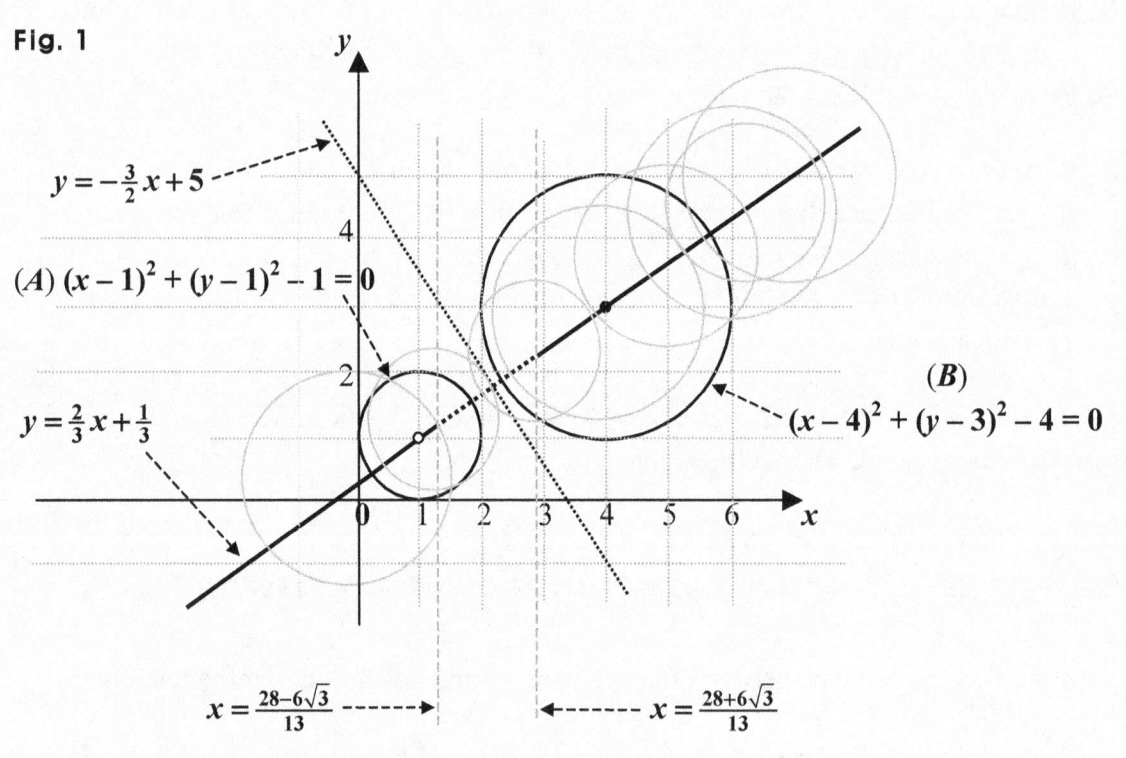

Fig. 1

$y = -\frac{3}{2}x + 5$

$(A) \ (x - 1)^2 + (y - 1)^2 - 1 = 0$

$y = \frac{2}{3}x + \frac{1}{3}$

(B)

$(x - 4)^2 + (y - 3)^2 - 4 = 0$

$x = \frac{28 - 6\sqrt{3}}{13}$

$x = \frac{28 + 6\sqrt{3}}{13}$

Note that in the graph above, the radii of the circles in gray are not accurate at all.

That is to say that the gray circles are no more than the ones for showing how the circles in each group can be put in a graph.

The key idea is that the centers of all the circles above are in the two rays and the small line segment.

In the graph above, the two circles A and B are in black, and each of all the circles is a linear combination of the two circles A and B.

Note however, for any value of k, the circle A, $x^2 + y^2 - 2x - 2y + 1 = 0$, which is equal to $(x - 1)^2 + (y - 1)^2 - 1 = 0$ cannot be made from the equation given in the problem.

The equation given can be put in $k(x^2 + y^2 - 2x - 2y + 1) + (x^2 + y^2 - 8x - 6y + 21) = 0$.

So for no value of k, the equation above can indicate the circle A.

However, the circle A can be obtained if we use such an equation as follows:

$m(x^2 + y^2 - 2x - 2y + 1) + n(x^2 + y^2 - 8x - 6y + 21) = 0$ where m and n are constants, and $m + n \neq 0$.

The equation above is not equivalent to the equation given, because it can indicate more circles that the equation given can.

In the equation, setting $m = 1$ and $n = 0$, we get the circle A, $x^2 + y^2 - 2x - 2y + 1 = 0$.

Suggestions or Solutions
To the Problem in the Example 4

See if there is any particular point the curve of the equation below passes through for any value of k, which is constant, and specify the particular point if any.

$$(k + 1)(x^2 + y^2) - 8(k + 1)x - 6(k + 1)y + 24k + 21 = 0.$$

First, we can put the equation the way below:

$$(k + 1)(x^2 + y^2) - 8(k + 1)x - 6(k + 1)y + 24k + 21$$
$$= k(x^2 + y^2) - 8kx - 6ky + 24k + x^2 + y^2 - 8x - 6y + 21$$
$$= k(x^2 + y^2 - 8x - 6y + 24) + (x^2 + y^2 - 8x - 6y + 21) = 0,$$ which can indicate a circle for a value of k.

Next, suppose two circles: $x^2 + y^2 - 8x - 6y + 24 = 0$ and $x^2 + y^2 - 8x - 6y + 21 = 0$ meet each other at (s, t).

Then, we get: $s^2 + t^2 - 8s - 6t + 24 = 0$, and $s^2 + t^2 - 8s - 6t + 21 = 0$.

So we get: $k(s^2 + t^2 - 8s - 6t + 24) + s^2 + t^2 - 8s - 6t + 21 = 0$.

So if the two circles above meet at a particular point, for any value of k, the equation given indicates a circle that includes the particular point. However, the two circles are concentric circles, so they do not meet each other. Therefore, there is no particular point shared by every circle indicated by the given equation for every value of k.

Let's see now, for what value of k, the equation can indicate a circle.

Simplifying it first, we get: $(k + 1)(x^2 + y^2) - 8(k + 1)x - 6(k + 1)y + 24k + 21 = 0$

$$\Rightarrow x^2 + y^2 - \frac{8(k+1)}{k+1}x - \frac{6(k+1)}{k+1}y + \frac{24k+21}{k+1} = x^2 + y^2 - 8x - 6y + \frac{24k+21}{k+1} = 0.$$

Then, first of all, since a denominator cannot be 0, we need to have $k \neq -1$.

If $k = -1$ in fact, the equation doesn't even hold. Let's now set: $c = \frac{24k+21}{k+1}$. Then, we get:

$$x^2 + y^2 - 8x - 6y + c = 0 \Rightarrow (x - 4)^2 + (y - 3)^2 - 25 + c = 0 \Rightarrow (x - 4)^2 + (y - 3)^2 = 25 - c.$$

Then, if the equation above indicates a circle, we need to have: $25 - c > 0$.

So we get: $25 - c = 25 - \frac{24k+21}{k+1} = \frac{25k+25-24k-21}{k+1} = \frac{k+4}{k+1} = 1 + \frac{3}{k+1} > 0 \Rightarrow \frac{3}{k+1} > -1$.

Now, since $(k + 1)$ can be negative, multiply both sides by $(k + 1)^2$ so that the direction of the inequality sign does not change.   Then, we get:

$3(k + 1) > -(k + 1)^2 \Rightarrow 3k + 3 > -k^2 - 2k - 1 \Rightarrow k^2 + 5k + 4 > 0 \Rightarrow (k + 4)(k + 1) > 0$.

So if $k > -1$ or $k < -4$, then for each value of k, the equation given can indicate a circle.

Next, let's find the extent of the radius in the equation $(x - 4)^2 + (y - 3)^2 = 25 - c$.

The radius squared is: $25 - c = 1 + \frac{3}{k+1}$ for $k > -1$ or $k < -4$.   So we get:

$k > -1 \Rightarrow 1 + \frac{3}{k+1} > 1 \Rightarrow 25 - c > 1$.

$k < -4 \Rightarrow 0 < 1 + \frac{3}{k+1} < 1 \Rightarrow 0 < 25 - c < 1$.

Therefore, the radius is greater than 1 for $k > -1$, but is between 0 and 1 for $k < -4$. Let's now, put in a graph some of the circles concentric.

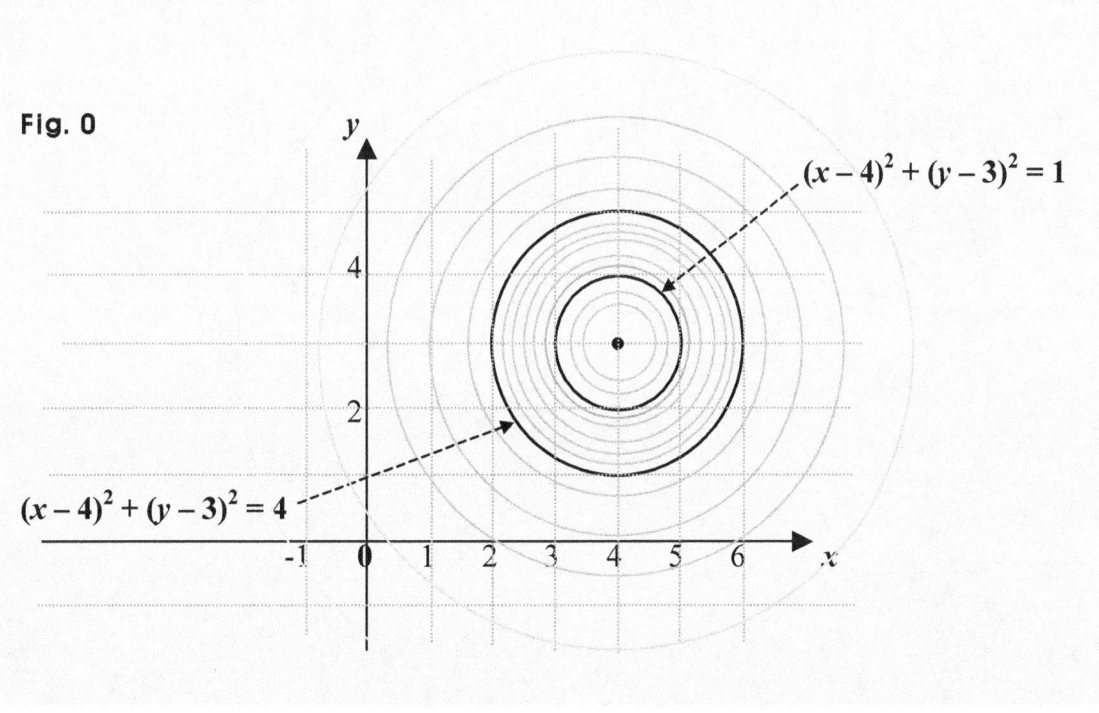

Fig. 0

$(x - 4)^2 + (y - 3)^2 = 1$

$(x - 4)^2 + (y - 3)^2 = 4$

In the graph, every circle in gray is a linear combination of the two circles in black.

Note that the smaller circle in black, $x^2 + y^2 - 8x - 6y + 24 = (x-4)^2 + (y-3)^2 - 1 = 0$ cannot be made from $k(x^2 + y^2 - 8x - 6y + 24) + (x^2 + y^2 - 8x - 6y + 21) = 0$ for any value of k.

However, the smaller one, too, can be obtained if we use such an equation as follows:

$m(x^2 + y^2 - 8x - 6y + 24) + n(x^2 + y^2 - 8x - 6y + 21) = 0$ where m and n are constants.

Setting $m = 1$ and $n = 0$, we get the smaller of the two.

Suggestions or Solutions
To the Problem in the Example 5

See if there is any particular point the curve of the equation below passes through for any value of k, which is constant, and specify the particular point if any.
$$(k + 1)(x^2 + y^2) - 2(2k + 3)x - 2(2k + 3)y - k + 17 = 0.$$

First, we can put the equation the way below:

$(k + 1)(x^2 + y^2) - 2(2k + 3)x - 2(2k + 3)y - k + 17$
$= k(x^2 + y^2) - 4kx - 4ky - k + x^2 + y^2 - 6x - 6y + 17$
$= k(x^2 + y^2 - 4x - 4y - 1) + (x^2 + y^2 - 6x - 6y + 17) = 0$, which can be an equation of a circle for each value of k.

Next, suppose two circles $x^2 + y^2 - 4x - 4y - 1 = 0$ and $x^2 + y^2 - 6x - 6y + 17 = 0$ meet at (s, t).

Then, we get: $s^2 + t^2 - 4s - 4t - 1 = 0$, and $s^2 + t^2 - 6s - 6t + 17 = 0$.

So we get: $k(s^2 + t^2 - 4s - 4t - 1) + s^2 + t^2 - 6s - 6t + 17 = 0$.

So if the two circles above meet at a particular point, for any value of k, the equation given indicates a circle that includes the particular point.

Te same is true for another particular point, too, if the two circles share the point.

So if the two circles meet at particular points, for any value of k, the equation given indicates a circle that includes the particular points.

Getting the centers and radii of two circles, we can see if they can meet each other.

$x^2 + y^2 - 4x - 4y - 1 = x^2 - 4x + 4 - 4 + y^2 - 4y + 4 - 4 - 1 = (x - 2)^2 + (y - 2)^2 - 9 = 0$
So the center is **(2, 2)**, and the radius is 3.

$x^2 + y^2 - 6x - 6y + 17 = x^2 - 6x + 9 - 9 + y^2 - 6y + 9 - 9 + 17$
$= (x - 3)^2 + (y - 3)^2 - 1 = 0.$ So the center is **(3, 3)**, and the radius is 1.

Suppose next, **D** is the distance between the two centers, and **T** is the difference between the radii. Then, we get:

$$D^2 = (\Delta x)^2 + (\Delta y)^2 = (3 - 2)^2 + (3 - 2)^2 = 1 + 1 = 2$$
$$T = 3 - 1 = 2 > D = \sqrt{2} \implies T > D.$$

So one of the two circles is inside the other, and the two circles do now meet each other.

Therefore, there is no particular point each and every circle indicated by the equation given includes for every value of **k**.

Fig. 0

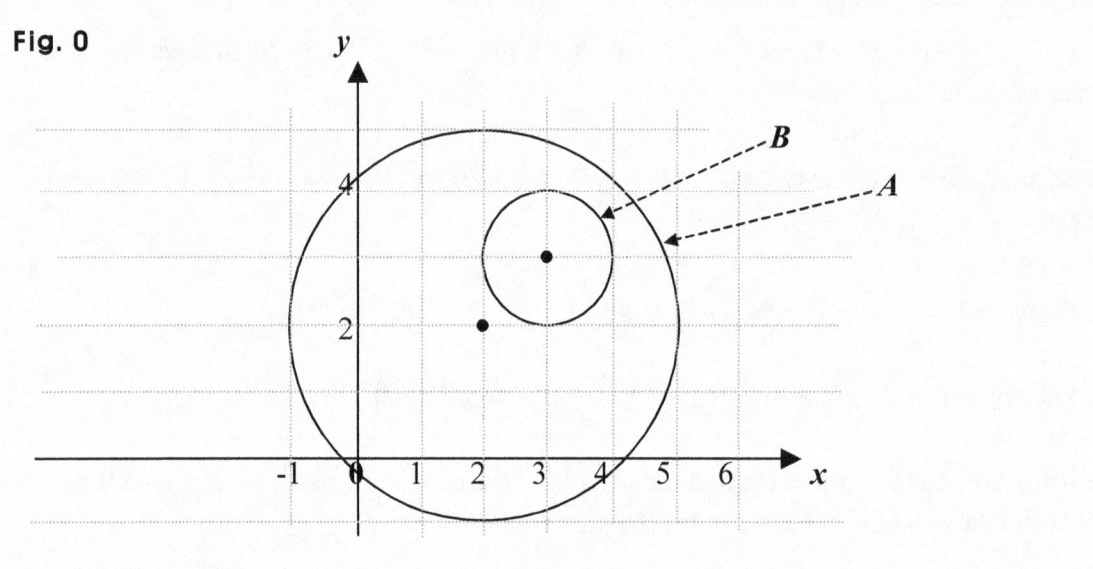

Now, let's find out for what value of **k**, the equation given can indicate a circle.

Then, simplifying the equation first, we get:

$$(k + 1)(x^2 + y^2) - 2(2k + 3)x - 2(2k + 3)y - k + 17 = 0$$

$$\Rightarrow x^2 + y^2 - \frac{2(2k+3)}{k+1}x - \frac{2(2k+3)}{k+1}y + \frac{17-k}{k+1} = 0.$$

Then, first of all, we need to have: **k ≠ -1** since a denominator cannot be 0.

Let's now, set: $a = \frac{2k+3}{k+1}$, and $b = \frac{17-k}{k+1}$. Then, we get:

$$x^2 + y^2 - 2ax - 2ay + b = 0$$

$$\Rightarrow x^2 - 2ax + a^2 - a^2 + y^2 - 2ay + a^2 - a^2 + b = (x-a)^2 + (y-a)^2 - 2a^2 + b = 0$$

$$\Rightarrow (x-a)^2 + (y-a)^2 = b - 2a^2.$$

So $b - 2a^2$ has to be positive. Thus, we need to have:

$$b - 2a^2 = \frac{17-k}{k+1} - \frac{2(2k+3)^2}{(k+1)^2} = \frac{(17-k)(k+1)-2(2k+3)^2}{(k+1)^2} > 0 \Rightarrow (17-k)(k+1) - 2(2k+3)^2 > 0$$

$$\Rightarrow -k^2 + 16k + 17 - 2(4k^2 + 12k + 9) = -9k^2 - 8k - 1 > 0 \Rightarrow 9k^2 + 8k + 1 < 0.$$

Meanwhile, $9k^2 + 8k + 1 = 9(k^2 + \frac{8}{9}k + \frac{16}{81} - \frac{16}{81}) + 1 = 9(k + \frac{4}{9})^2 - \frac{16}{9} + 1 = 9(k + \frac{4}{9})^2 - \frac{7}{9}.$

So we get: $9(k + \frac{4}{9})^2 - \frac{7}{9} < 0 \Rightarrow (k + \frac{4}{9})^2 - \frac{7}{81} < 0.$

Thus, we get: $(k + \frac{4}{9})^2 - \frac{7}{81} = \{(k + \frac{4}{9}) + \frac{\sqrt{7}}{9}\}\{(k + \frac{4}{9}) - \frac{\sqrt{7}}{9}\} = (k + \frac{4+\sqrt{7}}{9})(k + \frac{4-\sqrt{7}}{9}) < 0.$

So we get: $-\frac{4+\sqrt{7}}{9} < k < -\frac{4-\sqrt{7}}{9}$. We have: $k \neq -1$, too, though.

However, we have: $-\frac{4+\sqrt{7}}{9} > -1$, also.

So k can't be -1 in the case where $-\frac{4+\sqrt{7}}{9} < k < -\frac{4-\sqrt{7}}{9}$.

Therefore, if $-\frac{4+\sqrt{7}}{9} < k < -\frac{4-\sqrt{7}}{9}$,

the equation $(k + 1)(x^2 + y^2) - 2(2k + 3)x - 2(2k + 3)y - k + 17 = 0$ where k is constant can indicate a circle.

Examples 9 in Circles

0.   Find a line passing through two points where the following two circles meet:

$x^2 + y^2 + 4x + 2y - 4 = 0$ and $x^2 + y^2 - 2x - 2y + 1 = 0$.

1.   See if there is any particular point the curve of each equation below passes through for any value of k, which is constant, and specify the particular point if any.

1.0. $k(x^2 + y^2) - (6k - 1)x - 2(2k + 1)y + 9k + 3 = 0$.

1.1. $k(x^2 + y^2) - (2k - 1)x - 2(3k - 1)y + 5k - 12 = 0$.

1.2. $k(x^2 + y^2) - (6k - 1)x - (4k + 1)y + 10k + 2 = 0$.

Suggestions or Solutions
To the Problem in the Example 0

Find the line passing through the two points where the following two circles meet:
$x^2 + y^2 + 4x + 2y - 4 = 0$ **and** $x^2 + y^2 - 2x - 2y + 1 = 0$.

$-(x^2 + y^2 + 4x + 2y - 4) + x^2 + y^2 - 2x - 2y + 1 = 0$

$\Rightarrow -4x - 2y + 4 - 2x - 2y + 1 = -6x - 4y + 5 = 0 \Rightarrow y = \frac{5}{4} - \frac{3}{2}x.$

Therefore, the line is: $y = \frac{5}{4} - \frac{3}{2}x.$

If not quite sure of the idea behind the processes above, follow the steps below:

Setting: $k = -1$ in the equation $k(x^2 + y^2 + 4x + 2y - 4) + (x^2 + y^2 - 2x - 2y + 1) = 0$, we can immediately get the equation of the line this problem is asking.   How come?

We have covered the idea in **Examples 8**.

Let's now, check to see if the two circles really meet at two points.

We can see it comparing the distance between the centers and the sum of or the difference between the radii.

And we can get the centers and radii putting the equations in the standard form.

$x^2 + y^2 + 4x + 2y - 4 = (x + 2)^2 - 4 + (y + 1)^2 - 1 - 4 = (x + 2)^2 + (y + 1)^2 - 9 = 0$

So the center is **(-2, -1)**, and the radius is 3.

$x^2 + y^2 - 2x - 2y + 1 = (x - 1)^2 + (y - 1)^2 - 2 + 1 = (x - 1)^2 + (y - 1)^2 - 1 = 0$

So the center is **(1, 1)**, and the radius is 1.

Suppose now, D is the distance between the centers, S is the sum of the radii, and T is the difference between the radii. Then, we get:

$$D^2 = (\Delta x)^2 + (\Delta y)^2 = \{1 - (-2)\}^2 + \{1 - (-1)\}^2 = 9 + 4 = 13, S = 3 + 1 = 4, \& \ T = 3 - 1 = 2.$$

Thus, we get: $S > D$, because S and D are nonnegative and $S^2 = 16 > D^2 = 13$.

Also, we get: $T < D$, because T and D are nonnegative and $T^2 = 4 < D^2 = 13$.

So we get: $T < D < S$.

Therefore, the two circles do meet each other at two points, and share the two points.

If not sure, refer to the Problem 0 in the Example 0 in **Examples 8**.

Now, let's set $k = -1$ in the equation $k(x^2 + y^2 + 4x + 2y - 4) + (x^2 + y^2 - 2x - 2y + 1) = 0$, and actually, find the equation of the line passing through the two points where the two given circles meet each other.

$$k = -1 \Rightarrow k(x^2 + y^2 + 4x + 2y - 4) + (x^2 + y^2 - 2x - 2y + 1) = -4x - 2y + 4 - 2x - 2y + 1$$
$$= -6x - 4y + 5 = 0 \Rightarrow y = \tfrac{5}{4} - \tfrac{3}{2}x.$$

Let's now, put in a graph the two circles and the line passing through the two points.

Fig. 0

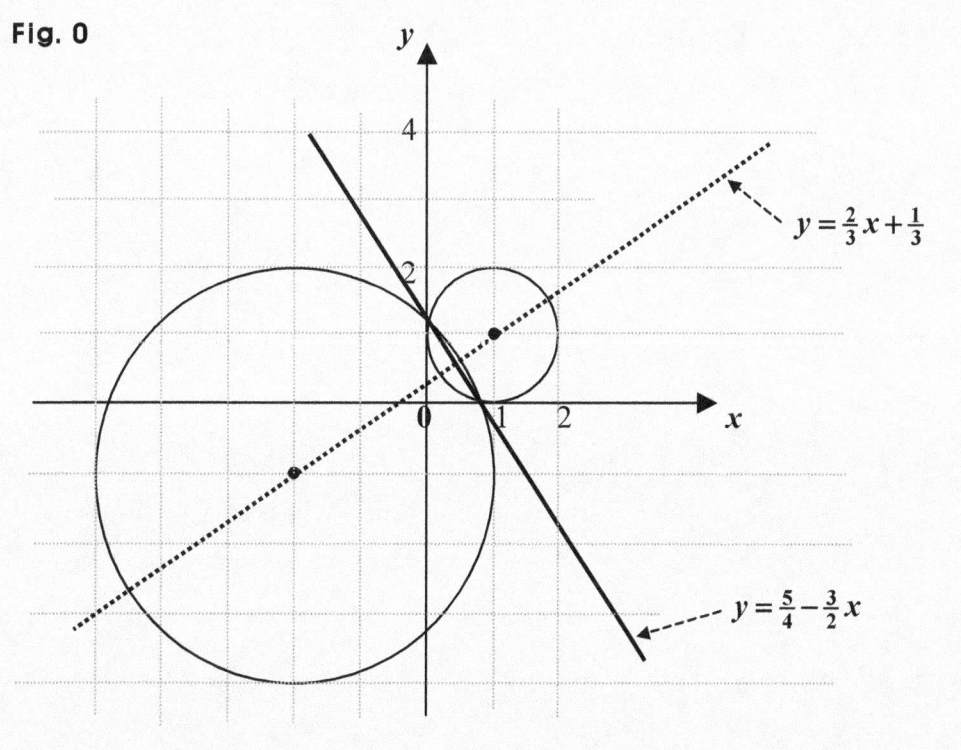

$$y = \tfrac{2}{3}x + \tfrac{1}{3}$$

$$y = \tfrac{5}{4} - \tfrac{3}{2}x$$

Of course, we can find first, the two points where the two circles meet, and then, get the equation of the line passing through the two points.

So we can get the line this problem is asking in such a way as above, too, but it will take some time to find such two points.

However, knowing how circles (equations for circles) work, we can get such a line in no time.

In short:

$-(x^2 + y^2 + 4x + 2y - 4) + x^2 + y^2 - 2x - 2y + 1 = 0$

$\Rightarrow -4x - 2y + 4 - 2x - 2y + 1 = -6x - 4y + 5 = 0 \Rightarrow y = \frac{5}{4} - \frac{3}{2}x.$

Therefore, the line is: $y = \frac{5}{4} - \frac{3}{2}x.$

Suggestions or Solutions
To the Problem 0 in the Example 1

See if there is any particular point the curve of the equation below passes through for any value of k, which is constant, and specify the particular point if any.
$$k(x^2 + y^2) - (6k - 1)x - 2(2k + 1)y + 9k + 3 = 0.$$

First, we can put the equation the way below:

$k(x^2 + y^2) - (6k - 1)x - 2(2k + 1)y + 9k + 3$

$= k(x^2 + y^2) - 6kx + x - 4ky - 2y + 9k + 3$

$= k(x^2 + y^2) - 6kx - 4ky + 9k + x - 2y + 3$

$= k(x^2 + y^2 - 6x - 4y + 9) + (x - 2y + 3) = 0$, which can indicate a circle for a value of k if $k \neq 0$.

Next, suppose a circle $x^2 + y^2 - 6x - 4y + 9 = 0$ meets a line $x - 2y + 3 = 0$ at (s, t).

Then, we get: $s^2 + t^2 - 6s - 4t + 9 = 0$, and $s - 2t + 3 = 0$.

So we get: $k(s^2 + t^2 - 6s - 4t + 9) + s - 2t + 3 = 0.$

So if the circle and the line share any particular point, for any value of k, other than 0, the equation given indicates a circle that includes the particular point.

Finding the points, we can get first: $x - 2y + 3 = 0 \Rightarrow x = 2y - 3$.

And next, we get: $x^2 + y^2 - 6x - 4y + 9 = (2y - 3)^2 + y^2 - 6(2y - 3) - 4y + 9$

$= 4y^2 - 12y + 9 + y^2 - 12y + 18 - 4y + 9 = 5y^2 - 28y + 36 = (5y - 18)(y - 2) = 0$

$\Rightarrow y = 2$ or $y = \frac{18}{5}$. Thus, $y = 2 \Rightarrow x = 2y - 3 = 1$, and $y = \frac{18}{5} \Rightarrow x = 2y - 3 = \frac{36}{5} - 3 = \frac{21}{5}$.

So for any value of k, other than 0, the equation given indicates a circle that includes the two points $(1, 2)$ and $(\frac{21}{5}, \frac{18}{5})$, which are therefore, the particular points.

If not quite sure of the idea behind the processes above, follow the steps below:

This problem is just about the same as the ones in **Examples 8**.
Let's see though, what's new here.

First, note that in math, a curve can be a line, too.   Next, we have a fact as follows:

Suppose m and n are constant, and the curve of $f(x, y) = 0$ meets that of $g(x, y) = 0$ at particular points.

Then, at the particular points, too, the curve of $mf(x, y) + ng(x, y) = 0$ meets the two curves of $f(x, y) = 0$ and $g(x, y) = 0$.

In other words, the three curves share the particular points.

And in particular, the equation $mf(x, y) + ng(x, y) = 0$ is called a linear combination of the two equations $f(x, y) = 0$ and $g(x, y) = 0$.

So now, getting back to the problem in this example, we want to see first, if the equation given can be such a linear combination. That is, we want to begin with extracting equations, of which the equation given is a linear combination.

$$k(x^2 + y^2) - (6k - 1)x - 2(2k + 1)y + 9k + 3 = k(x^2 + y^2) - 6kx + x - 4ky - 2y + 9k + 3$$
$$= k(x^2 + y^2) - 6kx - 4ky + 9k + x - 2y + 3 = k(x^2 + y^2 - 6x - 4y + 9) + (x - 2y + 3) = 0.$$

Thus, we can see two equations, one indicates a circle $x^2 + y^2 - 6x - 4y + 9 = 0$, and the other indicates a line $x - 2y + 3 = 0$.

So if C is the circle above, and L is the line above, the curve of the equation given is a linear combination of C and L, so we can use the fact below:

Suppose a circle $x^2 + y^2 + ax + by + c = 0$ and a line $dx + ey + f = 0$ meet each other at particular points, m and n are constant, and $m \neq 0$. Then, for each pair of the values of m and n, the equation $m(x^2 + y^2 + ax + by + c) + n(dx + ey + f) = 0$ indicates a circle that includes the particular points, and has its center in a line passing through the center of a circle $x^2 + y^2 + ax + by + c = 0$ and perpendicular to a line $dx + ey + f = 0$.

So let's now, see if the circle C and the line L share any point.

To begin with, we can get: $x - 2y + 3 = 0 \Rightarrow x = 2y - 3$.

Then, we get: $x^2 + y^2 - 6x - 4y + 9 = (2y - 3)^2 + y^2 - 6(2y - 3) - 4y + 9$

$= 4y^2 - 12y + 9 + y^2 - 12y + 18 - 4y + 9 = 5y^2 - 28y + 36 = (5y - 18)(y - 2) = 0$

$\Rightarrow y = 2$ or $y = \frac{18}{5}$.

So we get: $y = 2 \Rightarrow x = 2y - 3 = 1$, and $y = \frac{18}{5} \Rightarrow x = 2y - 3 = \frac{36}{5} - 3 = \frac{21}{5}$.

Therefore, the circle C meets the line L at two points $(1, 2)$ and $(\frac{21}{5}, \frac{18}{5})$.

It is not the case however, the equation $k(x^2 + y^2) - (6k - 1)x - 2(2k + 1)y + 9k + 3 = 0$ indicates a circle for each and every value of k. How come?

$k(x^2 + y^2) - (6k - 1)x - 2(2k + 1)y + 9k + 3 = k(x^2 + y^2 - 6x - 4y + 9) + (x - 2y + 3) = 0$.

So if $k = 0$, we don't get a circle but just a line.

Thus, other than 0, for each and every value of k, the equation given indicates a circle that includes the two points $(1, 2)$ and $(\frac{21}{5}, \frac{18}{5})$.

That is, the equation given represents a group of circles sharing altogether the two points.

Let's now, put in a graph the circle C and the line L, together with several circles in the group. To begin with, putting the circle C in the standard form, we get:

$x^2 + y^2 - 6x - 4y + 9 = x^2 - 6x + 9 + y^2 - 4y + 4 - 4 = (x - 3)^2 + (y - 2)^2 - 4 = 0$.

So the center of C is $(3, 2)$, and the radius is 2.

Note however. that the circle C cannot be indicated by the equation given.

We can indicate though, the circle C, too, if we use an equation as follows:

$m(x^2 + y^2 - 2x - 6y + 5) + n(x + 2y - 12) = 0$ where m and n are constants.

So in the equation above, setting $m = 1$ and $n = 0$, we get the circle C.

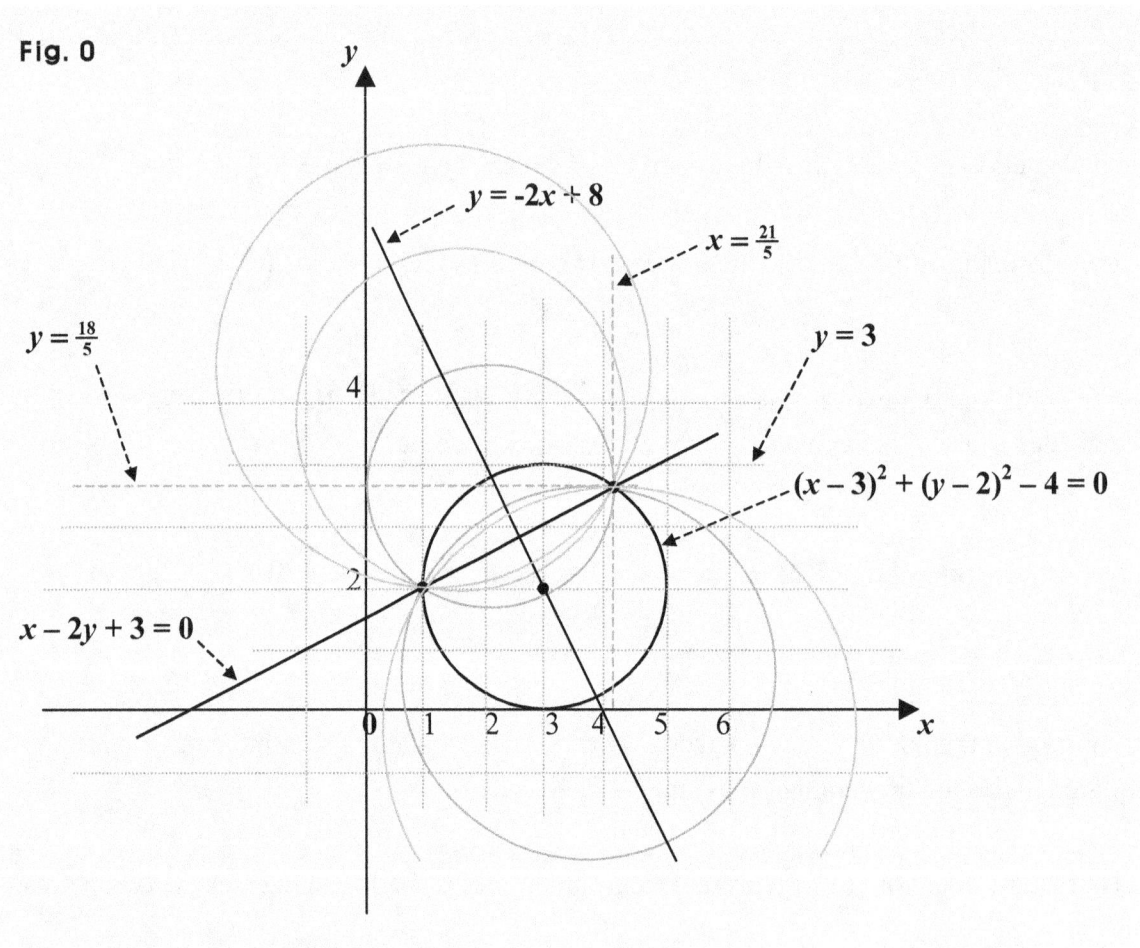

Fig. 0

The centers of all the circles above are in the line $y = -2x + 8$, which is perpendicular to the line $x - 2y + 3 = 0$, and connects the center of the circle $(x - 3)^2 + (y - 2)^2 = 4$ and the midpoint between the two particular points.

Suggestions or Solutions
To the Problem 1 in the Example 1

See if there is any particular point the curve of the equation below passes through for any value of k, which is constant, and specify the particular point if any.
$$k(x^2 + y^2) - (2k - 1)x - 2(3k - 1)y + 5k - 12 = 0.$$

First, we can put the equation the way below:

$$k(x^2 + y^2) - (2k - 1)x - 2(3k - 1)y + 5k - 12$$
$$= k(x^2 + y^2) - 2kx + x - 6ky + 2y + 5k - 12$$
$$= k(x^2 + y^2) - 2kx - 6ky + 5k + x + 2y - 12$$
$$= k(x^2 + y^2 - 2x - 6y + 5) + (x + 2y - 12) = 0,$$ which can indicate a circle for a value of k
if $k \neq 0$.

Next, suppose a circles $x^2 + y^2 - 2x - 6y + 5 = 0$ meets a line $x + 2y - 12 = 0$ at (s, t).

Then, we get: $s^2 + t^2 - 2s - 6t + 5 = 0$, and $s + 2t - 12 = 0$.

So we get: $k(s^2 + t^2 - 2s - 6t + 5) + s + 2t - 12 = 0$.

So if the circle and the line share any particular point, for any value of k, other than 0, the equation given indicates a circle that includes the particular point.
So let's now, find the points.

To begin with: $x + 2y - 12 = 0 \Rightarrow x = 12 - 2y$.

And next, we get: $x^2 + y^2 - 2x - 6y + 5 = (12 - 2y)^2 + y^2 - 2(12 - 2y) - 6y + 5$

$$= 144 - 48y + 4y^2 + y^2 - 24 + 4y - 6y + 5 = 5y^2 - 50y + 125 = 0$$

$$\Rightarrow y^2 - 10y + 25 = (y - 5)^2 = 0 \Rightarrow y = 5 \Rightarrow x = 12 - 2y = 2.$$

Therefore, $(2, 5)$ is the particular point that is shared by each and every circle indicated by the equation given for every value of k.

Fig. 0

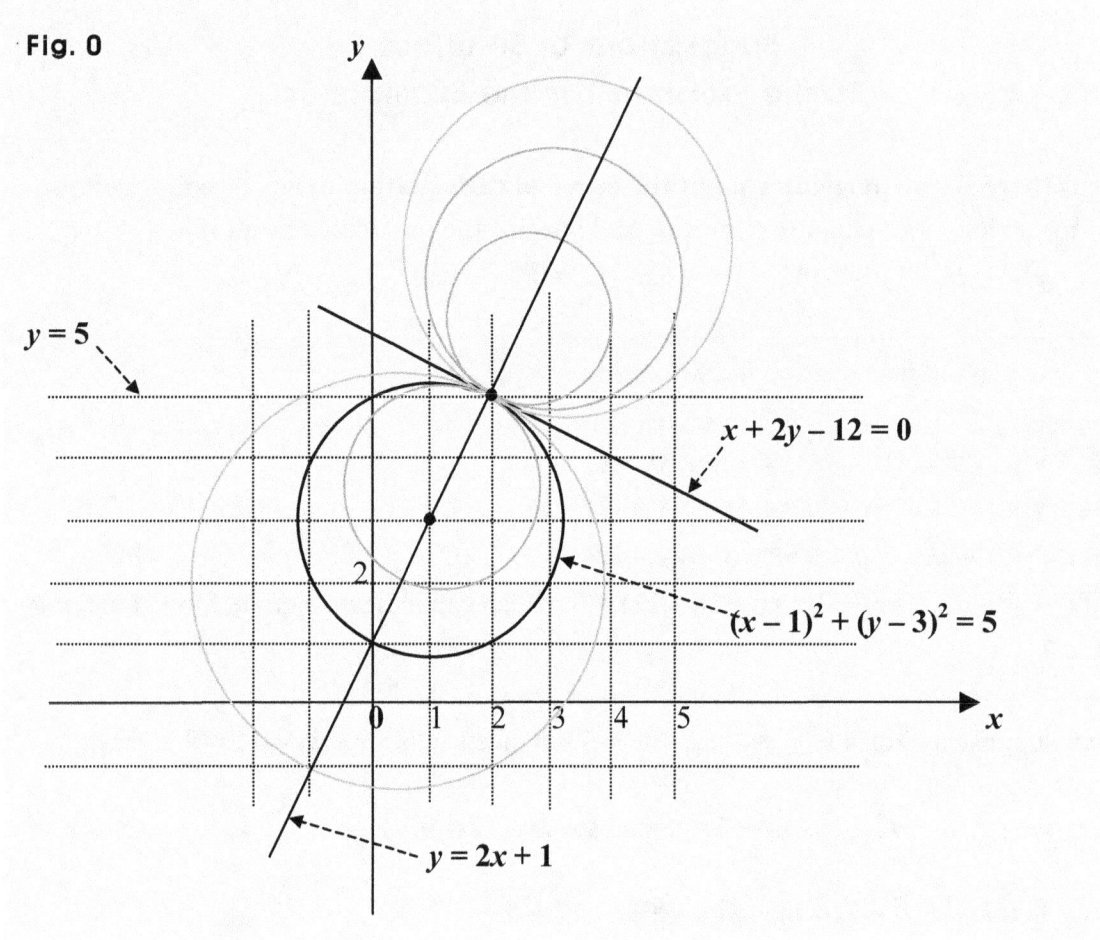

$y = 5$

$x + 2y - 12 = 0$

$(x-1)^2 + (y-3)^2 = 5$

$y = 2x + 1$

Suggestions or Solutions
To the Problem 2 in the Example 1

See if there is any particular point the curve of the equation below passes through for any value of k, which is constant, and specify the particular point if any.
$$k(x^2 + y^2) - (6k - 1)x - (4k + 1)y + 10k + 2 = 0.$$

To begin with, we can put the equation the way below:

$$k(x^2 + y^2) - (6k - 1)x - (4k + 1)y + 10k + 2 = k(x^2 + y^2) - 6kx + x - 4ky - y + 10k + 2$$

$$= k(x^2 + y^2) - 6kx - 4ky + 10k + x - y + 2 = k(x^2 + y^2 - 6x - 4y + 10) + (x - y + 2) = 0.$$

So the given equation can be an equation of a circle for a value of k if $k \neq 0$.

Next, suppose a circles $x^2 + y^2 - 6x - 4y + 10 = 0$ meets a line $x - y + 2 = 0$ at (s, t).

Then, we get: $s^2 + t^2 - 6s - 4t + 10 = 0$, and $s - t + 2 = 0$.

So we get: $k(s^2 + t^2 - 6s - 4t + 10) + s - t + 2 = 0$.

So if the circle and the line share any particular point, for any value of k, other than 0, the equation given indicates a circle that includes the particular point.
Finding the particular point, we get:

$$x - y + 2 = 0 \Rightarrow x = y - 2 \Rightarrow x^2 + y^2 - 6x - 4y + 10 = (y - 2)^2 + y^2 - 6(y - 2) - 4y + 10$$

$$= y^2 - 4y + 4 + y^2 - 6y + 12 - 4y + 10 = 2y^2 - 14y + 26 = 0 \Rightarrow y^2 - 7y + 13 = 0.$$

However, we have:

$$y^2 - 7y + 13 = y^2 - 7y + (\tfrac{7}{2})^2 - (\tfrac{7}{2})^2 + 13 = (y - \tfrac{7}{2})^2 - \tfrac{49}{4} + 13 = (y - \tfrac{7}{2})^2 + \tfrac{3}{4} > 0.$$

Thus, the circle does not meet the line, so there is no particular point that can be shared by every circle indicated by the given equation for every nonzero k.

That is, there is no particular point that can be shared by each and every circle indicated by the given equation for every value of k.

If not quite sure of the idea behind the processes above, follow the steps below:

For this problem, too, we can use the fact below:

Suppose m and n are constant, and at particular points, the curve of $f(x, y) = 0$ meets the curve of $g(x, y) = 0$.

Then, at the particular points, too, the curve of $mf(x, y) + ng(x, y) = 0$ meets the two curves of $f(x, y) = 0$ and $g(x, y) = 0$.

In other words, the three curves share the particular points.

And in particular, the equation $mf(x, y) + ng(x, y) = 0$ is called a linear combination of the two equations $f(x, y) = 0$ and $g(x, y) = 0$.

So we want to check to see if the equation given can be such a linear combination. That is, we get to extract equations, which the equation given is a linear combination of.

$$k(x^2 + y^2) - (6k - 1)x - (4k + 1)y + 10k + 2 = k(x^2 + y^2) - 6kx + x - 4ky - y + 10k + 2$$

$$= k(x^2 + y^2) - 6kx - 4ky + 10k + x - y + 2 = k(x^2 + y^2 - 6x - 4y + 10) + (x - y + 2) = 0.$$

So we can see two equations, one indicates a circle $x^2 + y^2 - 6x - 4y + 10 = 0$, and the other indicates a line $x - y + 2 = 0$.

So suppose C is the circle above, and L is the line above. Then, the curve of the equation given is a linear combination of C and L, so we can use such a fact as follows:

Suppose a circle $x^2 + y^2 + ax + by + c = 0$ meets a line $dx + ey + f = 0$ at a particular point, m and n are constant, and $m \neq 0$. Then, for each pair of the values of m and n, the equation $m(x^2 + y^2 + ax + by + c) + n(dx + ey + f) = 0$ indicates another circle that includes the particular point, and has its center in a line passing through the center of a circle $x^2 + y^2 + ax + by + c = 0$ and perpendicular to a line $dx + ey + f = 0$.

Let's see now, if the circle C meets the line L.

To begin with, we can get: $x - y + 2 = 0 \Rightarrow x = y - 2$.

So next, we get: $x^2 + y^2 - 6x - 4y + 10 = (y - 2)^2 + y^2 - 6(y - 2) - 4y + 10$

$= y^2 - 4y + 4 + y^2 - 6y + 12 - 4y + 10 = 2y^2 - 14y + 26 = 0$. So we get: $y^2 - 7y + 13 = 0$.

However: $y^2 - 7y + 13 = y^2 - 7y + (\frac{7}{2})^2 - (\frac{7}{2})^2 + 13 = (y - \frac{7}{2})^2 - \frac{49}{4} + 13 = (y - \frac{7}{2})^2 + \frac{3}{4} > 0$.

In other words, we have: $y^2 - 7y + 13 \neq 0$. Therefore, we can see that there is no solution to the equation $y^2 - 7y + 13 = 0$, so the circle does not meet the line.

We can check the same using the discriminant, too, of course.

Finding the discriminant of $au^2 + bu + c = 0$, we get: $b^2 - 4ac$, and if the discriminant is negative, the equation has no root, so it has no solution.

However, the equation given can represent many circles. That is, other than 0, for each of some values of k, the equation given still indicates a circle. That's because we have:

$k(x^2 + y^2) - (6k - 1)x - (4k + 1)y + 10k + 2 = k(x^2 + y^2 - 6x - 4y + 10) + (x - y + 2) = 0$.

So if $k \neq 0$, we can get a circle. And if $k = 0$, of course, we just get a line: $x - y + 2 = 0$.

Let's now, put in a graph the circle C and the line L.

To begin with, putting the circle C in the standard form, we get:

$x^2 + y^2 - 6x - 4y + 10 = x^2 - 6x + 9 - 9 + y^2 - 4y + 4 - 4 + 10$

$= (x - 3)^2 + (y - 2)^2 - 3 = 0 \Rightarrow (x - 3)^2 + (y - 2)^2 = 3$.

So the center of C is at $(3, 2)$, and the radius is $\sqrt{3}$.

Fig. 0

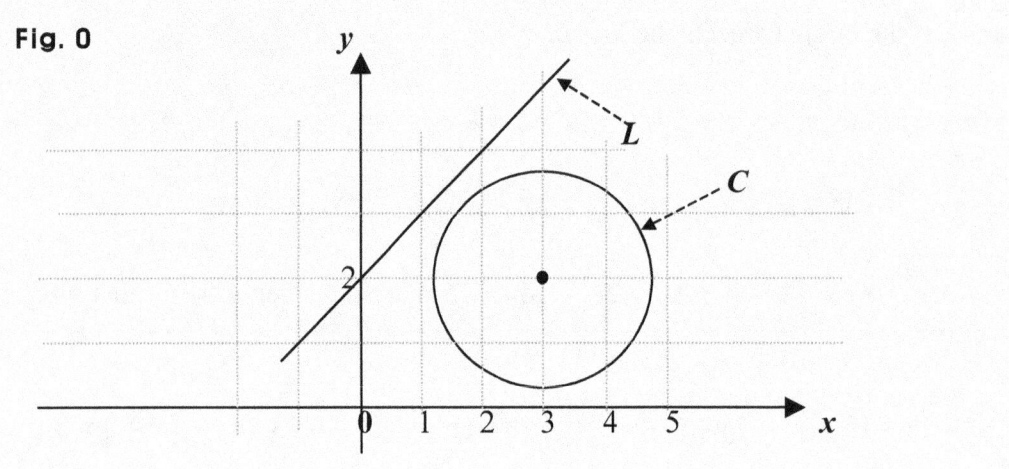

Now, for <u>some</u> values of k, the equation $k(x^2 + y^2) - (6k - 1)x - (4k + 1)y + 10k + 2 = 0$ indicates a circle.   So it is not the case where it indicate a circle for every value of k.

For what value of k then, can the given equation indicate a circle?

Let's first, put the given equation in the standard form so that we can see the radius.

To begin with, $k(x^2 + y^2) - (6k - 1)x - (4k + 1)y + 10k + 2 = 0$

$\Rightarrow \ x^2 + y^2 - \frac{6k-1}{k}x - \frac{4k+1}{k}y + \frac{10k+2}{k} = 0 \Rightarrow x^2 + y^2 - \frac{2(6k-1)}{2k}x - \frac{2(4k+1)}{2k}y + \frac{10k+2}{k} = 0.$

And next, setting: $a = \frac{6k-1}{2k}$, $b = \frac{4k+1}{2k}$, and $c = \frac{10k+2}{k}$, we get: $x^2 + y^2 - 2ax - 2by + c = 0$

$\Rightarrow (x - a)^2 + (y - b)^2 + c - a^2 - b^2 = 0 \Rightarrow (x - a)^2 + (y - b)^2 = a^2 + b^2 - c.$

Thus, $a^2 + b^2 - c$ has to be positives since a radius squared is positive.

So we get: $a^2 + b^2 - c = \frac{(6k-1)^2}{4k^2} + \frac{(4k+1)^2}{4k^2} - \frac{10k+2}{k} = \frac{(6k-1)^2 + (4k+1)^2 - 4k(10k+2)}{4k^2} > 0.$

Thus, $(6k - 1)^2 + (4k + 1)^2 - 4k(10k + 2) = 36k^2 - 12k + 1 + 16k^2 + 8k + 1 - 40k^2 - 8k$

$= 12k^2 - 12k + 2 = 2(6k^2 - 6k + 1) > 0 \Rightarrow 6k^2 - 6k + 1 > 0.$

So we need to have: $6k^2 - 6k + 1 = 6(k^2 - k) + 1 = 6(k^2 - k + \frac{1}{4} - \frac{1}{4}) + 1$

$= 6(k - \frac{1}{2})^2 - \frac{3}{2} + 1 = 6(k - \frac{1}{2})^2 - \frac{1}{2} > 0 \Rightarrow (k - \frac{1}{2})^2 - \frac{1}{12} > 0.$

Meanwhile, we have a factorization identity where $P^2 - Q^2 = (P + Q)(P - Q)$.

So we get: $\{(k - \frac{1}{2}) + \sqrt{\frac{1}{12}}\}\{(k - \frac{1}{2}) - \sqrt{\frac{1}{12}}\} > 0 \Rightarrow \{(k - \frac{1}{2}) + \frac{1}{2\sqrt{3}}\}\{(k - \frac{1}{2}) - \frac{1}{2\sqrt{3}}\} > 0$

$\Rightarrow \{(k - \frac{1}{2}) + \frac{\sqrt{3}}{6}\}\{(k - \frac{1}{2}) - \frac{\sqrt{3}}{6}\} > 0 \Rightarrow (k - \frac{3 - \sqrt{3}}{6})(k - \frac{3 + \sqrt{3}}{6}) > 0$

$\Rightarrow k > \frac{3 + \sqrt{3}}{6} \approx 0.789$ or $k < \frac{3 - \sqrt{3}}{6} \approx 0.211.$

Therefore, when $k > \frac{3 + \sqrt{3}}{6}$ or $k < \frac{3 - \sqrt{3}}{6}$, the given equation where:

$k(x^2 + y^2) - (6k - 1)x - (4k + 1)y + 10k + 2 = 0$ can indicate a circle.

The centers of all the circles that can be indicated by the equation above are in a line, which is perpendicular to the line L, which is: $x - y + 2 = 0$, and also, passes through the center of the circle C, which is: $x^2 + y^2 - 6x - 4y + 10$, which is: $(x - 3)^2 + (y - 2)^2 = 3$.

Suppose N is the perpendicular line above. Then, N passes through all the centers.

So let's now, get the line N.   To begin with, we have a fact below:

Given the slope and a point in a line, we can readily find the line.

So if p is the slope, and (s, t) is a point in a line, the line is: $y - t = p(x - s)$.

Now, the line N is perpendicular to the line L, $x - y + 2 = 0$, where the slope is 1.

So the slope of N is **-1** because the product of the slopes of two lines perpendicular to each other is -1, and the slope of L is 1.

Next, the line N includes the center of the circle C, and the center of C is **(3, 2)**.

So the line N is: $y - 2 = -(x - 3)$, which equals: $y = -x + 5$, in which the centers of all the circles that are linear combinations of C and L are located.

And yet, not the entire line can be the location where all the centers can be put.

That's because we have the limitation on the constant k in the equation given.

So let's find in what part of the line N all the centers can be put.

To begin with, we have put the given equation in such a from as follows:

$(x - a)^2 + (y - b)^2 = a^2 + b^2 - c$ where $a = \frac{6k-1}{2k}$, $b = \frac{4k+1}{2k}$, and $c = \frac{10k+2}{k}$.

So each of the centers is at (a, b).

And finding the extent of a, we can see the extent of the x-coordinate of the center.

We have: $a = \frac{6k-1}{2k} = 3 - \frac{1}{2k}$ for $k > \frac{3+\sqrt{3}}{6}$ or $k < \frac{3-\sqrt{3}}{6}$. So to begin with, we can get:

$k > \frac{3+\sqrt{3}}{6} \Rightarrow 2k > \frac{3+\sqrt{3}}{3} \Rightarrow 0 < \frac{1}{2k} < \frac{3}{3+\sqrt{3}} = \frac{3(3-\sqrt{3})}{(3+\sqrt{3})(3-\sqrt{3})} = \frac{3-\sqrt{3}}{2} \Rightarrow 0 < \frac{1}{2k} < \frac{3-\sqrt{3}}{2}$.

How come we get not only $\frac{1}{2k} < \frac{3}{3+\sqrt{3}}$ but $0 < \frac{1}{2k}$, too?

That's because k is positive since $k > \frac{3+\sqrt{3}}{6}$, so we get: $\frac{1}{2k} > 0$, too.

Thus, we get: $0 < \frac{1}{2k} < \frac{3-\sqrt{3}}{2} \Rightarrow 0 > -\frac{1}{2k} > -\frac{3-\sqrt{3}}{2} \Rightarrow 3 > 3 - \frac{1}{2k} > 3 - \frac{3-\sqrt{3}}{2} = \frac{3+\sqrt{3}}{2}$.

So we get: $3 > a > \frac{3+\sqrt{3}}{2} \approx 2.366$.

Next, let's move on to $k < \frac{3-\sqrt{3}}{6}$. Then, we get: $k < \frac{3-\sqrt{3}}{6} \Rightarrow 2k < \frac{3-\sqrt{3}}{3}$.

So we get: $\frac{1}{2k} > \frac{3}{3-\sqrt{3}}$ or $\frac{1}{2k} \leq 0$. How come we get: $\frac{1}{2k} \leq 0$, too, though?

That's because we have: $k < \frac{3-\sqrt{3}}{6}$, so k can be ≤ 0.

Now, beginning with $\frac{1}{2k} \leq 0$, we get: $\frac{1}{2k} \leq 0 \Rightarrow -\frac{1}{2k} \geq 0 \Rightarrow 3 - \frac{1}{2k} \geq 3 \Rightarrow a \geq 3$.

Next, we get: $\frac{1}{2k} > \frac{3}{3-\sqrt{3}} = \frac{3(3+\sqrt{3})}{(3-\sqrt{3})(3+\sqrt{3})} = \frac{3+\sqrt{3}}{2} \Rightarrow \frac{1}{2k} > \frac{3+\sqrt{3}}{2} \Rightarrow -\frac{1}{2k} < -\frac{3+\sqrt{3}}{2}$

$\Rightarrow 3 - \frac{1}{2k} < 3 - \frac{3+\sqrt{3}}{2} = \frac{3-\sqrt{3}}{2} \Rightarrow a < \frac{3-\sqrt{3}}{2} \approx 0.634$.

Therefore, we get: $a \geq 3$ or $a < \frac{3-\sqrt{3}}{2} \approx 0.634$.

Thus, we now have: $3 > a > \frac{3+\sqrt{3}}{2} \approx 2.366$, $a \geq 3$, or $a < \frac{3-\sqrt{3}}{2} \approx 0.634$.

Putting threads together, we have: $a > \frac{3+\sqrt{3}}{2} \approx 2.366$ or $a < \frac{3-\sqrt{3}}{2} \approx 0.634$.

Now, since a is the x-coordinate of each center, all the centers are in two rays as follows:

$y = -x + 5$ for $x > \frac{3+\sqrt{3}}{2} \approx 2.366$ or $x < \frac{3-\sqrt{3}}{2} \approx 0.634$.

Note that the circle C, $x^2 + y^2 - 6x - 4y + 10 = 0$, that is, $(x-3)^2 + (y-2)^2 = 3$ cannot be made by the equation given.   That's because we have:

$k(x^2 + y^2) - (6k-1)x - (4k+1)y + 10k + 2 = k(x^2 + y^2 - 6x - 4y + 10) + (x - y + 2) = 0$.

So if $k \neq 0$, we can get a circle. And if $k = 0$, of course, we just get a line: $x - y + 2 = 0$.

However, the circle C can be obtained if we use such an equation as follows:

$m(x^2 + y^2 - 6x - 4y + 10) + n(x - y + 2) = 0$ where m and n are constants.

So setting: $m = 1$ and $n = 0$, we get the circle C.

Let's now, put in a graph some circles each of which is a linear combination of the circle C and the line L, together with the circle C and the line L.

Note that in the graph above, the radii of the circles in gray are not accurate at all. That is, the gray circles are no more than the ones for showing how the circles can be put in a graph. The key idea is that the centers of all the circles are on the two rays.

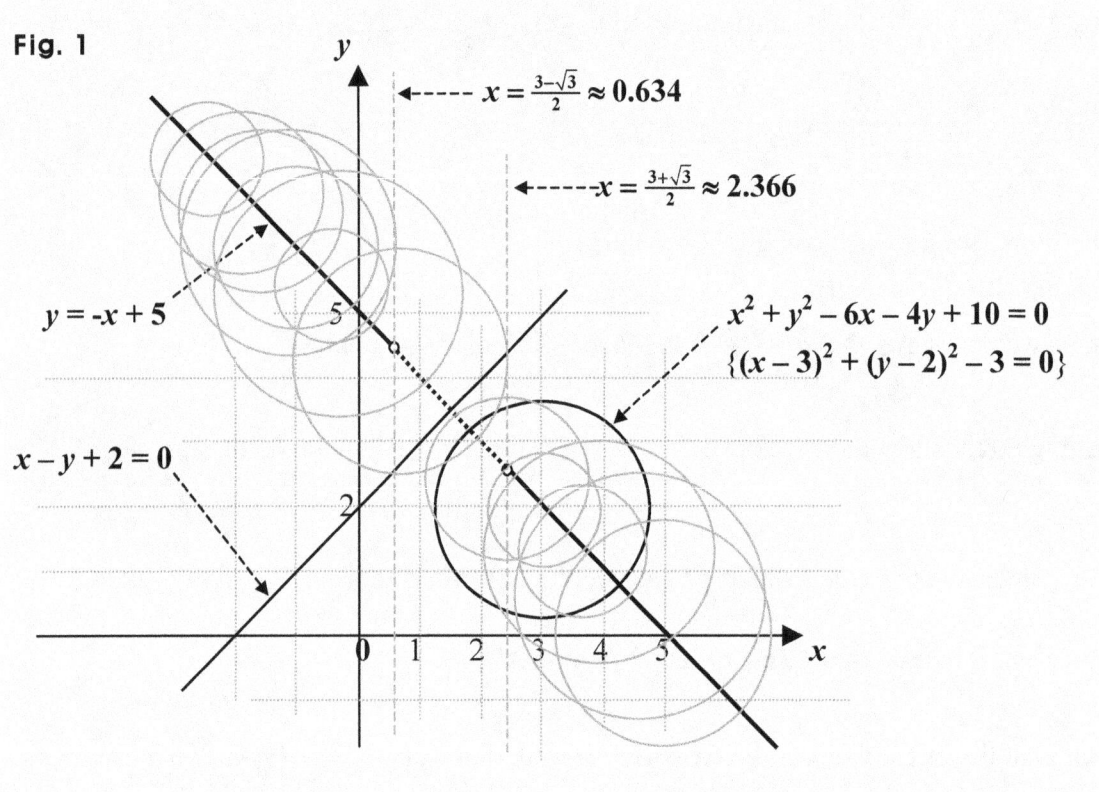

Fig. 1

$y = -x + 5$

$x - y + 2 = 0$

$x = \frac{3-\sqrt{3}}{2} \approx 0.634$

$x = \frac{3+\sqrt{3}}{2} \approx 2.366$

$x^2 + y^2 - 6x - 4y + 10 = 0$

$\{(x-3)^2 + (y-2)^2 - 3 = 0\}$

Now, what if one of the curves in such a linear combination as above is not a circle?

Suppose for instance, the equation given is as follows:

$k(x^2 + y^2) - 2(k-1)x - 2(3k-1)y + 12k - 1 = 0$.

Then, extracting the curves, we get:

$k(x^2 + y^2) - 2(k-1)x - 2(3k-1)y + 12k - 1$

$= k(x^2 + y^2) - 2kx + 2x - 6ky + 2y + 12k - 1$

$= k(x^2 + y^2) - 2kx - 6ky + 12k + 2x + 2y - 1$

$= k(x^2 + y^2 - 2x - 6y + 12) + (2x + 2y - 1) = 0$.

So one is: $x^2 + y^2 - 2x - 6y + 12 = 0$, and the other is a line: $2x + 2y - 1 = 0$.

And next, putting the first of the two above in the standard form, we get:

$$x^2 + y^2 - 2x - 6y + 12 = x^2 - 2x + 1 - 1 + y^2 - 6y + 9 - 9 + 12$$

$= (x - 1)^2 + (y - 3)^2 + 2 = 0$. So we get: $(x - 1)^2 + (y - 3)^2 = \text{-}2$, which however, does not indicate a circle. So does the equation given not indicate a circle for any value of k?

It is not necessarily the case. It can still be the case where the equation can indicate a circle for some value of k.

So let's check to see if there is any value for k so that the equation indicates a circle. Then, to begin with, we can put the equation this way:

$$k(x^2 + y^2) - 2(k - 1)x - 2(3k - 1)y + 12k - 1 = 0$$

$$\Rightarrow x^2 + y^2 - \frac{2(k-1)}{k}x - \frac{2(3k-1)}{k}y + \frac{12k-1}{k} = 0.$$

And next, setting: $a = \frac{k-1}{k}$, $b = \frac{3k-1}{k}$, and $c = \frac{12k-1}{k}$, we get: $x^2 + y^2 - 2ax - 2by + c = 0$.

Then, we get: $x^2 + y^2 - 2ax - 2by + c = x^2 - 2ax + a^2 - a^2 + y^2 - 2by + b^2 - b^2 + c$

$= (x - a)^2 + (y - b)^2 + c - (a^2 + b^2) = 0 \Rightarrow (x - a)^2 + (y - b)^2 = a^2 + b^2 - c.$

Now, if the equation $(x - a)^2 + (y - b)^2 = a^2 + b^2 - c$ indicates a circle, $a^2 + b^2 - c$ has to be positive. If it indicates a circle, we need to have:

$$a^2 + b^2 - c = \left(\frac{k-1}{k}\right)^2 + \left(\frac{3k-1}{k}\right)^2 - \frac{12k-1}{k} = \frac{(k-1)^2 + (3k-1)^2 - (12k-1)k}{k^2} > 0.$$

Then, first, we need to have: $k \neq 0$ since a denominator cannot be 0.

Next, we need to have: $(k - 1)^2 + (3k - 1)^2 - (12k - 1)k > 0.$

Expanding the left hand side and simplifying the result, we get:
$$k^2 - 2k + 1 + 9k^2 - 6k + 1 - 12k^2 + k = \text{-}2k^2 - 7k + 2 > 0 \Rightarrow 2k^2 + 7k - 2 < 0.$$

Next, setting: $2k^2 + 7k - 2 = 0$, we get: $k = \frac{-7 \pm \sqrt{49 - 4 \cdot 2 \cdot (-2)}}{4} = \frac{-7 \pm \sqrt{65}}{4}$.

So we get: $\frac{-7 - \sqrt{65}}{4} < k < \frac{-7 + \sqrt{65}}{4}$, since the parabola $y = 2x^2 + 7x - 2$ is concave-up.

Therefore, we need to have: $\frac{-7 - \sqrt{65}}{4} < k < \frac{-7 + \sqrt{65}}{4}$, yet k cannot be 0.

So we need to have: $\frac{-7 - \sqrt{65}}{4} < k < 0$, or $0 < k < \frac{-7 + \sqrt{65}}{4}$.

Thus, the given equation $k(x^2 + y^2) - 2(k - 1)x - 2(3k - 1)y + 12k - 1 = 0$ does indicate a circle if $\frac{-7 - \sqrt{65}}{4} < k < 0$, or $0 < k < \frac{-7 + \sqrt{65}}{4}$.

Now, we have: $k(x^2 + y^2) - 2(k - 1)x - 2(3k - 1)y + 12k - 1$
$= k(x^2 + y^2 - 2x - 6y + 12) + (2x + 2y - 1) = 0$.

So the equation can represent a group of circles each of which is a linear combination of the two curves as follows: $x^2 + y^2 - 2x - 6y + 12 = 0$, and $2x + 2y - 1 = 0$.

More specifically, it can represent two groups of circles:

One is for the case where $\frac{-7 - \sqrt{65}}{4} < k < 0$, and the other is for the case: $0 < k < \frac{-7 + \sqrt{65}}{4}$.

So next, let's check to see if the two curves meet each other.

Solving the system of $x^2 + y^2 - 2x - 6y + 12 = 0$, and $2x + 2y - 1 = 0$, we can get first:

$2x + 2y - 1 = 0 \Rightarrow y = -x + \frac{1}{2}$. So next, we get:

$x^2 + y^2 - 2x - 6y + 12 = x^2 + (-x + \frac{1}{2})^2 - 2x - 6(-x + \frac{1}{2}) + 12 = 2x^2 + 3x + \frac{37}{4} = 0$.

Checking to see if there is any root, we can check to see if the discriminant ≥ 0.

The discriminant is: $3^2 - 4 \cdot 2 \cdot \frac{37}{4} = 9 - 74$, which is negative.

So there is no root, and thus, the two curves don't meet.

So there is no particular point every circle indicated by the given equation includes for every value allowed for **k**.

What if the two curves did meet at particular points, though?

Then of course, all the circles in the groups stated above would meet altogether at the particular points. That is, each and every circle indicated by the given equation for every value allowed for **k** would pass through the particular points.

Examples A in Circles

Constructing a curve in math, we can put it in a graph. Putting it in a graph though, we need the equation of it. So constructing it, we want to find the equation of it if we are not given the equation, of course. And finding the equation, it is said that we get it, too.

0.   Assuming that a line is: $2x + 3y - 3 = 0$, construct:

0.0.   A circle that is tangent to the line, and has a radius of 2 and a center where the x-coordinate is 1.

0.1.   A circle that is tangent to the line, and has a radius of 2 and a center where the y-coordinate is 1.

1.   Construct a circle that is tangent to a line $x = a$, where a is constant, and has a radius of 2 and a center where the y-coordinate is 1.

2.   Construct a circle that is tangent to a line $y = b$, where b is constant, and has a radius of 2 and a center where the x-coordinate is 1.

Suggestions or Solutions
To the Problem 0 in the Example 0

Construct a circle that is tangent to a line $2x + 3y - 3 = 0$, and has a radius of 2 and a center where the x-coordinate is 1.

Suppose S is the circle to be constructed, A is the line $2x + 3y - 3 = 0$, B is a line parallel to A, and the distance between A and B is 2, which is the radius of S.

Then, B can be put in $y = -\frac{2}{3}x + c$ where c is constant and $c \neq 1$, for B is parallel to A.

Meanwhile, we get: $2x + 3y - 3 = 0 \Rightarrow y = -\frac{2}{3}x + 1$.

Suppose now, t is constant, and $(1, t)$ is a point in B. Then, the center of S can be put in $(1, -\frac{2}{3} + c)$, the distance from which to A is the radius of S.

So S can be for now, put in $(x-1)^2 + \{y - (-\frac{2}{3} + c)\}^2 = 4$.

The distance from a point (u, v) to a line $px + qy + r = 0$ is: $\frac{|pu + qv + r|}{\sqrt{p^2 + q^2}}$.

So the distance from the center to A is: $\frac{|2 \cdot 1 + 3(-\frac{2}{3} + c) - 3|}{\sqrt{2^2 + 3^2}} = \frac{|3c - 3|}{\sqrt{13}}$, which is 2. Thus, we get:

$\frac{3|1-c|}{\sqrt{13}} = 2 \Rightarrow \frac{9(1-c)^2}{13} = 4 \Rightarrow 9(1-c)^2 = 52 \Rightarrow 1 - c = \pm\frac{\sqrt{52}}{3} \Rightarrow c = 1 \pm \frac{2\sqrt{13}}{3} = \frac{3 \pm 2\sqrt{13}}{3}$.

So we get: $-\frac{2}{3} + c = -\frac{2}{3} + \frac{3 \pm 2\sqrt{13}}{3} = \frac{1 \pm 2\sqrt{13}}{3}$.

Therefore, the circle S is as follows: $(x-1)^2 + (y - \frac{1 \pm 2\sqrt{13}}{3})^2 = 4$.

If not quite sure of the idea behind the processes above, follow the steps below:

Suppose A is the line $2x + 3y - 3 = 0$, B is a line parallel to the line A, and the distance between A and B is 2.

Then, if the center of a circle is in the line B, and the circle is tangent to the line A, then the radius is 2. Meanwhile, $2x + 3y - 3 = 0 \Rightarrow y = -\frac{2}{3}x + 1$.

Let's now, put in a graph the ideas above.

Fig. 0

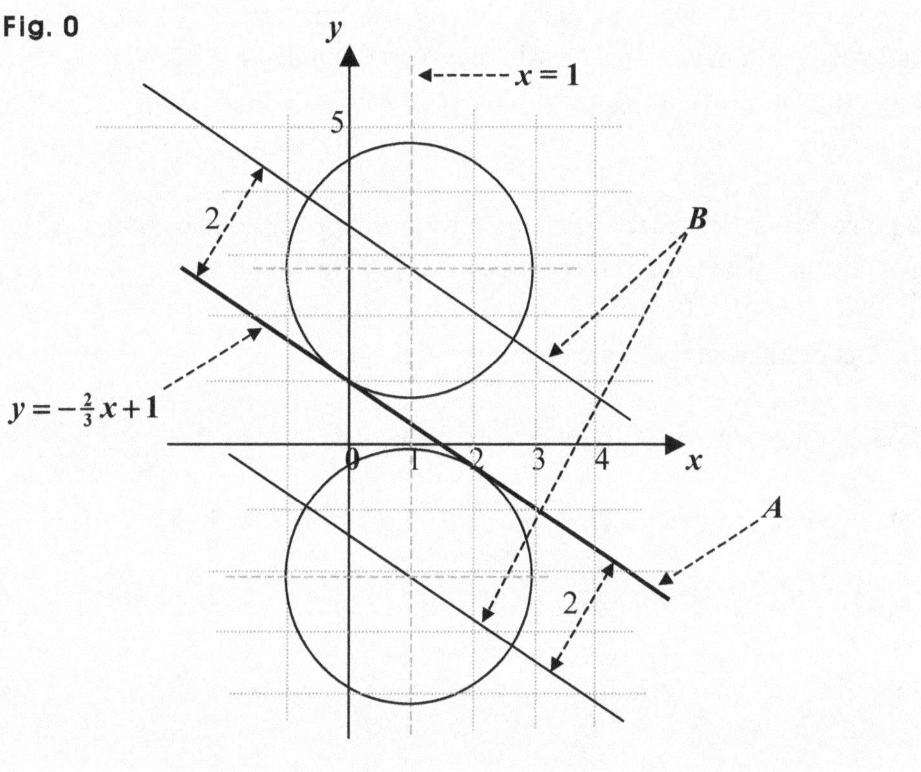

Now, we can see that there are two circles that can be the solution to this problem.

Given the center and radius, we can find the circle.
We know that the radius is 2, but the x-coordinate only of the center is given.
So we want to find the y-coordinate of the center.
Anyway, where do we want to look for the center?

The center is in the line **B**. That is, one of the points in the line **B** is the center.
So we want to find the line **B**, first.   How then, do we find the line **B**?

The line **B** is parallel to and 2 away from the line **A**.
Then, where in the line **B**, is the center?

We have a fact that at the center, the x-coordinate is 1.
So using the fact, we can find exactly where the center is in the line **B**.

Then, using the center, together with the radius, which is 2, we can find the circle.

Now, going back to the graph above, we can see that the two lines A and B are parallel to each other, so B can be put in $y = -\frac{2}{3}x + c$ where c is constant and $c \neq 1$, and of course, we want to find c.   How to find it, though? That is, how do we find the line B?

The line B has a point that is the center of the circle we want. So suppose now, that S is the circle to be constructed. Then, the center of the circle S is in the line B.

We know that the x-coordinate at the center is 1.

So we can suppose that the center is $(1, t)$ where t is a constant.

Then, the point $(1, t)$ belongs to the line B, so we can put it into B, which is $y = -\frac{2}{3}x + c$.

Thus, we get: $t = -\frac{2}{3} + c$, which is the y-coordinate at the center of the circle S.

So the center of the circle S can be put this way: $(1, -\frac{2}{3} + c)$. How do we find c, then?

We can find c using a fact stated earlier, and relevant to the radius. What fact?

The fact is that the distance from A to the center of S is the radius, which is 2. How then, do we use such a fact?

The fact is talking about the radius, so let's now, move on to the radius.

We know the center of S is in the line B, and the circle S is tangent to the line A.

So the radius of S is the distance from the center to the line A. We have a tool for such a distance, and the tool is a template, called a formula for a distance from a point to a line:

- The distance from a point (u, v) to a line $px + qy + r = 0$ is: $\frac{|pu+qv+r|}{\sqrt{p^2+q^2}}$.

Now, the line A is: $2x + 3y - 3 = 0$, and the center of S is $(1, -\frac{2}{3} + c)$. Thus, the distance from the center to A is: $\frac{|2\cdot1+3(-\frac{2}{3}+c)-3|}{\sqrt{2^2+3^2}} = \frac{|3c-3|}{\sqrt{13}}$, which is the radius of S, which is 2.

So we get: $\frac{3|1-c|}{\sqrt{13}} = 2 \Rightarrow \frac{9(1-c)^2}{13} = 4 \Rightarrow 9(1-c)^2 = 52 \Rightarrow 1-c = \pm\frac{\sqrt{52}}{3} \Rightarrow c = 1 \pm \frac{2\sqrt{13}}{3} = \frac{3\pm2\sqrt{13}}{3}$.

Now, we know that the center of S is $(1, -\frac{2}{3} + c)$.

So the center is $(1, -\frac{2}{3} + \frac{3\pm2\sqrt{13}}{3})$, which means we get two circles. More specifically:

First, if $c = \frac{3+2\sqrt{13}}{3}$, we get: $-\frac{2}{3} + c = -\frac{2}{3} + \frac{3+2\sqrt{13}}{3} = \frac{1+2\sqrt{13}}{3} \approx 2.737$.

Next, if $c = \frac{3-2\sqrt{13}}{3}$, we get: $-\frac{2}{3} + c = -\frac{2}{3} + \frac{3-2\sqrt{13}}{3} = \frac{1-2\sqrt{13}}{3} \approx -2.070$.

Now, since the radius is 2, using the standard form, we can see that the circle S is:

$(x-1)^2 + (y - \frac{1+2\sqrt{13}}{3})^2 = 4$ or $(x-1)^2 + (y - \frac{1-2\sqrt{13}}{3})^2 = 4$.

Of course, we can put the equations above this way, too: $(x-1)^2 + (y - \frac{1\pm2\sqrt{13}}{3})^2 = 4$.

Fig. 1

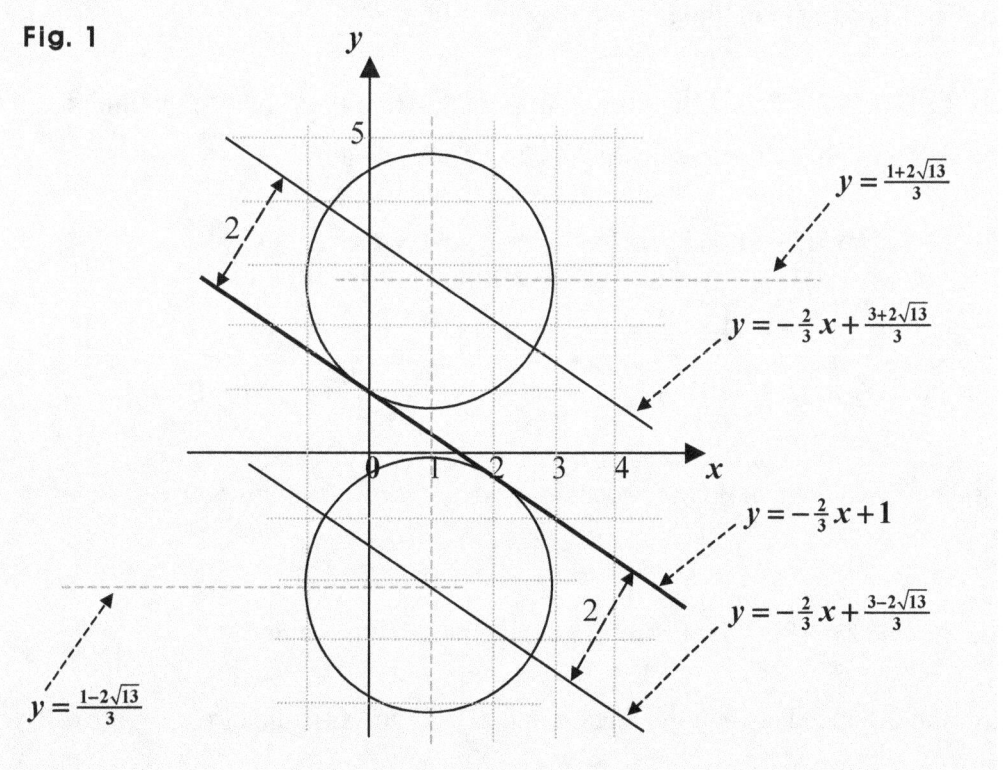

Suggestions or Solutions
To the Problem 1 in the Example 0

Construct a circle that is tangent to a line $2x + 3y - 3 = 0$, and has a radius of 2 and a center where the y-coordinate is 1.

Suppose first, C is the circle to be constructed, A is the line $2x + 3y - 3 = 0$, B is a line parallel to the line A, and the distance between the lines A and B is 2, which is the radius of C. Then, the center of the circle C is in the line B.

Next, putting the line A in the standard form, we get: $y = -\frac{2}{3}x + 1$, so the line B can be put in $y = -\frac{2}{3}x + c$, where c is constant and $c \neq 1$. Suppose now, that $(s, 1)$ is a point in B.

Then, $(s, 1)$ is the center of C, and we get: $1 = -\frac{2}{3}s + c$ since B is: $y = -\frac{2}{3}x + c$.

So we get: $1 = -\frac{2}{3}s + c \Rightarrow s = \frac{3(c-1)}{2}$, and thus, the center of C can be put in $(\frac{3(c-1)}{2}, 1)$.

So the circle C can be for now, put in $(x - \frac{3(c-1)}{2})^2 + (y-1)^2 = 4$.

The distance from a point (u, v) to a line $px + qy + r = 0$ is: $\frac{|pu+qv+r|}{\sqrt{p^2+q^2}}$.

The line A is: $2x + 3y - 3 = 0$, and therefore, the distance from the center to the line A is: $\frac{|2 \cdot \frac{3(c-1)}{2} + 3 \cdot 1 - 3|}{\sqrt{2^2+3^2}} = \frac{|3(c-1)|}{\sqrt{13}}$, which is the radius, which is 2. Thus, we get:

$\frac{3|1-c|}{\sqrt{13}} = 2 \Rightarrow \frac{9(1-c)^2}{13} = 4 \Rightarrow 9(1-c)^2 = 52 \Rightarrow 1 - c = \pm\frac{\sqrt{52}}{3} \Rightarrow c = 1 \pm \frac{2\sqrt{13}}{3} = \frac{3 \pm 2\sqrt{13}}{3}$.

So we get: $\frac{3(c-1)}{2} = \frac{3(\frac{3 \pm 2\sqrt{13}}{3} - 1)}{2} = \pm\sqrt{13}$.

Therefore, the circle C is: $(x + \sqrt{13})^2 + (y-1)^2 = 4$, or $(x - \sqrt{13})^2 + (y-1)^2 = 4$.

If not quite sure of the idea behind the processes above, follow the steps below:

Suppose A is the line $2x + 3y - 3 = 0$, and B is a line parallel to the line A.

Suppose also, the distance between the lines A and B is 2, and C is the circle we are after.

Then, the center of the circle C is in the line B, and the radius is the same as the distance between A and B. Meanwhile, $2x + 3y - 3 = 0 \Rightarrow y = -\frac{2}{3}x + 1$, which is the line A.

Now, putting in a graph the ideas above, we can put it the way below:

Fig. 0

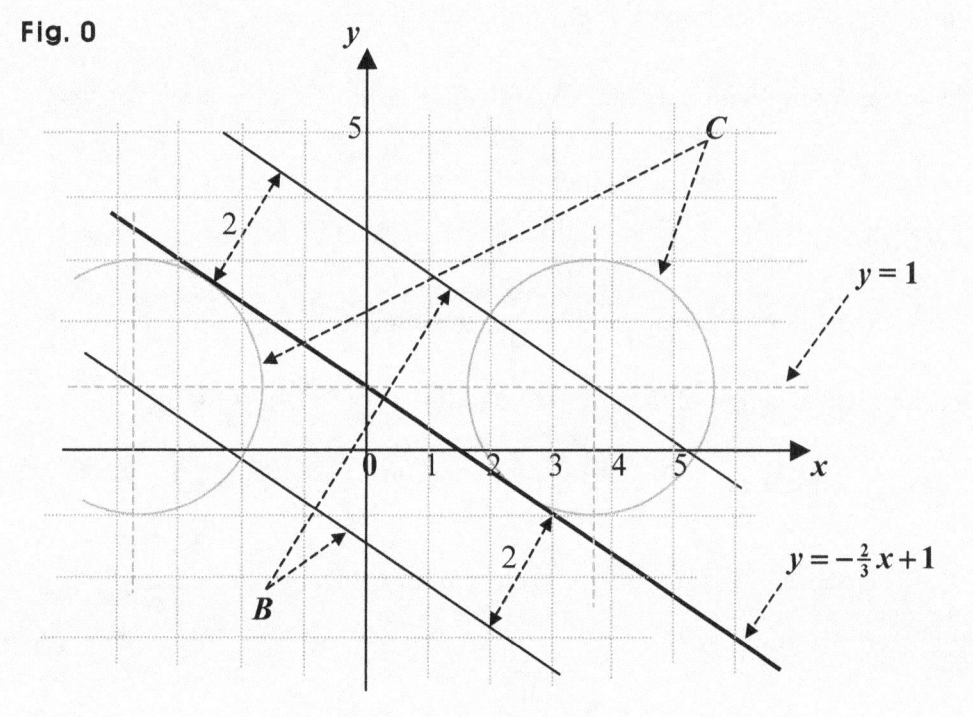

Now, we can see that there are two circles that can be the solution to this problem.

• Given the center and radius of a circle, we can find the circle.

We know that the radius is 2, but the y-coordinate only at the center is given.
So we want to find the x-coordinate of the center.
Anyway, whereabouts is the center?

The center is somewhere in the line B. So it is for sure that the line B passes through the center. So we want to find the line B, first.   How then, do we find it?

We know the line A.   So what about it?

The line A is parallel to, and is 2 away from the line B. And also, the line A is tangent to the circle we want. Then, what point in the line B, is the center?

We have a fact that at the center, the y-coordinate is 1. So using the fact, along with the line A, we can find the exact location of the center in the line B.

Then, using the center, along with the radius, which is 2, we can find the circle we want.

Now, the line A is: $y = -\frac{2}{3}x + 1$, and is parallel to the line B, so the line B can be put in an equation: $y = -\frac{2}{3}x + c$, where the slope is the same as A's, and c is constant and $\neq 1$.

Next, we know that the y-coordinate at the center of the circle C is 1.

So we can suppose that the center of the circle C is $(s, 1)$, where s is a constant.

Also, we know the center of C, that is, $(s, 1)$ is in the line B, which is: $y = -\frac{2}{3}x + c$.

So we get: $1 = -\frac{2}{3}s + c \Rightarrow s = \frac{3(c-1)}{2}$, and thus, the center of C can be put this way: $(\frac{3(c-1)}{2}, 1)$. How then, do we find c?

The distance from the line A to the center of the circle C is the radius, which is 2. And we have yet to use the fact above, so let's now, move on to the radius.

First, the center of the circle C is in the line B, and the circle C is tangent to the line A, so the radius of the circle C is the distance from the center to the line A. And next, there is a template, where the distance from a point (u, v) to a line $px + qy + r = 0$ is: $\frac{|pu + qv + r|}{\sqrt{p^2 + q^2}}$.

The line A is: $2x + 3y - 3 = 0$, the center of the circle C is: $(s, 1) = (\frac{3(c-1)}{2}, 1)$, so the distance from the center of C to A is: $\frac{|2 \cdot \frac{3(c-1)}{2} + 3 \cdot 1 - 3|}{\sqrt{2^2 + 3^2}} = \frac{|3(c-1)|}{\sqrt{13}}$, which is the radius, which is 2.

So we get: $\frac{3|1-c|}{\sqrt{13}} = 2 \Rightarrow \frac{9(1-c)^2}{13} = 4 \Rightarrow 9(1-c)^2 = 52 \Rightarrow 1 - c = \pm\frac{\sqrt{52}}{3} \Rightarrow c = 1 \pm \frac{2\sqrt{13}}{3} = \frac{3 \pm 2\sqrt{13}}{3}$.

Then, we get: $c = \frac{3 \pm 2\sqrt{13}}{3} \Rightarrow (\frac{3(c-1)}{2}, 1) = (\frac{3c-3}{2}, 1) = (\frac{3 \pm 2\sqrt{13}-3}{2}, 1) = (\frac{\pm 2\sqrt{13}}{2}, 1) = (\pm\sqrt{13}, 1)$.

Therefore, the center of the circle C is $(\sqrt{13}, 1)$ or $(-\sqrt{13}, 1)$, so we can get two circles.

Now, since the radius is 2, using the standard form, we can see that the circle C is:

$(x-\sqrt{13})^2 + (y-1)^2 = 4$ or $(x+\sqrt{13})^2 + (y-1)^2 = 4$. And we have: $\sqrt{13} \approx 3.606$.

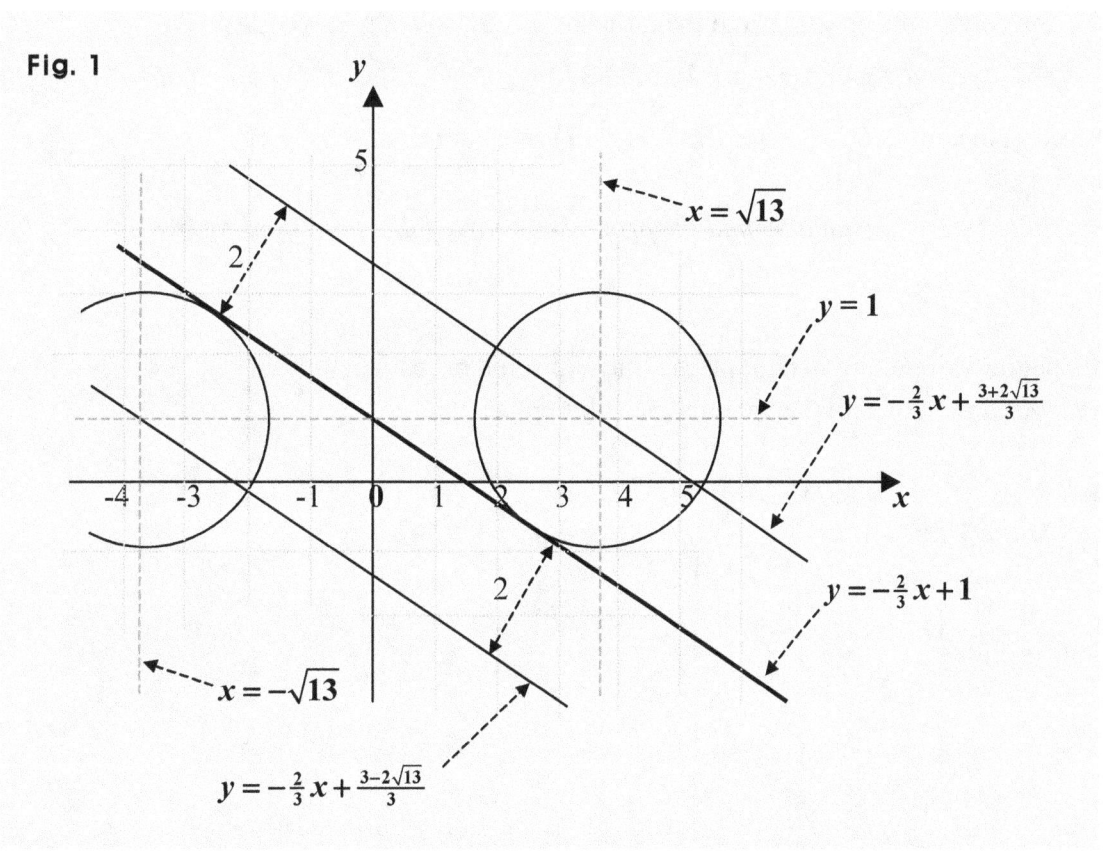

Fig. 1

Suggestions or Solutions
To the Problem in the Example 1

Construct a circle that is tangent to a line $x = a$, where a is constant, and has a radius of 2 and a center where the y-coordinate is 1.

Suppose C is the circle to be constructed, A is the line: $x = a$, B is another line: $x = c$, where c is a constant and $\neq a$, and the distance from A to B is 2, which is the radius of C.

Then, the center of C is $(c, 1)$, and the radius is $|c - a|$, so we get:

$|c - a| = 2 \Rightarrow c - a = \pm 2 \Rightarrow c = a + 2$ or $a - 2$.

Therefore, the circle C is: $(x - a + 2)^2 + (y - 1)^2 = 2^2$ or $(x - a - 2)^2 + (y - 1)^2 = 2^2$.

If not quite sure of the idea behind the processes above, follow the steps below:

Let's begin with putting in a graph the line and some probable circles.
Then, we can see better whereabouts the solution is.

Fig. 0

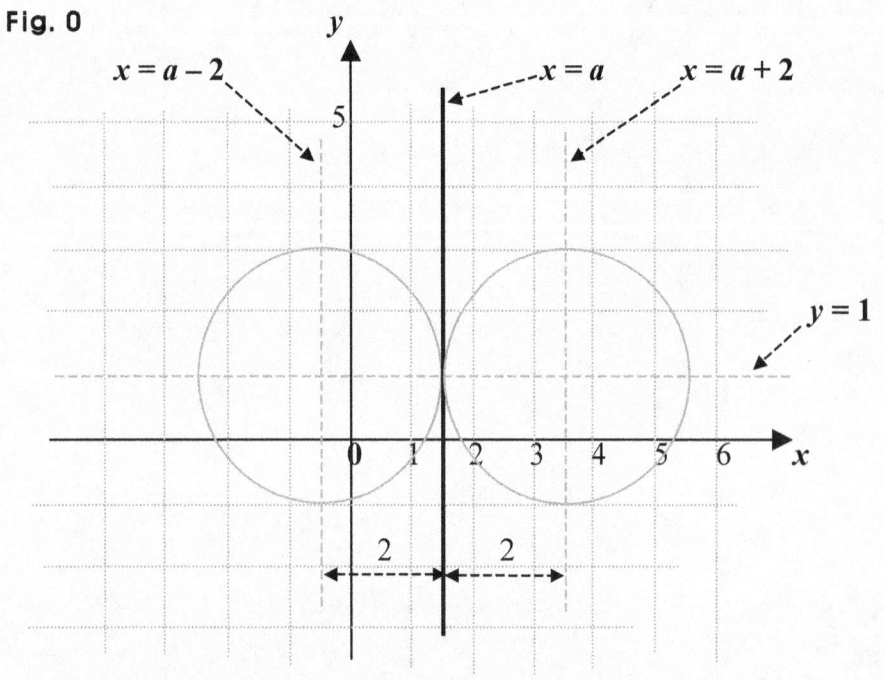

Examples A in Circles 223

Now, we can see that there can be such two circles as the one this problem is asking.

Given the center and radius of a circle, we can find the circle.

We know that the radius is 2, but the y-coordinate only of the center is given, and is 1. So we want to find the x-coordinate of the center.

Suppose C is the circle to be constructed, A is the line $x = a$, B is a line parallel to the line A, and the distance between the two lines A and B is 2.

Then, the center of the circle C is in the line B.

The line B is parallel to the line A, which is: $x = a$, so the line B can be put in an equation where $x = c$, where c is a constant, and $\neq a$. In the line $x = c$ in the x-y system, every value of x is c for every value of y. In short, x is c for all y.

So the center of the circle C is $(c, 1)$, for the center is in B and the y-coordinate is 1 there. Besides, the radius of the circle C is 2, so the circle C can be put this way, for now: $(x - c)^2 + (y - 1)^2 = 2^2$.

Let's now, move on to the radius to find c.

First, the distance between the two lines A and B is the radius of the circle C.

Next, looking at the graph, we can immediately see that the distance between A and B is $|c - a|$.

So this time, we don't have to use the formula for the distance from a point to a line.

Let's see however, if the formula still works in this case, too.

The distance from a point (u, v) to a line $px + qy + r = 0$ is $\frac{|pu+qv+r|}{\sqrt{p^2+q^2}}$.

The line A is: $x = a$, that is, $1 \cdot x + 0 \cdot y - a = 0$, and the center is $(c, 1)$.

So the distance from the line A to the center, that is, the radius is: $\frac{|1 \cdot c + 0 \cdot 1 - a|}{\sqrt{1^2 + 0^2}} = |c - a|$.

Now, since the radius is 2, we get: $|c - a| = 2 \Rightarrow c - a = \pm 2 \Rightarrow c = a + 2$ or $a - 2$.

So first, if $c = a + 2$, we get: $(x - c)^2 + (y - 1)^2 = 2^2 \Rightarrow (x - a - 2)^2 + (y - 1)^2 = 2^2$.

Next, if $c = a - 2$, we get: $(x - c)^2 + (y - 1)^2 = 2^2 \Rightarrow (x - a + 2)^2 + (y - 1)^2 = 2^2$.

Thus, the circle C is: $(x - a - 2)^2 + (y - 1)^2 = 2^2$ or $(x - a + 2)^2 + (y - 1)^2 = 2^2$, which is a circle that is tangent to a line: $x = a$, and has a radius of 2 and a center in which the y-coordinate is 1.

In short:

Suppose C is the circle to be constructed, A is the line: $x = a$, B is another line: $x = c$, where c is a constant and $\neq a$, and the distance from A to B is 2, which is the radius of C.

Then, the center of C is $(c, 1)$, and the radius is $|c - a|$, so we get:

$|c - a| = 2 \Rightarrow c - a = \pm 2 \Rightarrow c = a + 2$ or $a - 2$.

Therefore, the circle C is: $(x - a + 2)^2 + (y - 1)^2 = 2^2$ or $(x - a - 2)^2 + (y - 1)^2 = 2^2$.

Suggestions or Solutions
To the Problem in the Example 2

Construct a circle that is tangent to a line $y = b$, where b is constant, and has a radius of 2 and a center where the x-coordinate is 1.

Suppose C is the circle to be constructed, A is the line $y = b$, B is a line $y = c$, where c is constant and $\neq b$, and the distance from A to B is 2, which is the radius of the circle C.

Then, the center of the circle C is $(1, c)$, and the radius of C is $|c - b|$, so we get:

$|c - b| = 2 \Rightarrow c - b = \pm 2 \Rightarrow c = b + 2$ or $b - 2$.

Therefore, the circle C is: $(x - 1)^2 + (y - b + 2)^2 = 2^2$ or $(x - 1)^2 + (y - b - 2)^2 = 2^2$.

If not quite sure of the idea behind the processes above, follow the steps below:

Let's begin with putting the problem in a graph. Then, we can see better whereabouts the solution is.

Fig. 0

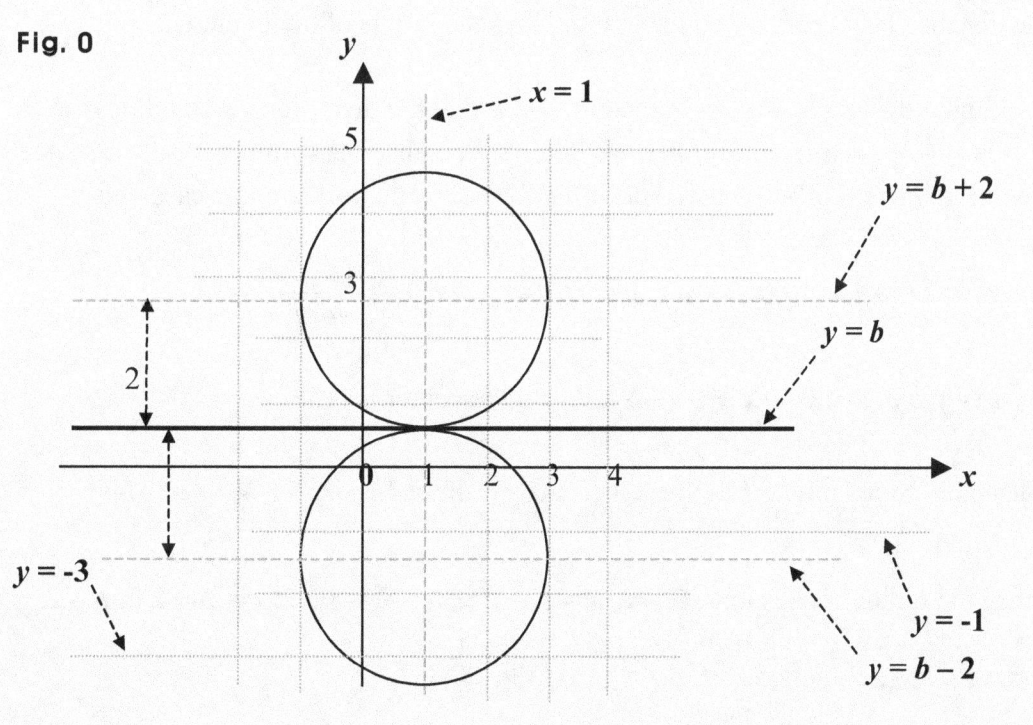

Now, to this problem, too, we can see that two circles can be the solution.

Suppose C is the circle we want, A is the line $y = b$, B is a line parallel to the line A, and the distance from the line A to the line B is 2, which is the radius of the circle C.

Then, the center of the circle C is in the line B, for the radius is the distance above.

Since the two lines A and B are parallel to each other, the line B can be indicated by an equation of a line $y = c$, where c is constant and $\neq b$.

In the line $y = c$ in the x-y system, every value of y is c for every value of x.
In short, y is c for all x.

Since the x-coordinate of the center of the circle C is 1, we can suppose the center of the circle C is $(1, t)$, where t is constant. Then, we want to find t.

The center $(1, t)$ is in the line B. So we get: $t = c$ since the line B is a line $y = c$.
Thus, the center of the circle C is $(1, c)$.

Let's now, move on to the radius.

First, the distance between the two lines A and B is the radius of the circle C.

Next, looking at the graph, we can immediately see that the distance from the line A to the line B is $|c - b|$. So this time, either, we don't have to use the formula for the distance from a point to a line. Let's see however, if the formula still works in this case, too.

• The distance from a point (u, v) to a line $px + qy + r = 0$ is: $\frac{|pu+qv+r|}{\sqrt{p^2+q^2}}$.

The line A is: $y = b$, that is, $0 \cdot x + 1 \cdot y - b = 0$, and the center is $(1, c)$.

So the distance from the line A to the center, that is, the radius is: $\frac{|0 \cdot 1 + 1 \cdot c - a|}{\sqrt{0^2 + 1^2}} = |c - b|$.

Now, since the radius of the circle C is 2, $|c - b| = 2 \Rightarrow c - b = \pm 2 \Rightarrow c = b + 2$ or $b - 2$.

How come $c - b = \pm 2$, though?

$|v|$ indicates the absolute value of v, that is, the magnitude of v.

So for instance, we get: $|w| = 1 \Rightarrow w = 1$ or -1, that is, $w = \pm 1$.

Now, since the radius of the circle C is 2 and the center is $(1, c)$, the circle C can be put this way, for now: $(x - 1)^2 + (y - c)^2 = 2^2$.

So first, if $c = b + 2$, we get: $(x - 1)^2 + (y - c)^2 = 2^2 \Rightarrow (x - 1)^2 + (y - b - 2)^2 = 2^2$.

Next, if $c = b - 2$, we get: $(x - 1)^2 + (y - c)^2 = 2^2 \Rightarrow (x - 1)^2 + (y - b + 2)^2 = 2^2$.

Thus, the circle C is: $(x - 1)^2 + (y - b - 2)^2 = 2^2$ or $(x - 1)^2 + (y - b + 2)^2 = 2^2$, which is a circle that is tangent to the line $y = b$ and has a radius of 2 and a center where the x-coordinate is 1.

We can put the circle C this way, too, of course: $(x - 1)^2 + (y - b \pm 2)^2 = 2^2$.

In short:

Suppose C is the circle to be constructed, A is the line $y = b$, B is a line $y = c$, where c is constant and $\neq b$, and the distance from A to B is 2, which is the radius of the circle C.

Then, the center of the circle C is $(1, c)$, and the radius of C is $|c - b|$, so we get:

$|c - b| = 2 \Rightarrow c - b = \pm 2 \Rightarrow c = b + 2$ or $b - 2$.

Therefore, the circle C is: $(x - 1)^2 + (y - b \pm 2)^2 = 2^2$.

Examples B in Circles

Constructing a curve in math, we can put it in a graph. Putting it in a graph though, we need the equation of it. So constructing it, we want to find the equation of it if we are not given the equation, of course. And finding the equation, it is said that we get it, too.

Assuming a line A is $y = 2x + 1$:

0. Construct a circle that is tangent to the line A and has a radius of 2 and a center where the x-coordinate is 1.

1. Construct every circle that is tangent to the line A, and has a radius of 2.

2. Find all the circles that have the same radii and are tangent to the line A.

Suggestions or Solutions
To the Problem in the Example 0

Construct a circle that is tangent to a line $y = 2x + 1$, and has a radius of 2 and a center where the x-coordinate is 1.

Suppose first, S is the circle we want, A is the line $y = 2x + 1$, B is a line $y = 2x + c$, where c is constant, but $\neq 1$, and the distance between A and B is 2, which is the radius of S.

Suppose next, $(1, t)$ is the center of S, and is a point in B, since the x-coordinate of the center is 1.

Then, we get: $t = 2 + c$, and the center of S can be put this way: $(1, 2 + c)$.

So the circle S can be put for now, this way: $(x-1)^2 + \{y - (2+c)\}^2 = 4$.

The distance from a point (u, v) to a line $px + qy + r = 0$ is: $\frac{|pu+qv+r|}{\sqrt{p^2+q^2}}$.

The line A is: $y = 2x + 1$, so A can be indicated by $2x - y + 1 = 0$, too.

So the distance from the center $(1, 2 + c)$ to the line A is: $\frac{|2\cdot1+(-1)(2+c)+1|}{\sqrt{2^2+(-1)^2}} = \frac{|2-2-c+1|}{\sqrt{4+1}} = \frac{|1-c|}{\sqrt{5}}$.

Thus, we get: $\frac{|1-c|}{\sqrt{5}} = 2 \Rightarrow \frac{(1-c)^2}{5} = 4 \Rightarrow (1-c)^2 = 20 \Rightarrow 1-c = \pm2\sqrt{5} \Rightarrow c = 1 \pm 2\sqrt{5}$.

Therefore, the circle S is: $(x-1)^2 + (y - 3 \pm 2\sqrt{5})^2 = 4$.

If not quite sure of the idea behind the processes above, follow the steps below:

It's a good idea to put the problem in a graph.
So let's begin with putting this problem in a graph.
Putting this problem in a graph, put in a graph the line given and some circles probable.
Then, you can see better where the solution can be around.

Fig. 0

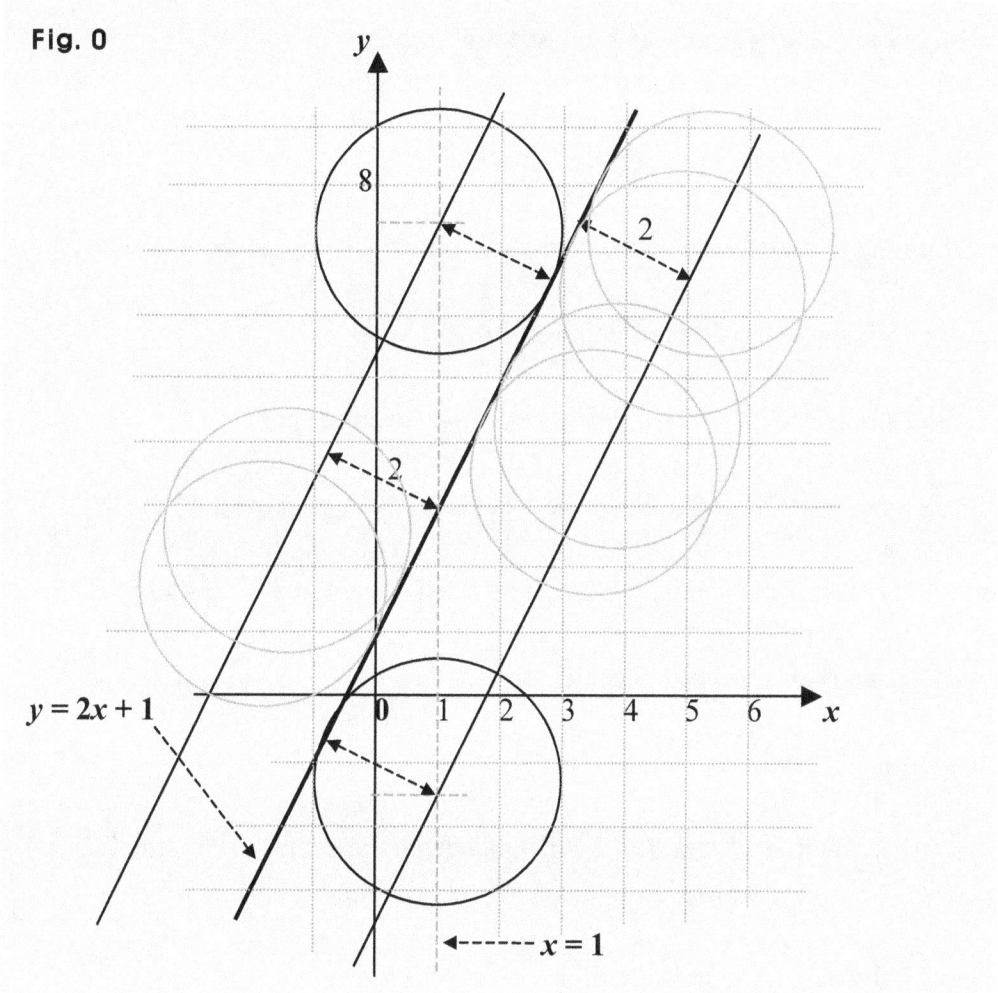

Now, in the graph above, is there any circle satisfying the conditions in the problem?

We can see that there are two such circles, and they are in black. How do we find them?

• Given the center and radius, we can readily find the circle. Then, the equation of it will be in the standard form.

We know that the radius is 2, but the x-coordinate only of the center is given.
So we want to find the y-coordinate of the center.

Suppose A is the line $y = 2x + 1$, and B is a line $y = 2x + c$, where c is constant and $\neq 1$.

Suppose also, the distance between the two lines A and B is 2.

Then, a circle tangent to the line A and centered at a point in the line B has a radius of 2, which is the distance between the two lines A and B.

So we want to find the value of c that makes the circle above have a radius of 2.

Suppose now, that S is the circle this problem is asking.

Then, we want the circle S to have a center where the x-coordinate is 1.

So suppose for now, that the center of the circle S is $(1, t)$ where t is constant.

Then, the center $(1, t)$ is in B, which is: $y = 2x + c$, since S is centered at a point in B.

So we get: $t = 2 + c$, and can say that the center is $(1, 2 + c)$.

Let's now, move on to the radius.

The distance between the lines A and B is the distance from the center of S to the line A, so the distance is the radius. And we know the center is $(1, 2 + c)$.

Besides, we have a formula for a distance from a point to a line:

• The distance from a point (u, v) to a line $px + qy + r = 0$ is: $\frac{|pu+qv+r|}{\sqrt{p^2+q^2}}$.

We can put the line A, $y = 2x + 1$, in this way, too: $2x - y + 1 = 0$.

Then, the distance, that is, the radius is as follows: $\frac{|2 \cdot 1 + (-1)(2+c)+1|}{\sqrt{2^2+(-1)^2}} = \frac{|2-2-c+1|}{\sqrt{4+1}} = \frac{|1-c|}{\sqrt{5}}$.

So the circle S can be for now, put this way: $(x-1)^2 + \{y-(2+c)\}^2 = \frac{(1-c)^2}{5}$.

Now, since the radius is 2 in the circle S, we get:

$\frac{|1-c|}{\sqrt{5}} = 2 \Rightarrow \frac{(1-c)^2}{5} = 4 \Rightarrow (1-c)^2 = 20 \Rightarrow 1-c = \pm 2\sqrt{5} \Rightarrow c = 1 \pm 2\sqrt{5}$.

Then, first, if $c = 1 + 2\sqrt{5}$, we get:

$$(x-1)^2 + \{y-(2+c)\}^2 = (x-1)^2 + (y-2-1-2\sqrt{5})^2 = 4.$$

And next, if $c = 1 - 2\sqrt{5}$, we get:

$$(x-1)^2 + \{y-(2+c)\}^2 = (x-1)^2 + (y-2-1+2\sqrt{5})^2 = 4.$$

Therefore, the circle S is as follows:

$$(x-1)^2 + (y-3-2\sqrt{5})^2 = 4 \text{ or } (x-1)^2 + (y-3+2\sqrt{5})^2 = 4.$$

Fig. 1

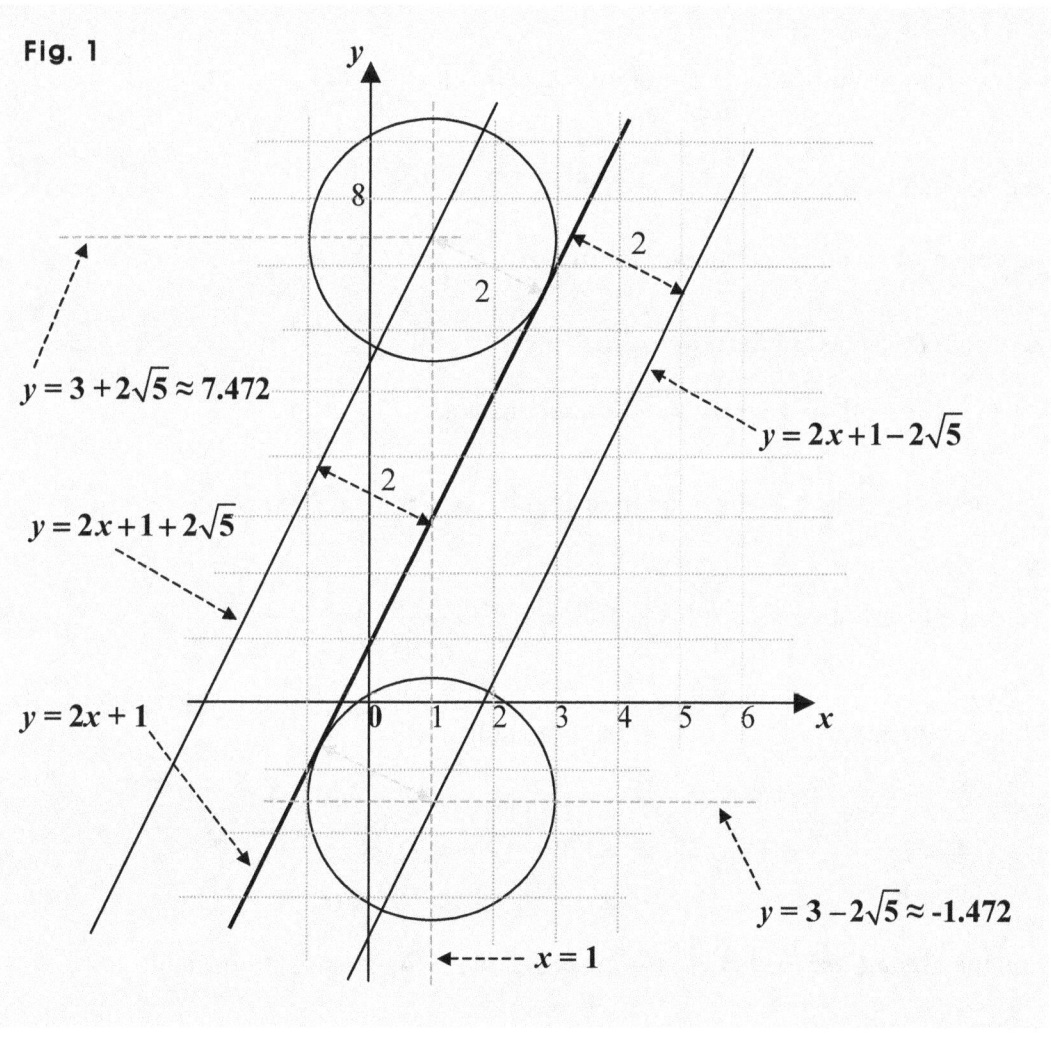

Suggestions or Solutions
To the Problem in the Example 1

Construct every circle that is tangent to a line $y = 2x + 1$, and has a radius of 2.

Suppose C is the circle we want, A is the line $y = 2x + 1$, B is a line $y = 2x + c$, where c is constant, but $\neq 1$, and the distance between the lines A and B is 2.

Then, the line B is parallel to the line A, and any circle tangent to the line A and centered at any point in the line B has a radius of 2.

The distance from a point (u, v) to a line $px + qy + r = 0$ is: $\frac{|pu+qv+r|}{\sqrt{p^2+q^2}}$.

Suppose now, that (s, t) is a point in B, which is: $y = 2x + c$.

Then, the center of C can be put in $(s, 2s + c)$.

Thus, the circle C can be put for now, in such a way as follows:

$(x-s)^2 + \{y-(2s+c)\}^2 = 4$, where s and c are constant.

Since the line A is $y = 2x + 1$, A can be indicated by $2x - y + 1 = 0$, too.

So the distance from the center to A is as follows: $\frac{|2s+(-1)(2s+c)+1|}{\sqrt{2^2+(-1)^2}} = \frac{|2s-2s-c+1|}{\sqrt{4+1}} = \frac{|1-c|}{\sqrt{5}}$.

The distance is the radius of the circle C, and the radius is 2.

So we get: $\frac{|1-c|}{\sqrt{5}} = 2 \Rightarrow \frac{(1-c)^2}{5} = 4 \Rightarrow (1-c)^2 = 20 \Rightarrow 1-c = \pm 2\sqrt{5} \Rightarrow c = 1 \pm 2\sqrt{5}$.

Therefore, the circle C is: $(x-s)^2 + (y-2s-1\pm2\sqrt{5})^2 = 4$, where s is constant.

If not quite sure of the idea behind the processes above, follow the steps below:

Putting the problem in a graph, that is, putting in a graph the line given and some circles probable, which are of radius 2, we can see better whereabouts the solution can be.

Fig. 0

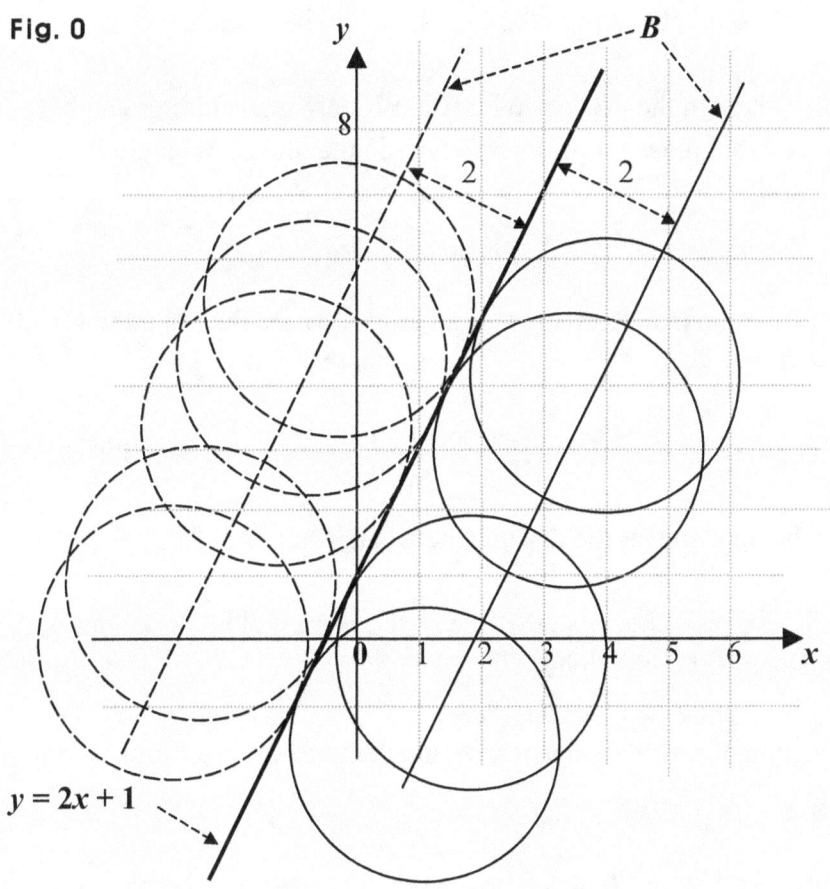

$y = 2x + 1$

Now, finding a circle, we need the center as well as the radius. The radius is 2, but the center is not specified. The information on the center is implicit, that is, hidden.
So we may want to begin with the center's whereabouts.   How then, do we get there?

The center in this case has much to do with the radius and the line tangent to the circle.

Suppose A is the line $y = 2x + 1$, and B is a line $y = 2x + c$, where c is constant, but $\neq 1$, and C is the circle this problem is asking.

Then, the line B is parallel to the line A. Thus, if a circle is tangent to the line A and centered at a point in the line B, the distance between the lines A and B is the radius.

Thus, every circle tangent to *A* and centered at a point in *B* has the same radius.

So if the distance between *A* and *B* is 2, we can see a lot of *C*s. That is, every circle centered at a point in the line *B* and tangent to the line *A* can be the circle *C*.

So the center of *C* is everywhere in the line *B*, and thus, infinitely many circles can be the circle *C*. Each radius is 2, but the center is everywhere in the line *B*. What do we do about such centers, then?

We can come up with an equation that can represent all circles where the radii are 2, and the centers are in the line *B*.

Thus, we want to find first, the line *B*, which is: $y = 2x + c$ where *c* is constant, but $\neq 1$.

So we want to find the value of *c* so that the distance between *A* and *B* is 2.

Then, no matter what point we may choose in the line *B*, the point will be the center of a circle of radius 2, which is tangent to the line *A*.

So we want to get an equation of a circle tangent to *A* and centered at an arbitrary point in the line *B*.

Then, the equation represents all circles that can be the circle *C*, and thus, is the solution to this problem.

So for instance, in the equation, which is in the standard form, the center will be specified by **(*s*, *t*)**, which is an arbitrary point in the line *B*.

Then, specifying only the *x*-coordinate, that is *s*, of a point in *B*, we can get a circle centered at the point and tangent to *A*.

That is, we have only to specify the *x*-coordinate at the center, and the *y*-coordinate will be determined, and then, we will get the circle.

Note however, that two lines can be the line *B*. One is above the line *A*, and below is the other. So we can get two circles at a time. One is above the line *A*, and below is the other.

Fig. 1

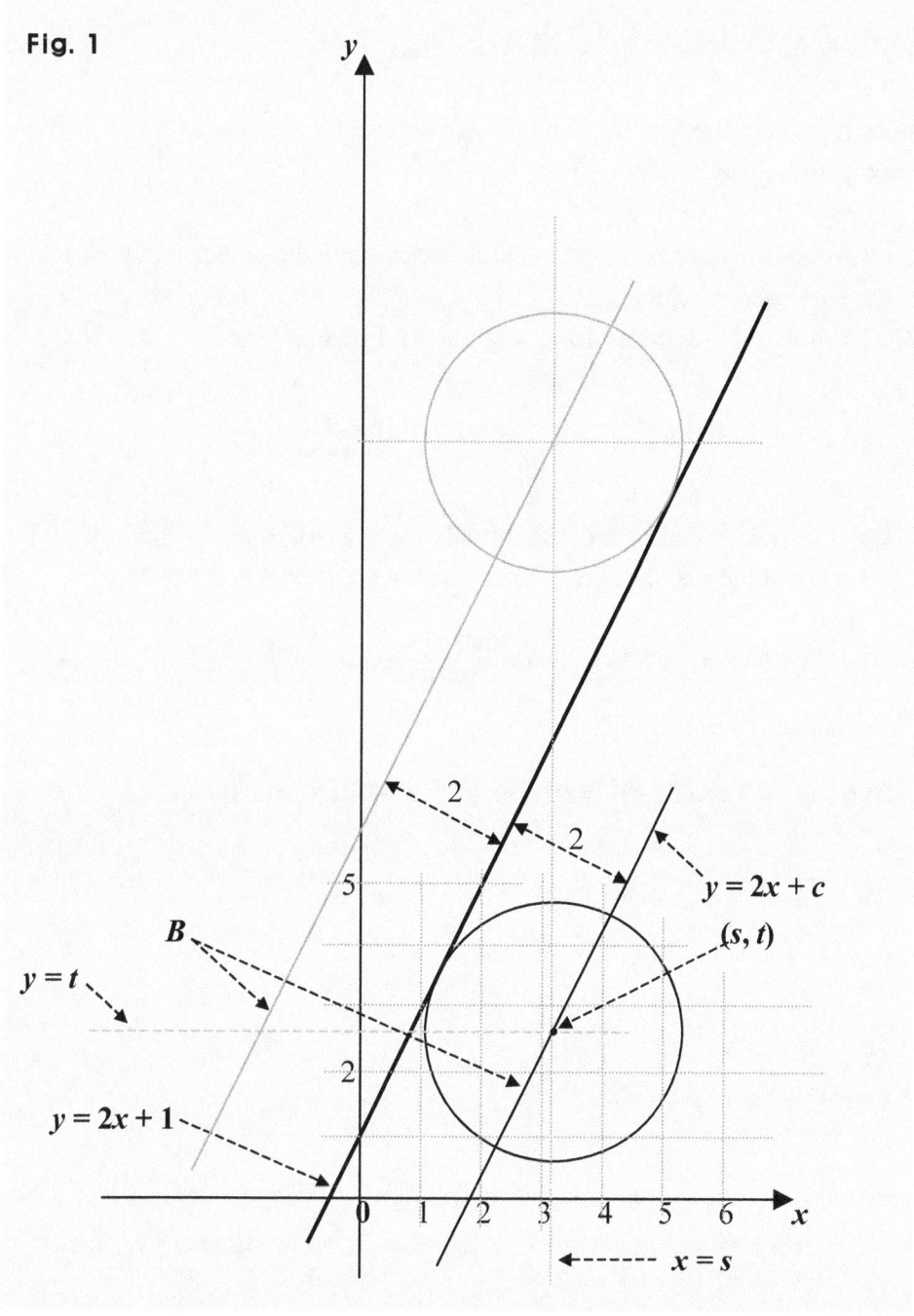

Suppose now, that a point **(s, t)** is in the line **B**.

Suppose also, that the point **(s, t)** is the center of a circle tangent to the line **A**.

Then, **s** is the **x**-coordinate of the center, so the center is in the line **x = s** as shown above. Of course, two centers are in the line since we can get two circles at a time.

Now, we are ready to find the value of c.

To begin with, since (s, t) is in the line B, which is: $y = 2x + c$, we get: $t = 2s + c$.
So the center can be put this way: $(s, 2s + c)$.

Next, the distance between the lines A and B is the distance from the center $(s, 2s + c)$ to
the line A, and is the very radius, which is 2.
Besides, we have a formula for a distance from a point to a line in a plane:

- The distance from a point (u, v) to a line $px + qy + r = 0$ is: $\frac{|pu+qv+r|}{\sqrt{p^2+q^2}}$.

Since the line A is: $y = 2x + 1$, A can be indicated by $2x - y + 1 = 0$, too.
The center is $(s, 2s + c)$, and the distance from the center to the line A is the radius.

So the radius can be put in such a way as follows: $\frac{|2s+(-1)(2s+c)+1|}{\sqrt{2^2+(-1)^2}} = \frac{|2s-2s-c+1|}{\sqrt{4+1}} = \frac{|1-c|}{\sqrt{5}}$.

Thus, we get: $\frac{|1-c|}{\sqrt{5}} = 2$ since the radius is 2.

So we get: $\frac{|1-c|}{\sqrt{5}} = 2 \Rightarrow \frac{(1-c)^2}{5} = 4 \Rightarrow (1-c)^2 = 20 \Rightarrow 1-c = \pm 2\sqrt{5} \Rightarrow c = 1 \pm 2\sqrt{5}$.

Then, first, if $c = 1 + 2\sqrt{5} \approx 5.472$, we get:
$$(x-s)^2 + \{y - (2s+c)\}^2 = (x-s)^2 + (y-2s-1-2\sqrt{5})^2 = 4.$$

Next, if $c = 1 - 2\sqrt{5} \approx -3.472$, we get:
$$(x-s)^2 + \{y - (2s+c)\}^2 = (x-s)^2 + (y-2s-1+2\sqrt{5})^2 = 4.$$

Therefore, the equation for circles we are after is as follows:
$(x - s)^2 + (y - 2s - 1 - 2\sqrt{5})^2 = 4$ or $(x - s)^2 + (y - 2s - 1 + 2\sqrt{5})^2 = 4$, where s is a
constant.

We can put together the two above the way below, too:
$(x - s)^2 + (y - 2s - 1 \pm 2\sqrt{5})^2 = 4$ where s is a constant.

Now, the equation above indicates every circle of radius 2 that is tangent to the line A.
All the centers are all the points in the line B.

So if we choose a point in the line **B**, a circle centered at the point and tangent to the line **A** gets determined. Choosing the center, we have only to specify the *x*-coordinate of a point in the line **B**. In fact, for each choice of the *x*-coordinate, which is *s*, we can get two circles of radius 2, each of which is tangent to the given line **A**, and is centered at each of the two points, $(s, 2s + 1 + 2\sqrt{5})$ and $(s, 2s + 1 - 2\sqrt{5})$.

By means of the equation above, we can determine twice as many circles as the number of the values of the constant *s*, which is the *x*-coordinate of the center, which is a point in either of the two lines $y = 2x + 1 + 2\sqrt{5}$ and $y = 2x + 1 - 2\sqrt{5}$, each of which can be the line **B**, of course.

Fig. 2

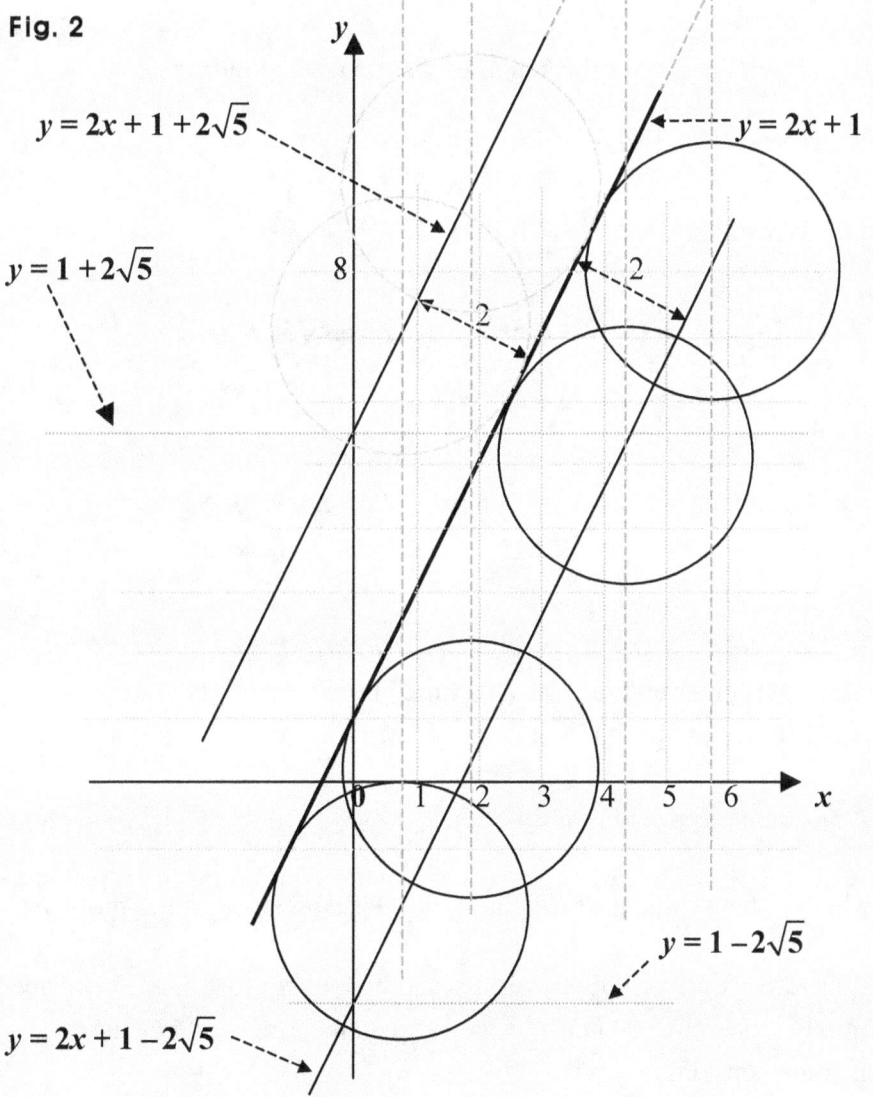

$y = 2x + 1 + 2\sqrt{5}$

$y = 1 + 2\sqrt{5}$

$y = 2x + 1$

$y = 1 - 2\sqrt{5}$

$y = 2x + 1 - 2\sqrt{5}$

Suggestions or Solutions
To the Problem in the Example 2

Find all the circles that have the same radii, and are tangent to a line $y = 2x + 1$.

Suppose A is the line $y = 2x + 1$, B is a line $y = 2x + c$ where c is constant and $c \neq 1$, r is the distance between the two lines A and B, and C is the circle to be constructed.

Suppose also, s and t are constant, and a point (s, t) is in B, and is the center of a circle tangent to A.

Then, we get: $(s, t) = (s, 2s + c)$, which is the center of C, and r is the radius of C.

The distance from a point (u, v) to a line $px + qy + r = 0$ is: $\frac{|pu+qv+r|}{\sqrt{p^2+q^2}}$.

So the distance from $(s, 2s + c)$ to A is: $\frac{|2s+(-1)(2s+c)+1|}{\sqrt{2^2+(-1)^2}} = \frac{|2s-2s-c+1|}{\sqrt{4+1}} = \frac{|1-c|}{\sqrt{5}}$.

Thus, we get: $\frac{|1-c|}{\sqrt{5}} = r \Rightarrow \frac{(1-c)^2}{5} = r^2 \Rightarrow (1-c)^2 = 5r^2 \Rightarrow 1-c = \pm r\sqrt{5} \Rightarrow c = 1 \pm r\sqrt{5}$.

Therefore, the circle C is: $(x-s)^2 + (y-2s-1 \pm r\sqrt{5})^2 = r^2$ where r and s are constant.

If not quite sure of the idea behind the processes above, follow the steps below:

Finding a circle, we need the center and radius. This time however, both the radius and center are not specified. So they are probably hidden. Where then, can they possibly be?

They must be somewhere in the problem description.   So let's break apart the problem.

To begin with, many, or rather infinitely many circles can be tangent to a line. So we are after not one but infinitely many circles. Thus, we can expect the radius and center to be expressed in terms of some constants.

Next, no matter what the radii may be, they all have to be the same.

The information on the centers has to be given somehow, too.

So this time, too, we may want to begin with the centers' whereabouts.

So putting in a graph the line given and some probable circles we can get:

Fig. 0

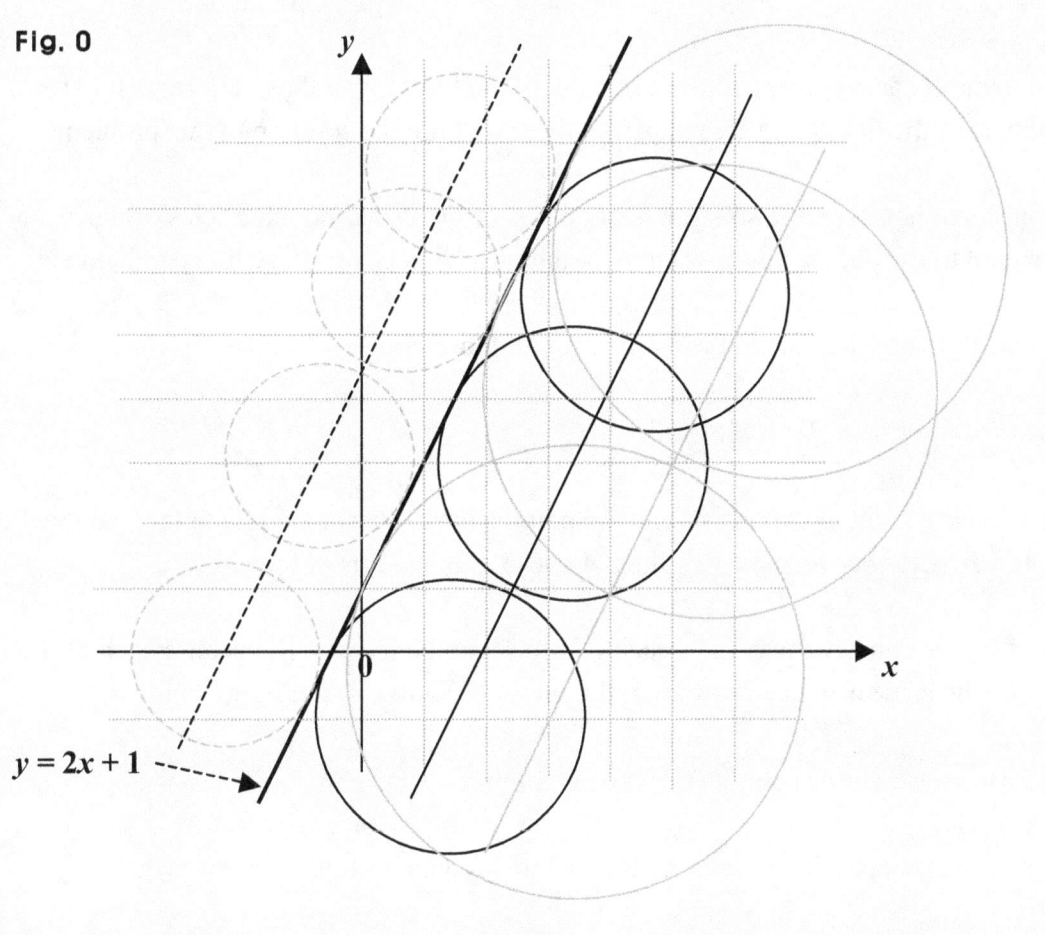

$y = 2x + 1$

The center in a circle has much to do with the radius and a line tangent to the circle.

So suppose A is the line $y = 2x + 1$, which the tangent line. Then, every circle tangent to the line A and centered at a point in a line parallel to the line A has the same radius.

That is, the line parallel to the line A is a particular distance away from the line A, and the particular distance is the same radius.

So suppose now, B is a line parallel to the line A.

In fact, not just one but infinitely many lines can be parallel to the line A.
Anyway, we can put the line B this way: $y = 2x + c$, where c is a constant, which is $\neq 1$.

For each value of c, the equation $y = 2x + c$ indicates a line parallel to the line A.

Thus, the line B can represent all lines parallel to the line A, and those circles that have their centers in the line B and are tangent to the line A are the solution to this problem.

In this problem however, the solution is not a line or lines as B but all circles of the same radii tangent to the line A. So we want to assume that B is an arbitrary line parallel to A.

Let's now, find the equation that can represent all those circles.

To begin with, we have the fact as follows:

• If we choose a point in the line B, and the point is the center of a circle tangent to the line A, the distance between the two lines A and B is the radius of the circle.

Then in fact, we can have two lines, each of which can be the line B, because each of the two can be the same distance away from the line A as shown in the figure below.

• So we can construct the solution the way as follows.

To begin with, we specify the value of the radius. We can set it to be r, for instance.

Then, we will get two values of c, because we can have two lines, each of which can be the line B since two lines can be the same distance away from the line A. One is above the line A, and the other is below the line A.

Next, specifying only the x-coordinate of a point in the line B, we can get two circles tangent to the line A. One is above the line A, and the other is below the line A.

Fig. 1

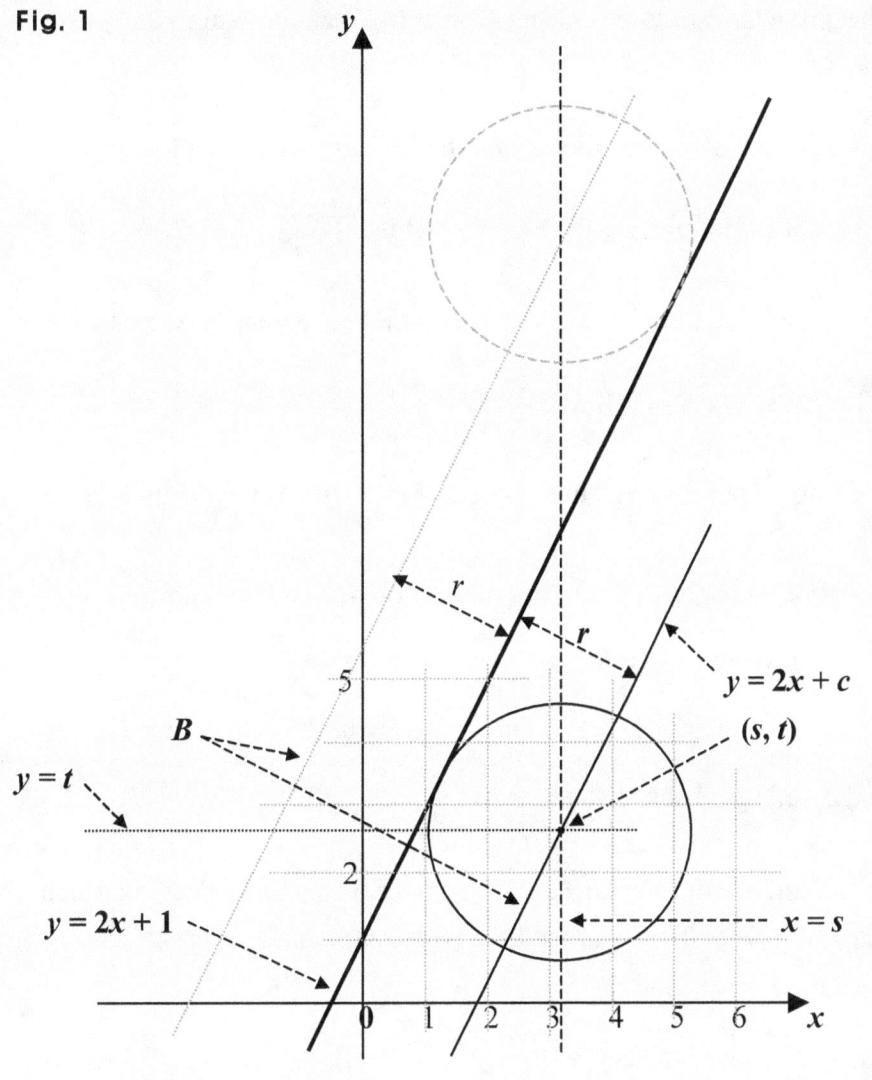

Suppose now, that *s* and *t* are constant, that a point *(s, t)* is in *B*, which is parallel to the line *A*, and that the point *(s, t)* is the center of a circle of radius *r*, which is tangent to *A*.

Then, since *(s, t)* is in the line *B*, which is: $y = 2x + c$, we get: $t = 2s + c$.

So the center can be put in *(s, 2s + c)*, for now. And we will take care of *c* shortly.

The distance between the lines *A* and *B* is the distance from the center *(s, 2s + c)* to the line *A*, so the distance is the radius *r*. Thus, we can put *c* in terms of the radius *r*.

Suppose now, C is the circle tangent to A and centered at (s, t), and let's find c. How then, can we find c?

We have a formula for such a distance as above, and the formula goes as follows:

- The distance from a point (u, v) to a line $px + qy + r = 0$ is: $\frac{|pu+qv+r|}{\sqrt{p^2+q^2}}$.

The line A is: $y = 2x + 1$, equivalent to: $2x - y + 1 = 0$. And the center is $(s, 2s + c)$.

So the distance, that is, the radius is as follows: $\frac{|2s+(-1)(2s+c)+1|}{\sqrt{2^2+(-1)^2}} = \frac{|2s-2s-c+1|}{\sqrt{4+1}} = \frac{|1-c|}{\sqrt{5}}$.

Therefore, the circle C can be put for now, this way: $(x - s)^2 + \{y - (2s + c)\}^2 = \frac{(1-c)^2}{5}$.

So the center of the circle C at $(s, 2s + c)$, and the radius is $\frac{|1-c|}{\sqrt{5}}$. And we know r is the radius of the circle C. So we get: $\frac{|1-c|}{\sqrt{5}} = r$.

Thus, we get: $\frac{|1-c|}{\sqrt{5}} = r \Rightarrow \frac{(1-c)^2}{5} = r^2 \Rightarrow (1-c)^2 = 5r^2 \Rightarrow 1-c = \pm r\sqrt{5} \Rightarrow c = 1 \pm r\sqrt{5}$.

Now that we've found c, let's get back to the line B.

The line B is: $y = 2x + c$, and we have: $c = 1 \pm r\sqrt{5}$. So we get two lines, each of which is a particular distance away from the line A, and the particular distance is r. One of the two is: $y = 2x + 1 + r\sqrt{5}$, and the other is: $y = 2x + 1 - r\sqrt{5}$.

Now, putting threads together, we have: $(x - s)^2 + \{y - (2s + c)\}^2 = r^2$, and $c = 1 \pm r\sqrt{5}$.

So the circle C is: $(x - s)^2 + (y - 2s - 1 \pm r\sqrt{5})^2 = r^2$, where s and r are constant.

The equation above represents two circles for each value of s, which is the x-coordinate of the center. One circle is: $(x - s)^2 + (y - 2s - 1 + r\sqrt{5})^2 = r^2$, and the other circle is: $(x - s)^2 + (y - 2s - 1 - r\sqrt{5})^2 = r^2$.

Now, choosing a value of the radius r, we get two parallel lines, each of which is r away from the line A. One of the two is: $y = 2x + 1 + r\sqrt{5}$, and the other is: $y = 2x + 1 - r\sqrt{5}$.

Then, choosing a point in either of the two lines, we can get two circles tangent to the line A. Each of the two has a radius of r, of course.

So choosing a point in either line, we choose the centers of the two circles. Choosing the centers, we have only to specify the *x*-coordinate at one of the two centers since the two centers have the same *x*-coordinates.

Fig. 2

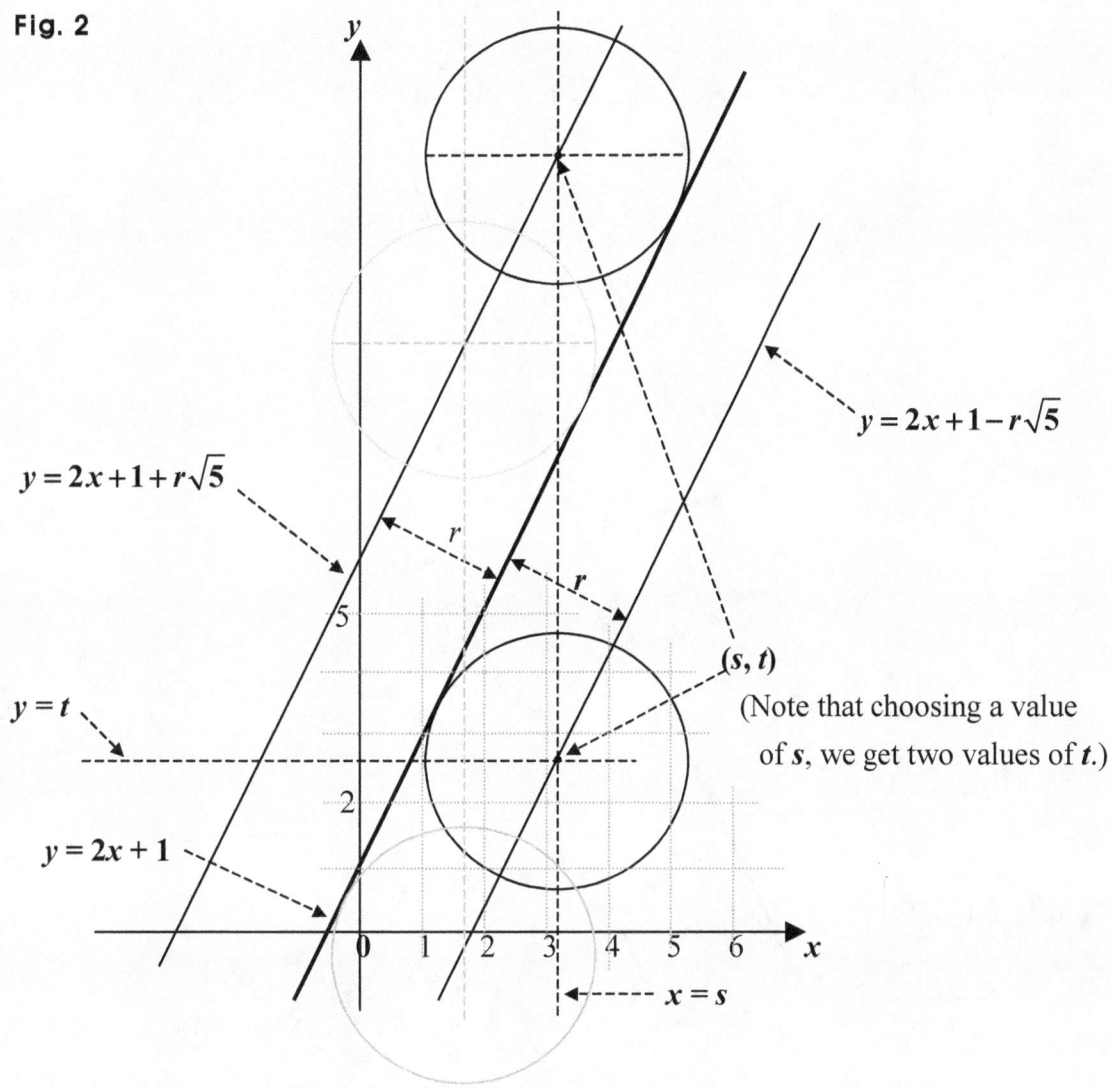

$y = 2x + 1 - r\sqrt{5}$

$y = 2x + 1 + r\sqrt{5}$

r

r

(s, t)

$y = t$

(Note that choosing a value of *s*, we get two values of *t*.)

$y = 2x + 1$

$x = s$

Assuming for instance, $r = \sqrt{5}$, we can get two lines, each of which is $\sqrt{5}$ away from the line A, $y = 2x + 1$. One of the two is: $y = 2x + 6$, and the other is: $y = 2x - 4$. Then, choosing a point in either of the two lines, we can get two circles tangent to the line A. In each of the two, the radius is $\sqrt{5}$, of course. So choosing a point in either line, we choose the centers of the two circles. Choosing the centers, we have only to specify the *x*-coordinate of one of the two centers since the two centers have the same *x*-coordinates.

Suppose for instance, we choose 2 for the *x*-coordinate of the center. Then, we get:

$$(x-s)^2+(y-2s-1\pm r\sqrt{5})^2=r^2 \text{ where } s=2, \text{ and } r=\sqrt{5}.$$

$$\Rightarrow (x-2)^2+(y-4-1\pm 5)^2=5 \Rightarrow (x-2)^2+(y-5\pm 5)^2=5.$$

So one circle is: $(x-2)^2+y^2=5$, and the other circle is: $(x-2)^2+(y-10)^2=5.$

Fig. 3

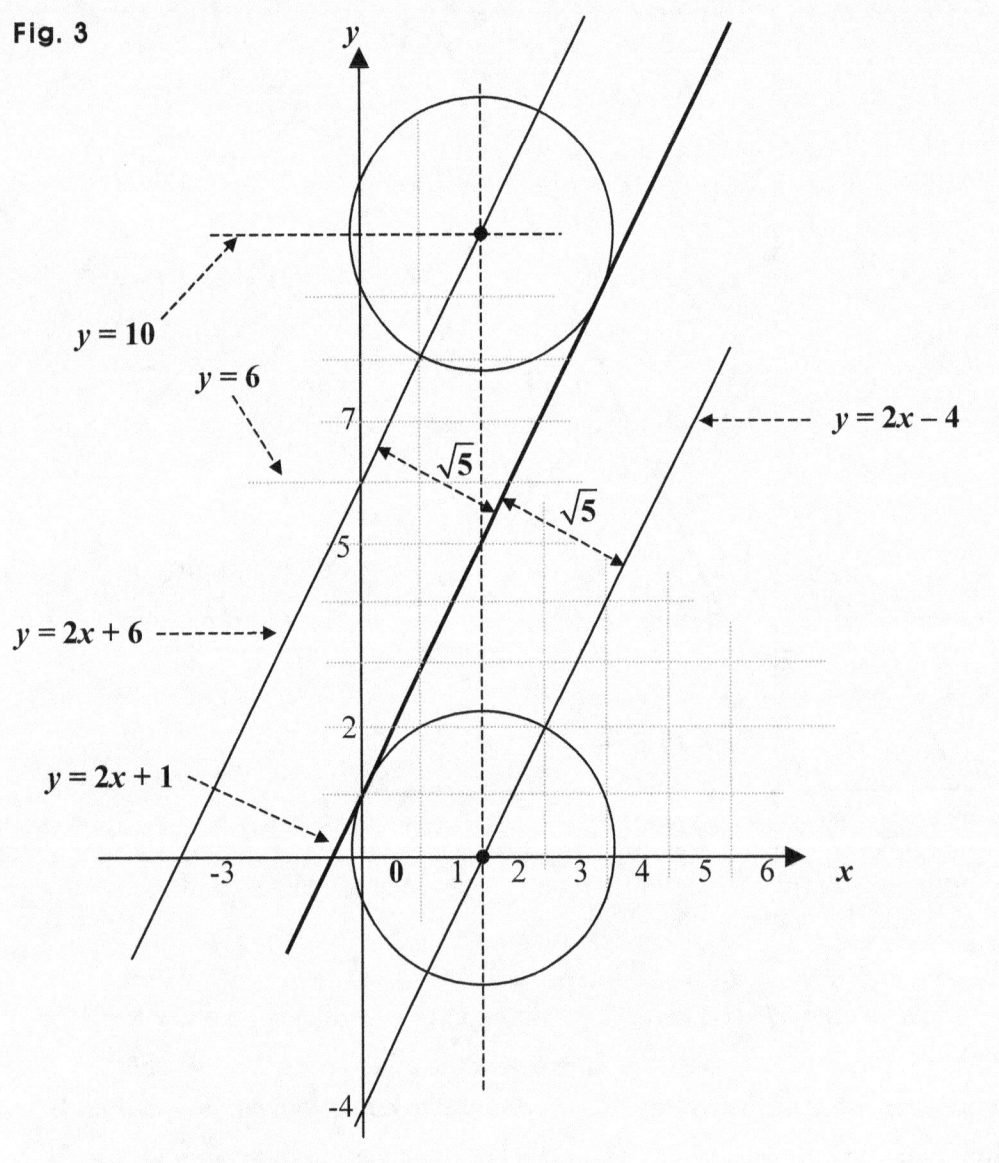

Specifying another value of *s*, we get another pair of circles, one of which is centered at a points in the line $y=2x-4$, and the other is centered at a point in the line $y=2x+6$.

Examples C in Circles

Constructing a curve in math, we can put it in a graph. Putting it in a graph though, we need the equation of it. So constructing it, we want to find the equation of it if we are not given the equation, of course. And finding the equation, it is said that we get it, too.

0. Find all the circles that have the same radii, and are tangent to a line $y = ax + b$ where a and b are constant.

1. Assuming a, b, c, and d are constant:

1.0. Construct a line tangent to a circle $(x - a)^2 + (y - b)^2 = c^2$.

1.1. Construct a line of slope d tangent to the circle above.

Suggestions or Solutions
To the Problem in the Example 0

Find all the circles that have the same radii, and are tangent to a line $y = ax + b$ where a and b are constant.

Suppose A is the line $y = ax + b$, B is a line $y = ax + c$, where c is constant and $c \neq b$, r is the distance between the two lines A and B, and C is the circle to be constructed.

Suppose also, that s and t are constant, and that a point (s, t) is in the line B, and the center of a circle tangent to the line A.

Then, we get: $(s, t) = (s, as + c)$, which is the center of C, and r is the radius of C.

The distance from a point (u, v) to a line $px + qy + r = 0$ is: $\frac{|pu+qv+r|}{\sqrt{p^2+q^2}}$.

So the distance from $(s, as + c)$ to the line A is: $\frac{|as+(-1)(as+c)+b|}{\sqrt{a^2+(-1)^2}} = \frac{|b-c|}{\sqrt{a^2+1}}$.

Thus, we get:

$$\frac{|b-c|}{\sqrt{a^2+1}} = r \Rightarrow \frac{(b-c)^2}{a^2+1} = r^2 \Rightarrow (b-c)^2 = r^2(a^2+1) \Rightarrow b-c = \pm r\sqrt{a^2+1} \Rightarrow c = b \pm r\sqrt{a^2+1}.$$

So the circle C is $(x-s)^2 + (y - as - b \pm r\sqrt{a^2+1})^2 = r^2$, where a, b, r, and s are constant.

If not quite sure of the idea behind the processes above, follow the steps below:

Finding circles, we need the centers and radii.
This problem, too, is saying that no matter what the radii may be, they all have to be the same.
The center in this problem has much to do with the radius and the line tangent to the circle.
So putting in a graph some circles of the same radii, along with a line tangent, we can see better the solution's whereabouts.

Fig. 0

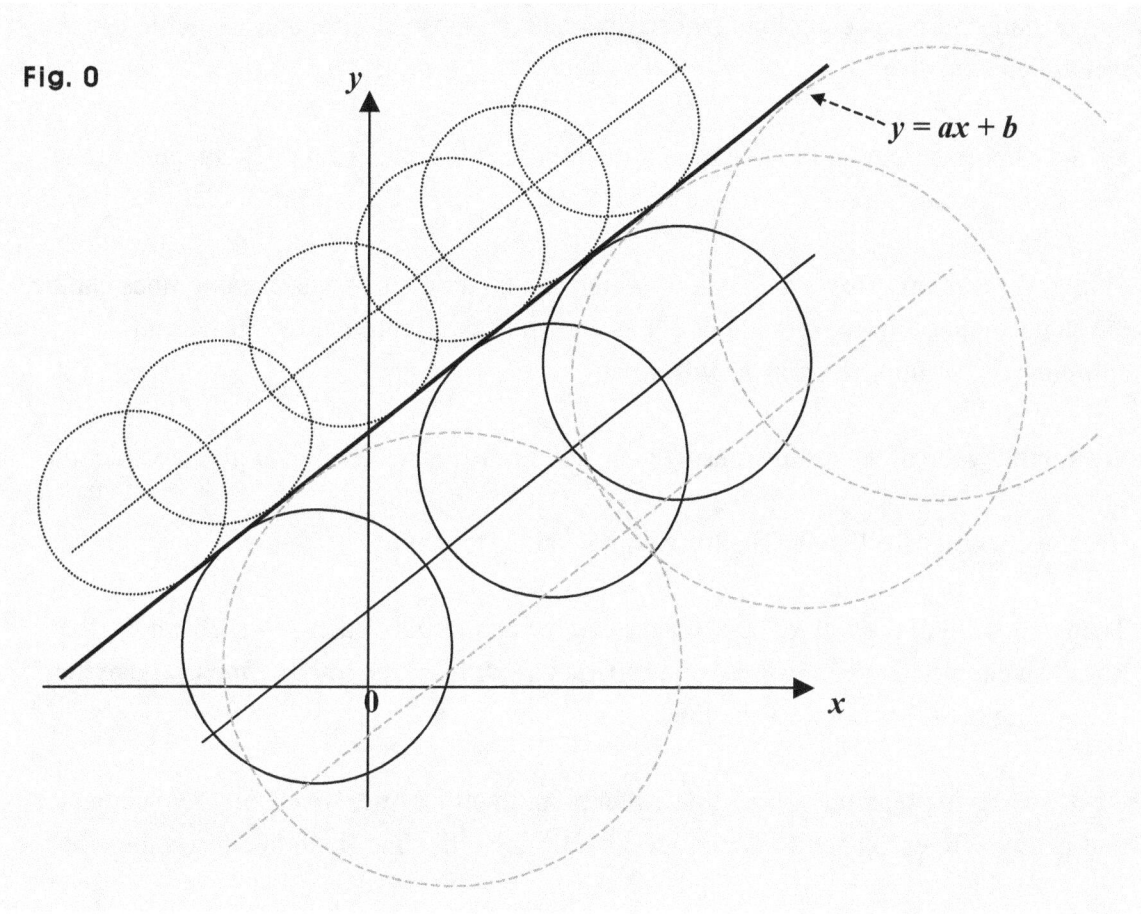

Suppose now, that *A* is the line *y = ax + b*.

Then, we can see that all the circles that have the same radii and are tangent to the line *A* have their centers in a line parallel to the line *A*. That is, the line parallel to the line *A* is a particular distance away from the line *A*, and the particular distance is the radius.

What then, is the line parallel to the line *A*?

It represents all lines parallel to *A*, and thus, is an arbitrary line parallel to the line *A*.

So suppose that *B* is a line parallel to the line *A*, *y = ax + b*.

Then, the line *B* can be put in *y = ax + c*, where *c* is a constant and *c ≠ b*.

And those circles that have their centers in the line *B* and are tangent to the line *A* are the solution to this problem.

So we want to find an equation that can represent all those circles, and we get to use the fact as follows:

• If we choose a point in the line **B**, and the point is the center of a circle tangent to the line **A**, the distance between the two lines **A** and **B** is the radius of the circle.

In fact, we can have two lines, each of which can be the line **B**, because two lines can be the same distance away from the line **A** as shown in the figure below. So we can approach the solution the way as follows.

We set the value of the radius first. We can set it to **r**, for instance, and give a value to **r**.

Next, we specify the line **A**. That is, we give values to **a** and **b**.

Then, we will get two values of **c**, for we can have two lines, each of which can be the line **B** since two lines can be the same distance away from the line **A**. One is above the line **A**, and the other is below the line **A**.

Next, specifying only the **x**-coordinate of a point in either line, we can get two centers, that is, two circles tangent to the line **A**. One is above the line **A**, and below is the other.

Fig. 1

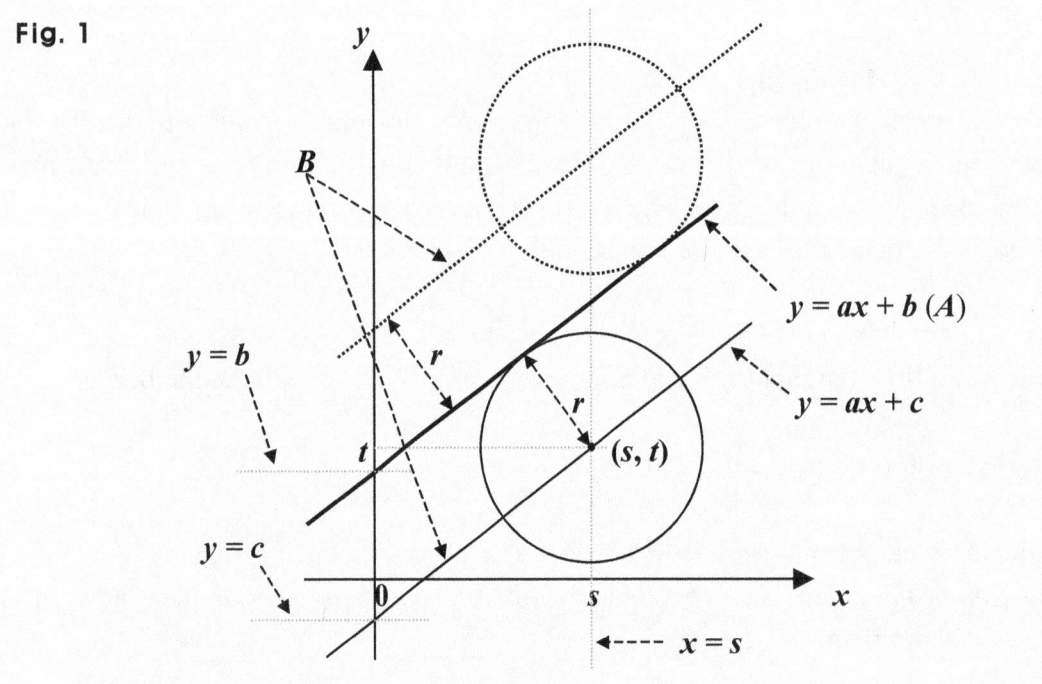

Suppose now, r, s, and t are constant, and C is a circle of radius r, which is centered at a point (s, t) in the line B, and tangent to the line A. That is, C is the circle we want.

Then first, since (s, t) is in the line B, which is: $y = ax + c$, we get: $t = as + c$.

Thus, the center can be put this way: $(s, as + c)$.

Next, the distance between the lines A and B is the distance from the center $(s, as + c)$ to the line A. And we know the distance is the very radius r. So let's now, find the radius r.

Then, we want to get first, the distance above. We have a formula for such a distance:

- The distance from a point (u, v) to a line $px + qy + r = 0$ is: $\frac{|pu+qv+r|}{\sqrt{p^2+q^2}}$.

The line is A, which is $y = ax + b$, which can be indicate by $ax - y + b = 0$, too.
The point is the center of the circle C, which is at $(s, as + c)$.

So the distance, that is, the radius is as follows: $\frac{|as+(-1)(as+c)+b|}{\sqrt{a^2+(-1)^2}} = \frac{|as-as-c+b|}{\sqrt{a^2+1}} = \frac{|b-c|}{\sqrt{a^2+1}}$.

Therefore, $(x - s)^2 + \{y - (as + c)\}^2 = \frac{(b-c)^2}{a^2+1}$ indicates the circle where the center is at (s, t) $= (s, as + c)$, and the radius is $\frac{|b-c|}{\sqrt{a^2+1}}$. And the circle is the circle C, of course.

Next, we know r is the radius of C, so we get: $\frac{|b-c|}{\sqrt{a^2+1}} = r$, which is of course, the distance between the two lines A and B, too.

Let's now, get the line B, which is parallel to and r away from the line A.

Finding the line B, we want to find c since B is: $y = ax + c$, and a is known since a is the slope of the line A, which is to be specified, and B is parallel to A.

Now, we have: $\frac{|b-c|}{\sqrt{a^2+1}} = r$, where a, b, and r are known, since those constants are given.

So we get:

$$\frac{|b-c|}{\sqrt{a^2+1}} = r \Rightarrow \frac{(b-c)^2}{a^2+1} = r^2 \Rightarrow (b-c)^2 = r^2(a^2+1) \Rightarrow b-c = \pm r\sqrt{a^2+1} \Rightarrow c = b \pm r\sqrt{a^2+1}.$$

Thus, two lines can be the line **B**, and each of the two is a particular distance away from the line **A**, and the particular distance is **r**, of course.

One line is $y = ax + b + r\sqrt{a^2 + 1}$, and the other is $y = ax + b - r\sqrt{a^2 + 1}$.

Now, putting threads together, we have:

$(x - s)^2 + \{y - (as + c)\}^2 = \frac{(b-c)^2}{a^2+1} = r^2$, and $c = b \pm r\sqrt{a^2 + 1}$.

So the circle **C** is $(x - s)^2 + (y - as - b \pm r\sqrt{a^2 + 1})^2 = r^2$, where **a**, **b**, **s**, and **r** are constant.

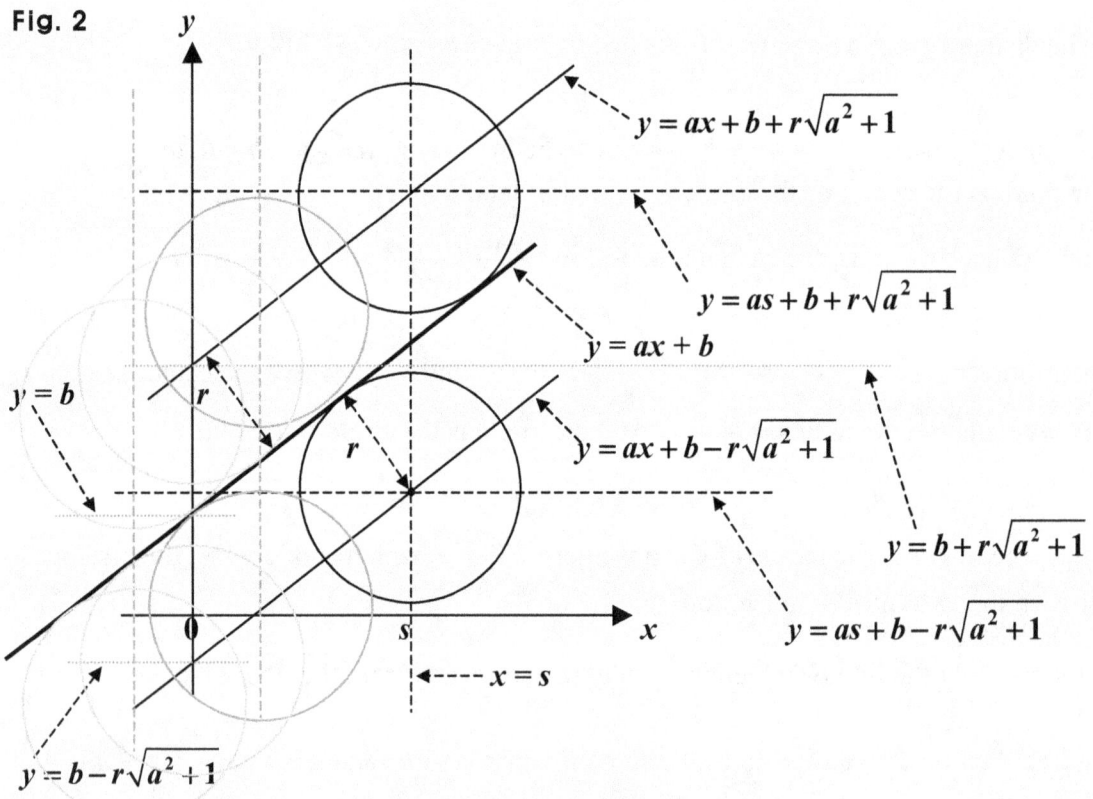

Fig. 2

Now, let's specify a specific group of circles that have the same radii, and are tangent to the line $y = ax + b$.

The equation for those circles is: $(x - s)^2 + (y - as - b \pm r\sqrt{a^2 + 1})^2 = r^2$.

First, we specify the line A and the radius. That is, we give values to a, b, and r.

Suppose for instance, the line A is $y = \frac{1}{2}x + 1$, and the radius r is $\frac{\sqrt{5}}{2}$.

Then, we can get two lines parallel to the line A, each of which is r away from the line A.

Then, the equation for circles above can represent two circles for each value of s, which is the x-coordinate at the center, which is in the line B.

One circle is: $(x - s)^2 + (y - \frac{1}{2}s + \frac{1}{4})^2 = \frac{5}{4}$, which is centered at $(s, \frac{1}{2}s - \frac{1}{4})$.

The other circle is: $(x - s)^2 + (y - \frac{1}{2}s - \frac{9}{4})^2 = \frac{5}{4}$, which is centered at $(s, \frac{1}{2}s + \frac{9}{4})$.

Thus, each of the two equations above represents a group of all circles where the centers are in one of two lines, which are parallel to and $\frac{\sqrt{5}}{2}$ away from the line A, $y = \frac{1}{2}x + 1$.

The two lines stated above can be the line B.

One of the two lines is as follows: $y = ax + b + r\sqrt{a^2 + 1} \Rightarrow y = \frac{1}{2}x + \frac{9}{4}$.

And the other line is as follows: $y = ax + b - r\sqrt{a^2 + 1} \Rightarrow y = \frac{1}{2}x - \frac{1}{4}$.

Then, having only to specify the x-coordinate at a point in either of the two lines, we get a circle tangent to the line A.

That is, giving a value to s in $(x - s)^2 + (y - \frac{1}{2}s + \frac{1}{4})^2 = \frac{5}{4}$ or in $(x - s)^2 + (y - \frac{1}{2}s - \frac{9}{4})^2 = \frac{5}{4}$, we get such a circle, since s is the x-coordinate of the point.

Suppose for instance, $s = 2$. That is, the x-coordinate of the point is 2.

Then, the circle we get is one of the two circles as follows:

One is: $(x - 2)^2 + (y - \frac{3}{4})^2 = \frac{5}{4}$, and the other is: $(x - 2)^2 + (y - \frac{13}{4})^2 = \frac{5}{4}$.

Note that for each value of r, the equation above represents all the circles that are tangent to the given line A, and have the same radii, which are the same as r, of course.

Fig. 3

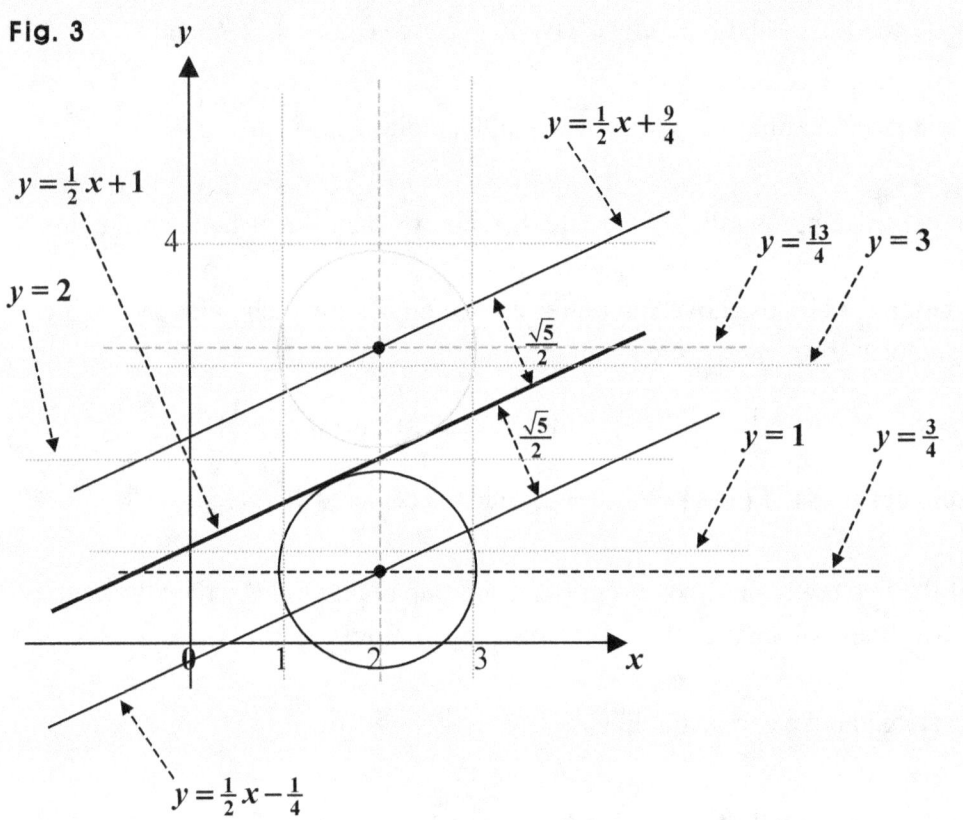

In short:

Suppose A is the line $y = ax + b$, B is a line $y = ax + c$, where c is constant and $c \neq b$, r is the distance between the two lines A and B, and C is the circle to be constructed.

Suppose also, that s and t are constant, and that a point (s, t) is in the line B, and the center of a circle tangent to the line A.

Then, we get: $(s, t) = (s, as + c)$, which is the center of C, and r is the radius of C.

The distance from a point (u, v) to a line $px + qy + r = 0$ is: $\frac{|pu + qv + r|}{\sqrt{p^2 + q^2}}$.

So the distance from $(s, as + c)$ to the line A is: $\frac{|as + (-1)(as + c) + b|}{\sqrt{a^2 + (-1)^2}} = \frac{|b - c|}{\sqrt{a^2 + 1}}$. Thus, we get:

$$\frac{|b-c|}{\sqrt{a^2+1}} = r \Rightarrow \frac{(b-c)^2}{a^2+1} = r^2 \Rightarrow (b-c)^2 = r^2(a^2+1) \Rightarrow b-c = \pm r\sqrt{a^2+1} \Rightarrow c = b \pm r\sqrt{a^2+1}.$$

So the circle C is $(x - s)^2 + (y - as - b \pm r\sqrt{a^2 + 1})^2 = r^2$, where a, b, r, and s are constant.

Suggestions or Solutions
To the Problem 0 in the Example 1

Construct a line tangent to a circle $(x - a)^2 + (y - b)^2 = c^2$, where a, b, and c are constant, and $c \neq 0$, of course.

Suppose B is a line tangent to the given circle at a point (s, t).

The center of the circle given is (a, b).

Suppose now, A is a line connecting the center (a, b) and the point (s, t).

Then, the line B includes the point (s, t), and is perpendicular to the line A.

The slope the line A is $\frac{b-t}{a-s}$, and the product of the slopes of the lines A and B is -1.

So the slope of the line B is $-\frac{1}{m} = -\frac{1}{\frac{b-t}{a-s}} = -\frac{a-s}{b-t}$.

Therefore, the line B is: $y - t = -\frac{a-s}{b-t}(x - s)$ since the line B includes (s, t).

Simplifying the equation above, we get:

$$y - t = -\tfrac{a-s}{b-t}(x - s) \Rightarrow (b - t)(y - t) = -(a - s)(x - s) \Rightarrow (a - s)(x - s) + (b - t)(y - t) = 0.$$

If not quite sure of the idea behind the processes above, follow the steps below:

Let's begin with putting in a graph the given circle and a line tangent to the circle.

It's a good idea to add to the graph a point indicating where the line is tangent to the circle. So we can see better what we can do about the problem.

Now, the center is at (a, b), and the radius is $|c|$. So assuming $c > 0$, putting in a graph the circle and a line, we can put them the way below:

Fig. 0

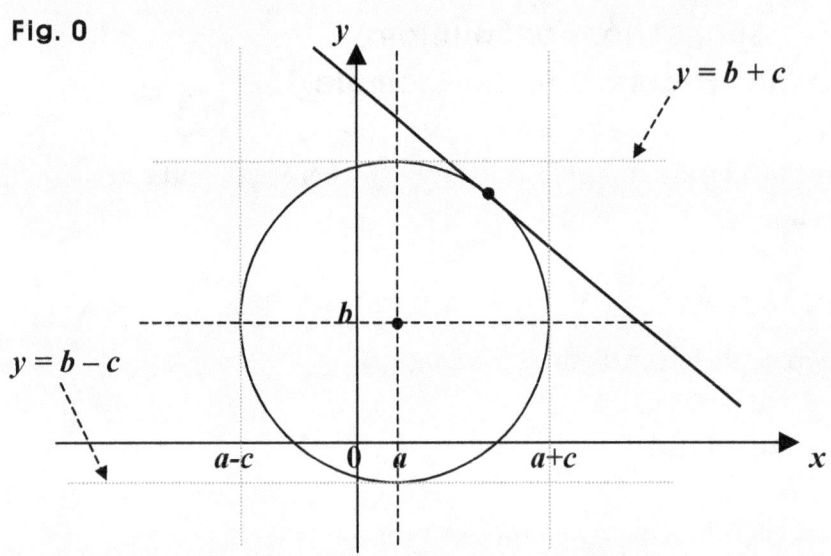

By the way, finding a line, what do we need to have?

Given a point in the line and the slope of the line, we can find the line.
Suppose for instance, a point **(u, v)** is in a line **L** in the **x-y** plane, and the slope is **w**.

Then, the line **L** is: $y - v = w(x - u)$.　(How come? Refer to **CONICS 1**.)

So we want to get a point and the slope in the line we are after in this problem.
What line are we after, though?

It is a line tangent to the circle $(x - a)^2 + (y - b)^2 = c^2$.
Where can a line be tangent to a circle, though?

A line can be tangent to a circle at a point, which is in the circle, of course.
So we may want to begin with a point where a line is tangent to a circle.

Suppose now, that:

B is the line we are after,
C is the circle $(x - a)^2 + (y - b)^2 = c^2$, which is given in the problem, and
s and **t** are constant, and **(s, t)** is the point where the line **B** is tangent to the circle **C**.

Then, the tangent point (s, t) is in the tangent line B, too. So next, what about the slope?

We can take advantage of a couple of facts as follows:

• A line tangent to a circle at a particular point is perpendicular to a line passing through the center of the circle and the particular point.

• Besides, the product of the slopes of two lines perpendicular to each other is -1.

Thus, finding the slope of the line connecting the center of the circle C and the tangent point (s, t), we can immediately get the slope of the line B tangent to the circle C.

Fig. 1

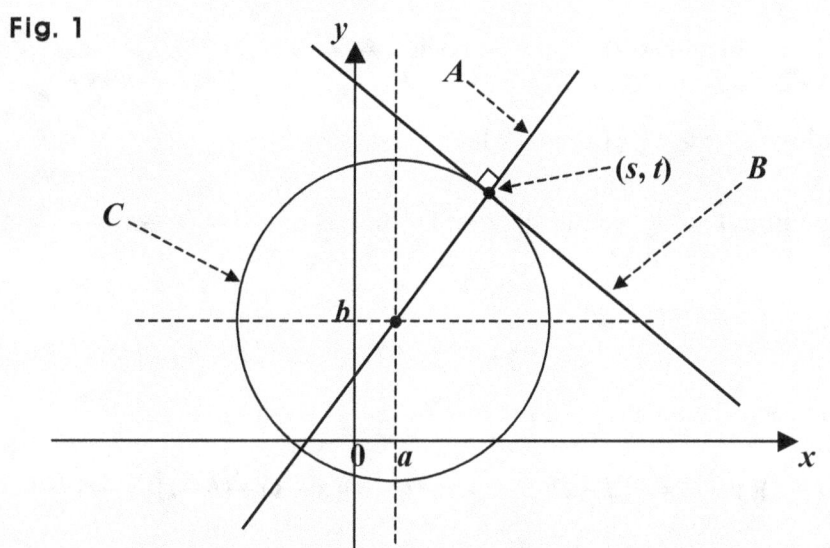

Suppose A is the line connecting the center (a, b) of the circle C and the point (s, t).
Then, the line A is perpendicular to the line B tangent to the circle C at the point (s, t).

Suppose next, m is the slope of the perpendicular line A.
Then, we can put the line A in an equation $y = mx + n$, where m and n are constant.

We know that the product of slopes of two lines perpendicular to each other is -1.
So we can put for now, the line B in an equation $y - t = -\frac{1}{m}(x - s)$ since B includes (s, t).

Fig. 2

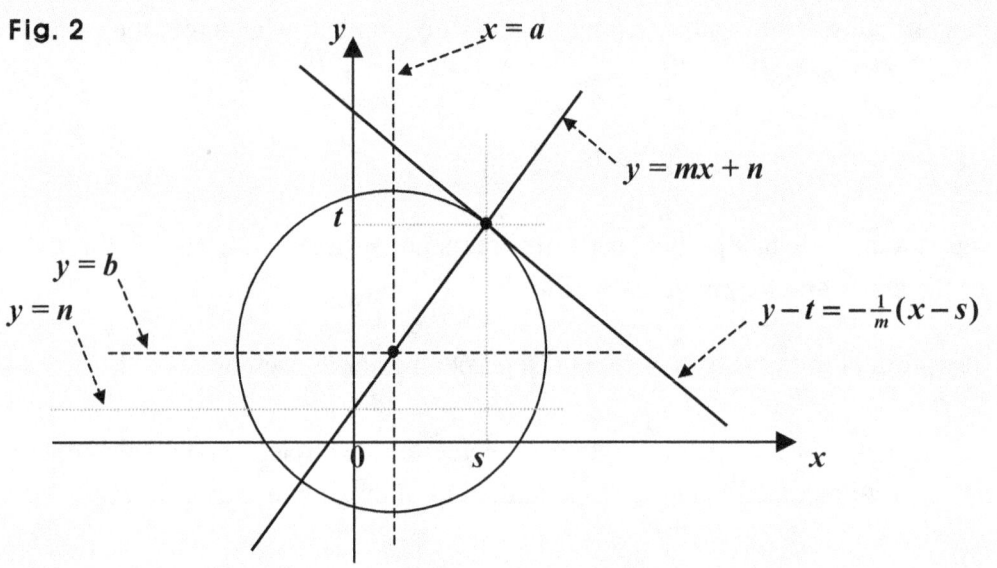

The line A passes through (a, b) and (s, t), so the slope $m = \frac{b-t}{a-s}$.

Thus, the line A is as follows: $y - b = \frac{b-t}{a-s}(x-a)$, where $\frac{b-t}{a-s}$ is the slope.

We know the slope of the line B is $-\frac{1}{m}$, so the slope of the line B is $-\frac{1}{\frac{b-t}{a-s}} = -\frac{a-s}{b-t}$.

Therefore, the line B is: $y - t = -\frac{a-s}{b-t}(x-s)$ since the line B passes through (s, t).

Now, we are going to simplify the equation of the line B.

$$y - t = -\frac{a-s}{b-t}(x-s) \Rightarrow (b-t)(y-t) = -(a-s)(x-s) \Rightarrow (a-s)(x-s) + (b-t)(y-t) = 0.$$

Therefore, the line B can also be indicated by $(a-s)(x-s) + (b-t)(y-t) = 0$, which is, of course, the same as $y - t = -\frac{a-s}{b-t}(x-s)$.

Thus, $(a-s)(x-s) + (b-t)(y-t) = 0$ is the line tangent to a circle centered at (a, b) at a point (s, t). Such a tangent line seems to have nothing to do with the radius, though.

What matters most in a line is the slope, yet a point it passes through matters, too.

In fact, the point (s, t) the tangent line passes through pertains the radius.

What if we want to make use of c, which is the radius, when we indicate the line B?

Then, we can put c^2 into the equation of the line B.

We don't just put it into the equation of B, of course.

The given circle C is: $(x - a)^2 + (y - b)^2 = c^2$, and the point (s, t) is in the circle C.

So we get: $(s - a)^2 + (t - b)^2 = c^2$.

We can now put $(s - a)^2 + (t - b)^2$ into the equation $(s - a)(x - s) + (t - b)(y - t) = 0$.

In fact, we will come up with $(s - a)^2$ and $(t - b)^2$ out of $(s - a)(x - s) + (t - b)(y - t) = 0$. So we've got to do some algebra.

$(s - a)(x - s) + (t - b)(y - t)$
$= (s - a)(x - s + a - a) + (t - b)(y - t + b - b)$
$= (s - a)(x - a - s + a) + (t - b)(y - b - t + b)$
$= (s - a)\{(x - a) - (s - a)\} + (t - b)\{(y - b) - (t - b)\}$
$= (s - a)(x - a) - (s - a)^2 + (t - b)(y - b) - (t - b)^2 = 0$.

So we get: $(s - a)(x - a) + (t - b)(y - b) = (s - a)^2 + (t - b)^2$, which is c^2.

Therefore, the equation of a line tangent at a point (s, t) to a circle of radius c centered at (a, b) is: $(s - a)(x - a) + (t - b)(y - b) = c^2$.

The equation above is in fact, equivalent to the equation $(s - a)(x - s) + (t - b)(y - t) = 0$.

Now, what if the center of the circle is at the origin $(0, 0)$?

Then, $a = 0$ and $b = 0$ since the center $(a, b) = (0, 0)$.

So we need to have: $a = 0$ and $b = 0$ in $(s - a)(x - a) + (t - b)(y - b) = c^2$.

Thus, we get: $sx + ty = c^2$, which is the line tangent at a point (s, t) to a circle of radius c centered at the origin $(0, 0)$.
Of course, we can use the equation $(s - a)(x - s) + (t - b)(y - t) = 0$, instead.

If $a = 0$ and $b = 0$, we get: $(s - a)(x - s) + (t - b)(y - t) = 0 \Rightarrow s(x - s) + t(y - t) = 0$
$\Rightarrow sx + ty = s^2 + t^2 = c^2 \Rightarrow sx + ty = c^2$. How come $s^2 + t^2 = c^2$, though?

The circle is: $x^2 + y^2 = c^2$, and the circle passes through (s, t). So we get: $s^2 + t^2 = c^2$.

Suggestions or Solutions
To the Problem 1 in the Example 1

Construct a line of slope of d tangent to a circle $(x - a)^2 + (y - b)^2 = c^2$, where a, b, c, and d are constant.

Suppose C is the circle given, and A is the line of slope d tangent to the circle C at a point $P(s, t)$, where s and t are constant.   Then, the line A is: $y - t = d(x - s)$.

Suppose also, B is a line perpendicular to the line A and connecting the center (a, b) and the point P. Then, B is a line of slope $-\frac{1}{d}$.

Besides, the line B has (a, b) and (s, t), so the slope of the line B can be put in $\frac{t-b}{s-a}$, too. Thus, we get: $-\frac{1}{d} = \frac{t-b}{s-a}$.

Meanwhile, (s, t) is in the circle C, too, so we get: $(s - a)^2 + (t - b)^2 = c^2$.

Thus, we get a system for s and t, where $-\frac{1}{d} = \frac{t-b}{s-a}$, and $(s - a)^2 + (t - b)^2 = c^2$.

Then, we get: $-\frac{1}{d} = \frac{t-b}{s-a} \Rightarrow s - a = -d(t - b) \Rightarrow (s - a)^2 + (t - b)^2 = d^2(t - b)^2 + (t - b)^2 = c^2$

$\Rightarrow (t - b)^2(1 + d^2) = c^2 \Rightarrow (t - b)^2 = \frac{c^2}{1+d^2} \Rightarrow t - b = \pm \frac{c}{\sqrt{1+d^2}}$.

So we now, have an equivalent system where $s - a = -d(t - b)$ and $t - b = \pm \frac{c}{\sqrt{1+d^2}}$.

Then first, we get: $t - b = \frac{c}{\sqrt{1+d^2}} \Rightarrow s - a = -d(t - b) = -\frac{cd}{\sqrt{1+d^2}} \Rightarrow s - a = -\frac{cd}{\sqrt{1+d^2}}$.

So we get: $s = a - \frac{cd}{\sqrt{1+d^2}}$, and $t = b + \frac{c}{\sqrt{1+d^2}}$.

Next, we get: $t - b = -\frac{c}{\sqrt{1+d^2}} \Rightarrow s - a = -d(t - b) = \frac{cd}{\sqrt{1+d^2}} \Rightarrow s - a = \frac{cd}{\sqrt{1+d^2}}$.

So we get: $s = a + \frac{cd}{\sqrt{1+d^2}}$, and $t = b - \frac{c}{\sqrt{1+d^2}}$.

Therefore, we get: $(s,t) = (a - \frac{cd}{\sqrt{1+d^2}}, b + \frac{c}{\sqrt{1+d^2}})$ or $(a + \frac{cd}{\sqrt{1+d^2}}, b - \frac{c}{\sqrt{1+d^2}})$.

Now, the line A is initially put in $y - t = d(x - s)$.   Therefore, the line A is as follows:

$y - (b + \frac{c}{\sqrt{1+d^2}}) = d\{x - (a - \frac{cd}{\sqrt{1+d^2}})\}$ or $y - (b - \frac{c}{\sqrt{1+d^2}}) = d\{x - (a + \frac{cd}{\sqrt{1+d^2}})\}$.

If not quite sure of the idea behind the processes above, follow the steps below:

Suppose C is the circle given, and A is the line we want to construct.

Then, A is the line of slope d, and is tangent to the circle C.

We can have two lines though, that are tangent to a circle and have the same slope.

So putting in a graph the ideas above, we can see better how the idea goes.

Fig. 0

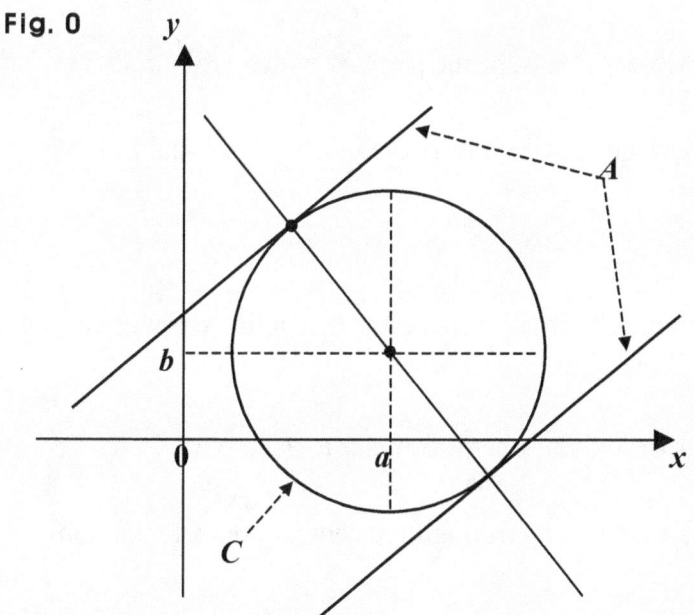

Now, we can see in the graph above, that two lines tangent to the circle C can have the same slope. So the two lines can be the line A, and thus, can be the solution.

So let's find the two lines. Finding a line though, what do we need?

• We need two points in the line, or we need a point in the line, together with the slope.

We know the tangent line A has a slope of d, so we need to get a point in the line A.

The solution to this problem is in fact, about points where lines are tangent to a circle.

A line and a circle can be tangent to each other at a point.

So at two points respectively, the circle C is tangent to two lines, which have the same slope, which is d. We may want to begin with therefore a point where a line is tangent to a circle. Getting the tangent point, since we know the slope, we can find the line tangent.

Suppose now, that the line A is tangent to the circle C at a point $P(s, t)$, where s and t are constant, of course.

Then, we want to find the point P, which is one of the points in the circle C, of course.

And finding the point P, we find the coordinates of the point P, which are s and t.

So we can expect that we are going to get a system of two equations for s and t.

To begin with, we can take advantage of a couple of facts as follows:

• A line tangent to a circle at a particular point is perpendicular to a line passing through the center of the circle and the particular point.

• Besides, the product of the slopes of two lines perpendicular to each other is -1.

Thus, we may want to begin with a line passing through the center and a tangent point.

So suppose now, that B is the line that is perpendicular to the tangent line A, and passes through the center (a, b) and the point $P(s, t)$.

Then, using the two lines A and B, we should be able to get a system of two equations for the two constants s and t.

First, the slope of the line A is d.

And next, the line A is tangent to the circle C at $P(s, t)$, and thus, has the point $P(s, t)$.

So the line A can be put in an equation as follows: $y - t = d(x - s)$.

Next, the line B is perpendicular to the line A, which is a line of slope d.

So B is a line of slope $-\frac{1}{d}$.

And the line **B** has the point **(s, t)**. So the line **B** can be put, for now, in $y - t = -\frac{1}{d}(x - s)$.

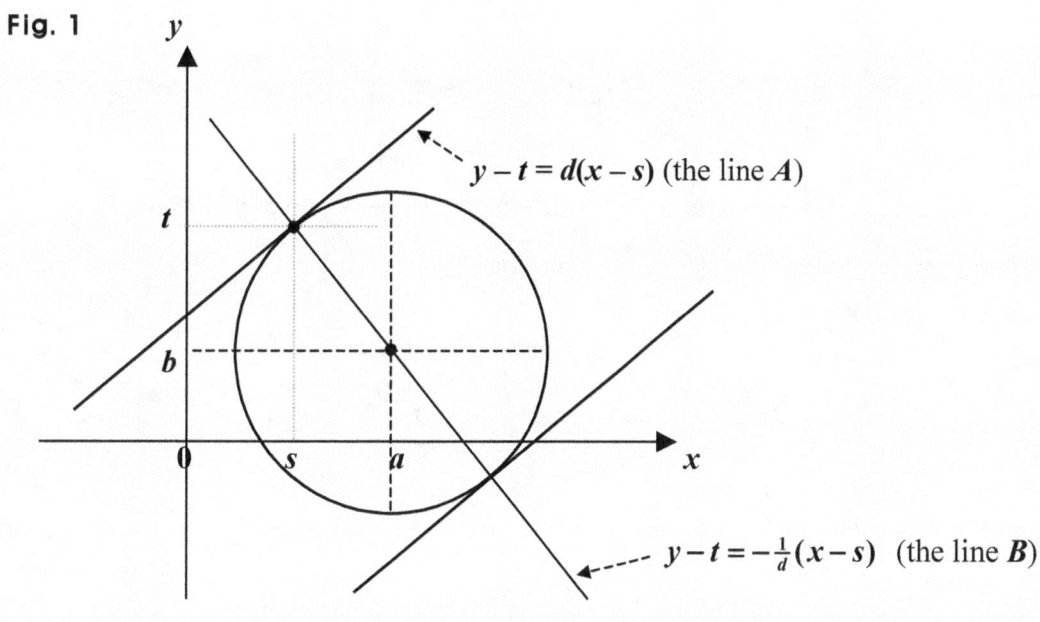

Fig. 1

$y - t = d(x - s)$ (the line A)

$y - t = -\frac{1}{d}(x - s)$ (the line B)

Now, the line **B** connects **(s, t)** and **(a, b)**, so its slope can be put in $\frac{t-b}{s-a}$, too.

Therefore, we can see that $-\frac{1}{d} = \frac{t-b}{s-a}$.

Meanwhile, the circle **C** is: $(x - a)^2 + (y - b)^2 = c^2$.

So since **(s, t)** is in the circle **C**, we get: $(s - a)^2 + (t - b)^2 = c^2$.

Thus, we get a system of two equations for **s** and **t** as follows:
$-\frac{1}{d} = \frac{t-b}{s-a}$, and $(s - a)^2 + (t - b)^2 = c^2$.

So we need to do some algebra.

To begin with, we get: $-\frac{1}{d} = \frac{t-b}{s-a} \Rightarrow s - a = -d(t - b)$.

Thus, we get: $(s-a)^2 + (t-b)^2 = c^2 \Rightarrow d^2(t-b)^2 + (t-b)^2 = c^2$

$\Rightarrow (t-b)^2(1+d^2) = c^2 \Rightarrow (t-b)^2 = \frac{c^2}{1+d^2} \Rightarrow t-b = \pm\frac{c}{\sqrt{1+d^2}}$.

Therefore, we now have: $s-a = -d(t-b)$ and $t-b = \pm\frac{c}{\sqrt{1+d^2}}$.

Then first, we get: $t-b = \frac{c}{\sqrt{1+d^2}} \Rightarrow s-a = -d(t-b) = -\frac{cd}{\sqrt{1+d^2}} \Rightarrow s-a = -\frac{cd}{\sqrt{1+d^2}}$.

So we get: $s = a - \frac{cd}{\sqrt{1+d^2}}$, and $t = b + \frac{c}{\sqrt{1+d^2}}$.

So next, we get: $t-b = -\frac{c}{\sqrt{1+d^2}} \Rightarrow s-a = -d(t-b) = \frac{cd}{\sqrt{1+d^2}} \Rightarrow s-a = \frac{cd}{\sqrt{1+d^2}}$.

So we get: $s = a + \frac{cd}{\sqrt{1+d^2}}$, and $t = b - \frac{c}{\sqrt{1+d^2}}$.

Therefore, we get: $(s,t) = (a - \frac{cd}{\sqrt{1+d^2}}, b + \frac{c}{\sqrt{1+d^2}})$ or $(a + \frac{cd}{\sqrt{1+d^2}}, b - \frac{c}{\sqrt{1+d^2}})$.

We know that the line A is initially indicated by $y - t = d(x - s)$.

So the line A can be either of the two lines as follows:

$y - (b + \frac{c}{\sqrt{1+d^2}}) = d\{x - (a - \frac{cd}{\sqrt{1+d^2}})\}$ and $y - (b - \frac{c}{\sqrt{1+d^2}}) = d\{x - (a + \frac{cd}{\sqrt{1+d^2}})\}$,

each of which is a line of slope d tangent to a circle $(x-a)^2 + (y-b)^2 = c^2$.

We can put the line A this way, too: $y - (b \pm \frac{c}{\sqrt{1+d^2}}) = d\{x - (a \mp \frac{cd}{\sqrt{1+d^2}})\}$.

Examples D in Circles

Constructing a curve in math, we can put it in a graph. Putting it in a graph though, we need the equation of it. So constructing it, we want to find the equation of it if we are not given the equation, of course. And finding the equation, it is said that we get it, too.

Suppose k is constant, A is $k(x + 1) - 2y + 4 = 0$, B is $2x + k(y - 3) - 10 = 0$, and G is the point where the curves of A and B meet each other.

Then, find the trace G makes as k varies in each of two cases as follows:

0. k is real.

1. $k \leq 1$.

Suggestions or Solutions
To the Problem in the Example 0

Assuming k is constant, A is $k(x + 1) - 2y + 4 = 0$, B is $2x + k(y - 3) - 10 = 0$, and G is the point where the curve of A meets that of B, find the trace G makes as k varies.

From A, we can get: $k(x + 1) - 2y + 4 = 0 \Rightarrow k(x + 1) + (4 - 2y) = 0$, which can be an equation of a line for each value of k.

Suppose a line $x + 1 = 0$ meets another line $4 - 2y = 0$ at (s, t).

Then, we get: $s + 1 = 0$, and $4 - 2t = 0$, so we get: $k(s + 1) + 4 - 2t = 0$.

So if the two lines above meet at a particular point, for any value of k, the equation given indicates a line that includes the particular point.

Finding the points, we get: $x + 1 = 0 \Rightarrow x = -1$, and $4 - 2y = 0 \Rightarrow y = 2$.
Thus, the two lines meet at **(-1, 2)**.
And the two lines can be put this way too: $x = -1$ and $y = 2$.
So all the lines indicated by the equation A meet each other altogether at **(-1, 2)**.

Next, from B, we can get $2x + k(y - 3) - 10 = 0 \Rightarrow k(y - 3) + 2x - 10 = 0$, which can be an equation of a line for each value of k.

Suppose a line $y - 3 = 0$ meets another line $2x - 10 = 0$ at (s, t).

Then, we get: $t - 3 = 0$, and $2s - 10 = 0$, so we get: $k(t - 3) + 2s - 10 = 0$.

So if the two lines above meet at a particular point, for any value of k, the equation given indicates a line that includes the particular point.

Finding the points, we get: $y - 3 = 0 \Rightarrow y = 3$, and $2x - 10 = 0 \Rightarrow x = 5$.
Thus, the two lines meet at **(5, 3)**.
And the two lines can be put this way too: $x = 5$ and $y = 3$.
So all the lines indicated by the equation B meet each other altogether at **(5, 3)**.

For each value of **k**, a couple of lines get determined, one of the two is from the equation **A**, and the other is from the equation **B**. We have:

A: $k(x + 1) - 2y + 4 = 0 \Rightarrow kx + k - 2y + 4 = 0 \Rightarrow y = \frac{k}{2}x + \frac{k+4}{2}$.

B: $2x + k(y - 3) - 10 = 0 \Rightarrow 2x + ky - 3k - 10 = 0 \Rightarrow y = -\frac{2}{k}x + \frac{3k+10}{k}$.

So the product of the slopes in each pair of the lines is -1, and therefore, the two lines in each pair are perpendicular to each other.

Thus, all the points at each of which the lines meet each other in a pair, form a circle.

The line segment between the two points **(-1, 2)** and **(5, 3)** is a diameter of the circle.

The center of the circle is the midpoint between the two endpoints **(-1, 2)** and **(5, 3)**.

So the center is $(\frac{-1+5}{2}, \frac{2+3}{2}) = (2, \frac{5}{2})$. Suppose now, **r** is the radius, and **d** is the diameter.

Then, we get: $d^2 = (\Delta x)^2 + (\Delta y)^2 = (-1 - 5)^2 + (2 - 3)^2 = 36 + 1 = 37$.

We have: $r = \frac{d}{2}$, too, so we get: $r^2 = \frac{d^2}{4} = \frac{37}{4}$, and the circle is: $(x - 2)^2 + (y - \frac{5}{2})^2 = \frac{37}{4}$.

However, because the equations **A** and **B** cannot indicate the two lines $x = -1$ and $y = 3$ for any value of **k**, the point **(-1, 3)** has to be excluded.

Therefore, the trace is $(x - 2)^2 + (y - \frac{5}{2})^2 = \frac{37}{4}$ where $x \neq -1$.

If not quite sure of the idea behind the processes above, follow the steps below:

To begin with, each of **A** and **B** indicates a group of lines, so the curves are lines. How come?

Since **k** is constant, we can choose one value for **k** at a time, so for each value of **k**, **A** indicates a line, and so does **B**. Thus, for each of **A** and **B**, we can get as many lines as the number of choices we make for **k**. So **A** and **B** each can be said to represent a group of lines.

Now, every time we choose a value for **k**, we can determine a couple of lines using the two equations **A** and **B**.

The two lines in each couple meet at a point if both are not parallel, of course.

Thus, we can determine as many points as the number of the choices for **k**.   So what?

Those many points are in a curve, which is the trace we are after in this problem.

So every one of such many points is **G**, which is therefore, an arbitrary point in the curve.

Thus, **G** can be put in (**s**, **t**), where **s** and **t** are constant. Then, as **k** varies, the values of **s** and **t** change, so points keep getting made. Thus, a curve gets made of the points, and the curve can be called the trace of **G**.   So let's now, find the trace that **G** makes.

To begin with, we have a fact as follows:

Suppose that **m** and **n** are constants, and that the curves of two equations $f(x, y) = 0$ and $g(x, y) = 0$ meet each other at particular points.

Then, the curve of $mf(x, y) + ng(x, y) = 0$ meets the curves of $f(x, y) = 0$ and $g(x, y) = 0$ at the particular points, too. That is, the three curves share the very particular points. In particular, the equation $mf(x, y) + ng(x, y) = 0$ is called a linear combination of the two equations $f(x, y) = 0$ and $g(x, y) = 0$.   Note that in math, a curve can be a line, too.

Now, in the fact above, the two equations *f* and *g* can indicate a line each.

So for this problem, we can use the fact as follows:

Suppose a line $ax + by + c = 0$ meets another line $dx + ey + f = 0$ at a particular point, and **m** and **n** are constant.

Then, another line $m(ax + by + c) + n(dx + ey + f) = 0$ passes through the particular point for each pair of values of **m** and **n**.

Let's now, start getting the trace.

We may want to first, check to see if each equation given in this problem can be such a linear combination as explained above.

So let's now, extract from each equation such curves as those of f and g explained above.

To begin with, for A, we get: $k(x + 1) - 2y + 4 = 0 \Rightarrow k(x + 1) + (4 - 2y) = 0$.

Thus, the curves are two lines $x + 1 = 0$ and $4 - 2y = 0$; in other words, $x = $ -1 and $y = 2$.

Next, for B, we get: $2x + k(y - 3) - 10 = 0 \Rightarrow k(y - 3) + 2x - 10 = 0$.

So the curves are two lines $y - 3 = 0$ and $2x - 10 = 0$; that is, $y = 3$ and $x = 5$.

Therefore, we can see that:

• All the lines the equation A represents meet each other altogether at a point **(-1, 2)**, because the two lines $x = $ -1 and $y = 2$ meet each other at the point **(-1, 2)**.

• All the lines the equation B represents meet each other altogether at a point **(5, 3)**, because the two lines $x = 5$ and $y = 3$ meet each other at the point **(5, 3)**.

Next, let's consider every point where a pair of lines for each value of k meet each other.

For each value of k, two lines get determined, one of them is from the equation A, and the other is from the equation B.
And we can notice that the two lines in each pair are perpendicular to each other.
How come?

For A, we get: $k(x + 1) - 2y + 4 = 0 \Rightarrow kx + k - 2y + 4 = 0 \Rightarrow y = \frac{k}{2}x + \frac{k+4}{2}$.

For B, we get: $2x + k(y - 3) - 10 = 0 \Rightarrow 2x + ky - 3k - 10 = 0 \Rightarrow y = -\frac{2}{k}x + \frac{3k+10}{k}$.

So the product of the slopes of the two lines A and B is $-\frac{2}{k} \cdot \frac{k}{2} = -1$, and therefore, the two lines in each pair for each value of k are perpendicular to each other.

Now, we can take advantage of the fact as follows:
• The arc angle for a half circle is $90°$.   What is the arc angle?

An arc angle is an angle made by two rays emitting from a point in a circle and limiting (or defining) an arc in the circle. So an arc angle is between $0°$ and $180°$.

For instance, the arc angle for a half circle is $90°$.

Fig. 0

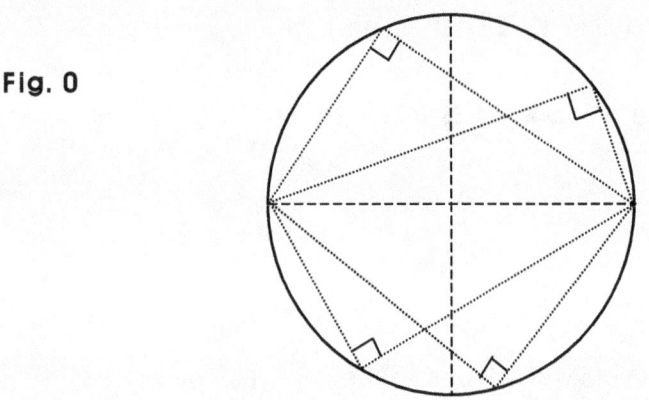

Let's now, put in a graph some pairs of the lines from the two equations.

The lines the equation *A* represents pass through the point **(-1, 2)**, and the lines the equation *B* represents pass through the point **(5, 3)**.

Fig. 1

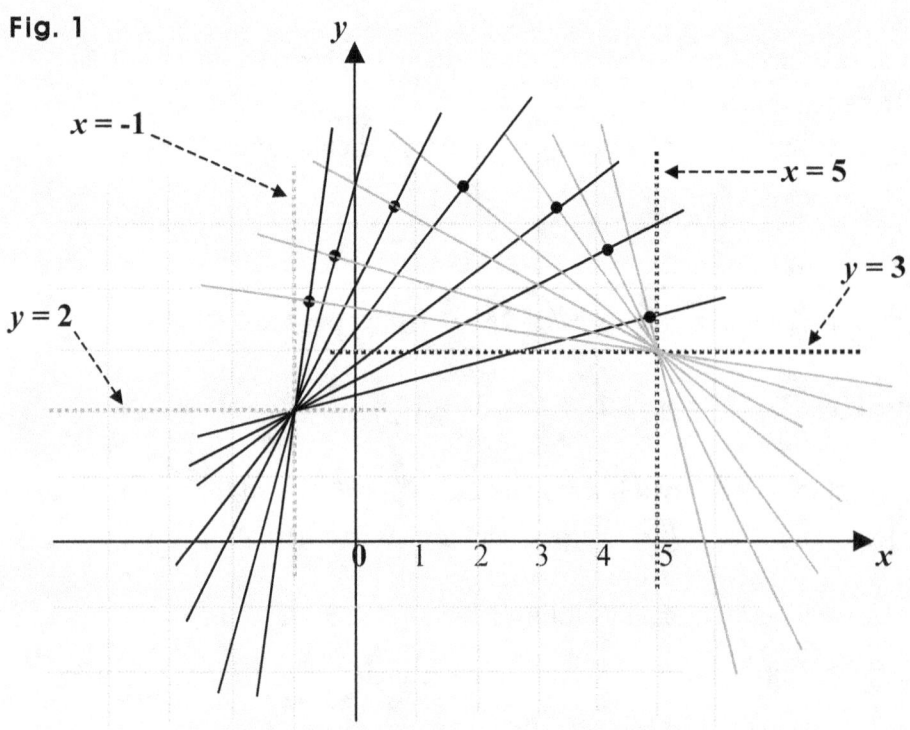

Notice that all the points at each of which two lines in each pair meet each other, are in a circle. And the line segment between the two points **(-1, 2)** and **(5, 3)** is a diameter of the circle. That's because two lines meeting at each and every point above are perpendicular to each other. And the arc angle for a half circle is 90°.

Fig. 2

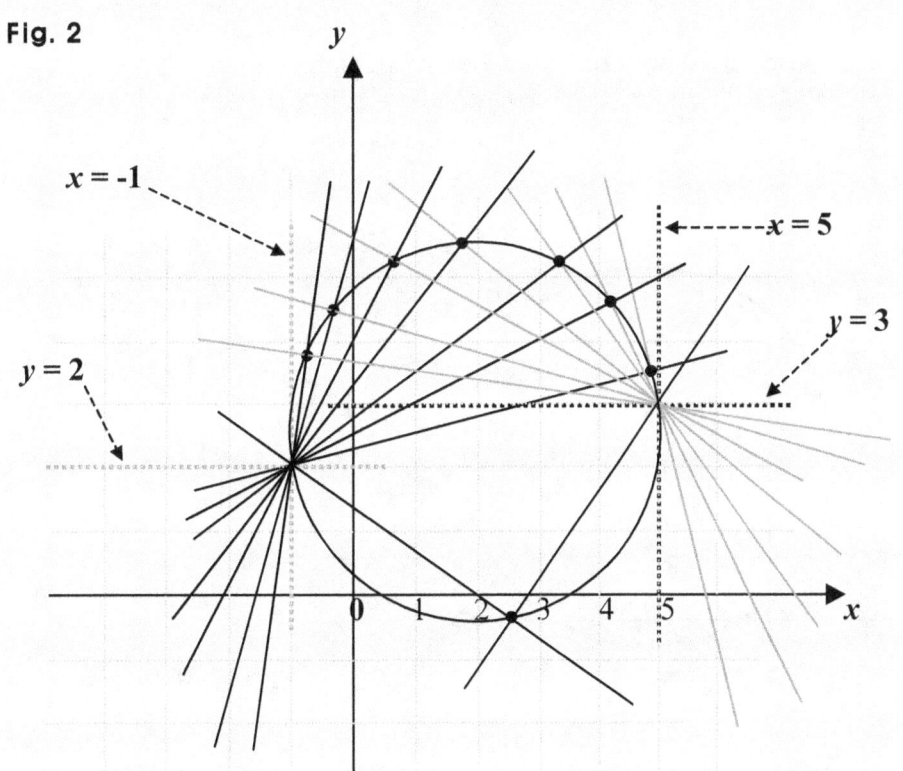

Now, let's find the circle. Finding a circle, what do we need to have?

We can use the standard form: $(x - a)^2 + (y - b)^2 = c^2$, where (a, b) is the center and c is the radius. So:

1. Given the center and the radius, we can directly use the standard form above.

2. Given three points in the circle, we can get a system of three equations for a, b, and c. And putting each of the three points into the standard equation above, we can get an equation. So we can get three equations for the three unknowns: a, b, and c. Solving the system, we get the values of a, b, and c.

In this case, we can readily find the center and the radius.
The center is the midpoint between the two endpoints **(-1, 2)** and **(5, 3)** of the diameter.

So the center is $(\frac{-1+5}{2}, \frac{2+3}{2}) = (2, \frac{5}{2})$. Suppose now, r is the radius, and d is the diameter. Then, we get: $d^2 = (\Delta x)^2 + (\Delta y)^2 = (-1 - 5)^2 + (2 - 3)^2 = 36 + 1 = 37$.

We have: $r = \frac{d}{2}$, too, so we get: $r^2 = \frac{d^2}{4} = \frac{37}{4}$, and the circle is: $(x - 2)^2 + (y - \frac{5}{2})^2 = \frac{37}{4}$.

Is the entire circle above the trace we've been after, then?

It's not quite the case, and we may want to take a look at a graph below.

Fig. 3

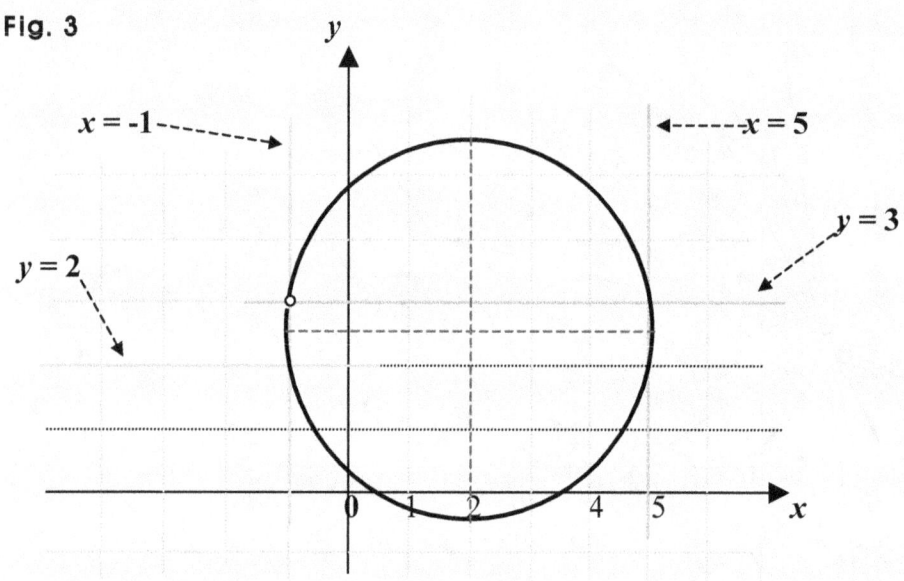

We can see that a particular point in the circle has to be excluded from the graph above, and the particular point is **(-1, 3)**. That's because the two lines $x = -1$ and $y = 3$ cannot be indicated by any of the two equations **A** and **B** for any value of k. The two equations are as follows:

For **A**, we get: $k(x + 1) - 2y + 4 = 0 \Rightarrow k(x + 1) + (4 - 2y) = 0$.

For **B**, we get: $2x + k(y - 3) - 10 = 0 \Rightarrow k(y - 3) + 2x - 10 = 0$.

Therefore, the trace we want is $(x - 2)^2 + (y - \frac{5}{2})^2 = \frac{37}{4}$ where $x \neq -1$.

Can the point **(-1, 3)** be in the circle, $(x - 2)^2 + (y - \frac{5}{2})^2 = \frac{37}{4}$, though?

Yes, it can. Putting the point **(-1, 3)** into the equation of the circle, we get:

$(-1 - 2)^2 + (3 - \frac{5}{2})^2 = 9 + \frac{1}{4} = \frac{37}{4}$.

Suggestions or Solutions
To the Problem in the Example 1

Suppose A is $k(x + 1) - 2y + 4 = 0$, B is $2x + k(y - 3) - 10 = 0$, in each of which k is constant, $k \leq 1$, and G is the point where the curve of A meets that of B. Then, find the trace G makes as k varies.

We have $k \leq 1$, and also:

For A, we get: $k(x + 1) - 2y + 4 = 0 \Rightarrow kx + k - 2y + 4 = 0 \Rightarrow y = \frac{k}{2}x + \frac{k+4}{2}$.

For B, we get: $2x + k(y - 3) - 10 = 0 \Rightarrow 2x + ky - 3k - 10 = 0 \Rightarrow y = -\frac{2}{k}x + \frac{3k+10}{k}$.

So first, the slope of every line by A is $\frac{k}{2}$, and we get: $k \leq 1 \Rightarrow \frac{k}{2} \leq \frac{1}{2}$.

Therefore, as k gets smaller, the line gets steeper, and eventually, gets nearly vertical.

So all the lines by the equation A range from a line $y = \frac{1}{2}x + \frac{5}{2}$ to a vertical line $x = $ -1, but the vertical line $x = $ -1 cannot be made because the equation A cannot indicate it.

Next, for each value of k, the slope of a line by the equation B is $-\frac{2}{k}$.

Thus, we get: $k \leq 1 \Rightarrow -k \geq -1 \Rightarrow (-\frac{1}{k} \leq$ -1 or $-\frac{1}{k} > 0) \Rightarrow (-\frac{2}{k} \leq$ -2 or $-\frac{2}{k} > 0)$.

So all the lines indicated by B range from a line $y = $ -2$x + 13$ to a vertical line $x = 5$, and then, from the vertical line $x = 5$ to a horizontal line $y = 3$, but the line $y = 3$ cannot be made since the equation B cannot indicate it.

Therefore, the trace is a part of the circle $(x - 2)^2 + (y - \frac{5}{2})^2 = \frac{37}{4}$.
However, the trace can be said to be composed of three curves.
One is the lower half of the circle, and the other two are two parts of the upper half.

$(y - \frac{5}{2})^2 = \frac{37}{4} - (x - 2)^2 \Rightarrow y - \frac{5}{2} = \pm\sqrt{\frac{37}{4} - (x - 2)^2}$.

So the lower half of the circle is: $y = \frac{5}{2} - \sqrt{\frac{37}{4} - (x - 2)^2}$, where $2 - \frac{\sqrt{37}}{2} \leq x \leq 2 + \frac{\sqrt{37}}{2}$.

The other two parts are above the line $y = \frac{5}{2}$, which includes a diameter of the circle. One is on the left in the upper half, and the other is on the right.

All the lines indicated by A range from a line $y = \frac{1}{2}x + \frac{5}{2}$ to a vertical line $x = -1$, but the vertical line $x = -1$ cannot be made, so the part on the left in the upper half is as follows:

$$y = \frac{5}{2} + \sqrt{\frac{37}{4} - (x-2)^2}, \text{ where } 2 - \frac{\sqrt{37}}{2} < x < -1.$$

For the part on the right in the upper half circle, we need to find where a line meets the circle above.

Suppose L is the line.
Then the line L is: $y = -2x + 13$, and the circle is: $(x-2)^2 + (y - \frac{5}{2})^2 = \frac{37}{4}$.

So first, we get: $y = -2x + 13 \Rightarrow y - \frac{5}{2} = -2x + 13 - \frac{5}{2} = -2x + \frac{21}{2}$.

Thus next, we get: $(x-2)^2 + (y-\frac{5}{2})^2 = (x-2)^2 + (-2x + \frac{21}{2})^2 = \frac{37}{4}$

$\Rightarrow x^2 - 4x + 4 + 4x^2 - 42x + \frac{441}{4} = 5x^2 - 46x + \frac{457}{4} = \frac{37}{4} \Rightarrow 5x^2 - 46x + \frac{420}{4} = 0$

$\Rightarrow 5x^2 - 46x + 105 = 0 \Rightarrow x^2 - \frac{46}{5}x + 21 = 0$.

Then first, L meets the circle at **(3, 5)**, so one of the two roots for the equation above is 5.

Next, suppose t is the other root. Then, we get: $5 + t = \frac{46}{5} \Rightarrow t = \frac{21}{5}$.

So the part on the right in the upper half circle is as follows:

$$y = \frac{5}{2} + \sqrt{\frac{37}{4} - (x-2)^2}, \text{ where } \frac{21}{5} \le x < 2 + \frac{\sqrt{37}}{2}.$$

Therefore, the trace is composed of the three curves as follows:

$y = \frac{5}{2} - \sqrt{\frac{37}{4} - (x-2)^2}$, where $2 - \frac{\sqrt{37}}{2} \le x \le 2 + \frac{\sqrt{37}}{2}$, which is the lower half circle.

$y = \frac{5}{2} + \sqrt{\frac{37}{4} - (x-2)^2}$, where $2 - \frac{\sqrt{37}}{2} < x < -1$, which is a part of the upper half circle.

$y = \frac{5}{2} + \sqrt{\frac{37}{4} - (x-2)^2}$, where $\frac{21}{5} \le x < 2 + \frac{\sqrt{37}}{2}$, which is a part of the upper half circle.

If not quite sure of the idea behind the processes above, follow the steps below:

In the example 0 above, we found the facts as follows:

For A, we get: $k(x+1) - 2y + 4 = 0 \Rightarrow kx + k - 2y + 4 = 0 \Rightarrow y = \frac{k}{2}x + \frac{k+4}{2}$.

For B, we get: $2x + k(y-3) - 10 = 0 \Rightarrow 2x + ky - 3k - 10 = 0 \Rightarrow y = -\frac{2}{k}x + \frac{3k+10}{k}$.

Every line by the equation A is a linear combination of two lines $x = -1$ and $y = 2$.
Every line by the equation B is a linear combination of two lines $x = 5$ and $y = 3$.

So all the lines the equation A represents meet each other altogether at $(-1, 2)$, and all the lines the equation B represents meet each other altogether at $(5, 3)$.

For each value of k, the product of slopes of the two lines by A and B is -1, and therefore, the two lines in each pair are perpendicular to each other.

All the points at each of which two lines in each pair meet each other, are in a circle.

The line segment between the two points $(-1, 2)$ and $(5, 3)$ is a diameter of the circle.

Now, let's start getting the trace in the case where $k \leq 1$.

To begin with, the slope of each line by A is $\frac{k}{2}$.   And we have: $k \leq 1$, so we get: $\frac{k}{2} \leq \frac{1}{2}$.

That is, for $k \leq 1$, slopes of all the lines by the equation A are less than or equal to $\frac{1}{2}$.
For instance, as k varies, we get:

$k = 1 \Rightarrow y = \frac{k}{2}x + \frac{k+4}{2} = \frac{1}{2}x + \frac{5}{2}$.

$k = 0 \Rightarrow y = \frac{0}{2}x + \frac{0+4}{2} = 2 \Rightarrow y = 2$.

$k = -1 \Rightarrow y = -\frac{1}{2}x + \frac{-1+4}{2} = -\frac{1}{2}x + \frac{3}{2}$.

$k = -100 \Rightarrow y = -\frac{100}{2}x + \frac{-100+4}{2} = -50x - 48$.

So we can see that as k gets smaller, the line gets steeper, and eventually, gets nearly vertical.

Thus, all the lines by the equation A range from a line $y = \frac{1}{2}x + \frac{5}{2}$ to a vertical line $x = $ -1, but the vertical line $x = $ -1 cannot be made because the equation A cannot indicate it.

Next, for B, we get: $2x + k(y-3) - 10 = 0 \Rightarrow 2x + ky - 3k - 10 = 0 \Rightarrow y = -\frac{2}{k}x + \frac{3k+10}{k}$.

So for each value of k, the slope of a line by the equation B is $-\frac{2}{k}$.

Since $k \le 1$, we get: $-k \ge -1 \Rightarrow (-\frac{1}{k} \le -1$ or $-\frac{1}{k} > 0) \Rightarrow (-\frac{2}{k} \le -2$ or $-\frac{2}{k} > 0)$.

Therefore, if $k \le 1$, we get: $-\frac{2}{k} \le$ **-2** or $-\frac{2}{k} >$ **0**.

That is, for $k \le 1$, slopes of all the lines by the equation B are less than or equal to -2, or are greater than 0.

Now, k is the denominator in $-\frac{2}{k}$, though.   Then, k cannot be 0 even if $k \le 1$, can it?

Yes, it can. Actually, the equation B is: $2x + k(y-3) - 10 = 0$, and not $y = -\frac{2}{k}x + \frac{3k+10}{k}$. So when $k = 0$, we get a vertical line, $x = 5$.

Let's see now, how the lines get generated by B as k varies.

$k = 1 \Rightarrow y = -\frac{2}{k}x + \frac{3k+10}{k} = $ -2x + 13.

$k = 0.1 \Rightarrow y = -\frac{2}{k}x + \frac{3k+10}{k} = -20x + \frac{0.3+10}{0.1} = -20x + 103$.

$k = 0.01 \Rightarrow y = -\frac{2}{k}x + \frac{3k+10}{k} = -200x + \frac{0.03+10}{0.01} = -200x + 1003$.

$k = 0 \Rightarrow 2x + k(y-3) - 10 = 0 \Rightarrow x = 5$.

$k = $ -1 $\Rightarrow y = -\frac{2}{k}x + \frac{3k+10}{k} = 2x + \frac{-3+10}{-1} = 2x - 7$.

$k = $ -2 $\Rightarrow y = -\frac{2}{k}x + \frac{3k+10}{k} = x + \frac{-6+10}{-2} = x - 2$.

$k = $ -100 $\Rightarrow y = -\frac{2}{k}x + \frac{3k+10}{k} = -\frac{2}{-100}x + \frac{-300+10}{-100} = \frac{1}{50}x + 2.9$.

Thus, we can see that:

As **k** gets smaller staying positive, the line gets steeper (i.e., the slope is negative with bigger magnitude), and eventually, the line gets vertical.

And then, as **k** gets smaller being negative, the line gets less steep (i.e., the slope is positive with smaller magnitude), and eventually, the line gets almost horizontal.

Therefore, all the lines indicated by the equation **B** range from a line $y = -2x + 13$ to a vertical line $x = 5$, and then, range from the vertical line $x = 5$ to a horizontal line $y = 3$, but the horizontal line $y = 3$ cannot be made since the equation **B** cannot indicate it.

Let's now, put in a graph some of the lines by the two equations, and see what the trace looks like.

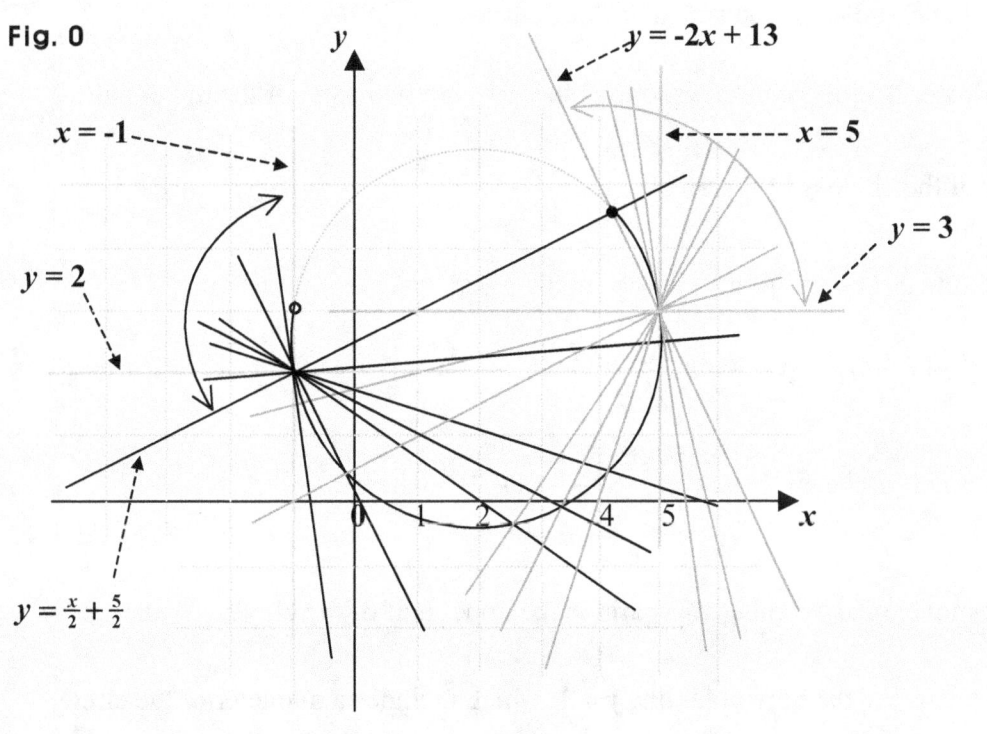

Fig. 0

Then, we can see that the trace is the part of the circle $(x-2)^2 + (y-\frac{5}{2})^2 = \frac{37}{4}$.
More specifically, the trace looks like the one in the figure below.

Fig. 1

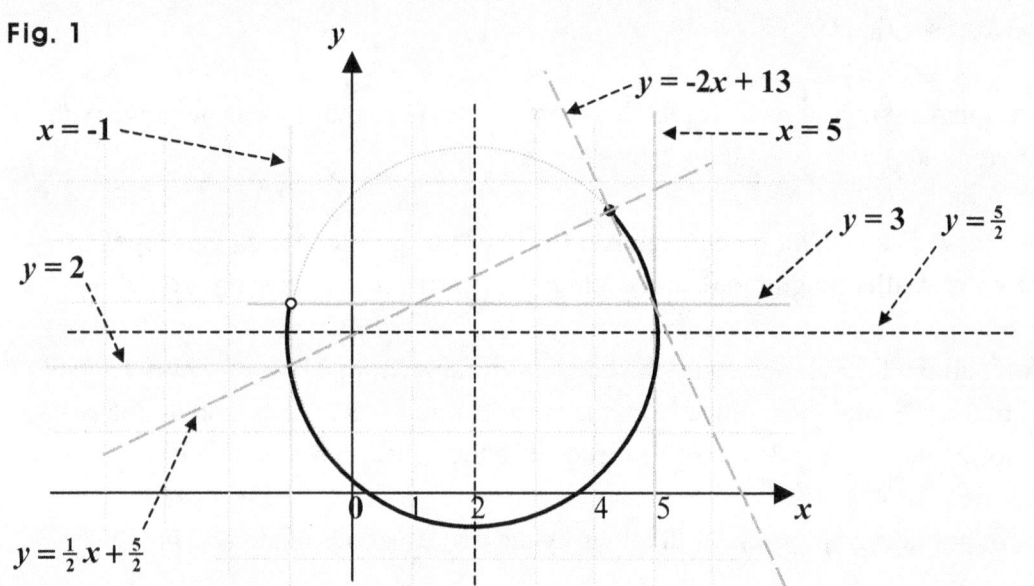

The trace can be said to be composed of three curves.

One is the lower half of the circle, and the other two are two parts of the upper half.

The center of the circle is $(2, \frac{5}{2})$, and the radius is $\frac{\sqrt{37}}{2}$.

Thus, the circle is: $(x-2)^2 + (y-\frac{5}{2})^2 = \frac{37}{4}$, and we get:

$$(y-\tfrac{5}{2})^2 = \tfrac{37}{4} - (x-2)^2 \Rightarrow y - \tfrac{5}{2} = \pm\sqrt{\tfrac{37}{4} - (x-2)^2}.$$

So the lower half of the circle is: $y = \frac{5}{2} - \sqrt{\frac{37}{4} - (x-2)^2}$, where $2 - \frac{\sqrt{37}}{2} \le x \le 2 + \frac{\sqrt{37}}{2}$.

Let's now, move on to the other two parts in the upper half of the circle.

Both parts are above the horizontal line $y = \frac{5}{2}$, which includes a diameter of the circle. One of the two is on the left in the upper half, and the other is on the right.

Now, all the lines by the equation A range from a line $y = \frac{1}{2}x + \frac{5}{2}$ to a vertical line $x = -1$, but the vertical line $x = -1$ cannot be made, for the equation A cannot indicate it.

The equation of the upper half circle is: $y = \frac{5}{2} + \sqrt{\frac{37}{4} + (x-2)^2}$ for $2 - \frac{\sqrt{37}}{2} \leq x \leq 2 + \frac{\sqrt{37}}{2}$.

So the part on the left is as follows: $y = \frac{5}{2} + \sqrt{\frac{37}{4} - (x-2)^2}$, where $2 - \frac{\sqrt{37}}{2} < x < -1$.

Next, for the other part, that is, the part on the right in the upper half circle, we need to know where a line and the circle meet each other.

Suppose L is the line.
Then, the line L is: $y = -2x + 13$, and of course, the circle is: $(x-2)^2 + (y - \frac{5}{2})^2 = \frac{37}{4}$.

So first, we get: $y = -2x + 13 \Rightarrow y - \frac{5}{2} = -2x + 13 - \frac{5}{2} = -2x + \frac{21}{2}$.

Thus next, we get: $(x-2)^2 + (y - \frac{5}{2})^2 = (x-2)^2 + (-2x + \frac{21}{2})^2 = \frac{37}{4}$

$\Rightarrow x^2 - 4x + 4 + 4x^2 - 42x + \frac{441}{4} = 5x^2 - 46x + \frac{457}{4} = \frac{37}{4} \Rightarrow 5x^2 - 46x + \frac{420}{4} = 0$

$\Rightarrow 5x^2 - 46x + 105 = 0 \Rightarrow x^2 - \frac{46}{5}x + 21 = 0$.

Now, being tangent to the circle, the line L meets the circle at one point, so the final quadratic equation above has a double root, and thus, can be put in a complete square.

On the other hand, if the line L meets the circle at two points, the final equation has two different roots.

Now, the line L is indicated by the equation B for $k = 1$, and we know all the lines the equation B represents meet each other altogether at **(5, 3)**, which is in the circle. So the line L meets the circle at the point **(5, 3)**, too.

So if the final quadratic equation above has two different roots, we can immediately see that one of the two is 5. What about the other root, then?

We can readily find the other root using the fact as follows:

- The sum of the two roots of the equation $ax^2 + bx + c = 0$ is $-\frac{b}{a}$.

Suppose the final equation has two different roots, and *t* is the other root.

Then, we get: $\frac{46}{5} = 5 + t \Rightarrow t = \frac{21}{5}$, which is not 5, and therefore, is the other root.

Meanwhile, we get: $x^2 - \frac{46}{5}x = x^2 - \frac{46}{5}x + (\frac{23}{5})^2 - (\frac{23}{5})^2 = (x - \frac{23}{5})^2 - (\frac{23}{5})^2$.

So we get: $x^2 - \frac{46}{5}x + 21 = (x - \frac{23}{5})^2 - (\frac{23}{5})^2 + 21 = (x - \frac{23}{5})^2 - \frac{529}{25} + \frac{525}{25}$

$= (x - \frac{23}{5})^2 - \frac{4}{25} = 0 \Rightarrow (x - \frac{23}{5})^2 = \frac{4}{25} \Rightarrow x - \frac{23}{5} = \pm\frac{2}{5} \Rightarrow x = \frac{23}{5} \pm \frac{2}{5} \Rightarrow x = 5$ or $\frac{21}{5}$.

So the part on the right in the upper half circle is as follows:

$y = \frac{5}{2} + \sqrt{\frac{37}{4} - (x - 2)^2}$, where $\frac{21}{5} \le x < 2 + \frac{\sqrt{37}}{2}$.

Therefore, the trace is composed of the three curves as follows:

$y = \frac{5}{2} - \sqrt{\frac{37}{4} - (x - 2)^2}$, where $2 - \frac{\sqrt{37}}{2} \le x \le 2 + \frac{\sqrt{37}}{2}$, which is the lower half circle.

$y = \frac{5}{2} + \sqrt{\frac{37}{4} - (x - 2)^2}$, where $2 - \frac{\sqrt{37}}{2} < x < -1$, which is a part of the upper half circle.

$y = \frac{5}{2} + \sqrt{\frac{37}{4} - (x - 2)^2}$, where $\frac{21}{5} \le x < 2 + \frac{\sqrt{37}}{2}$, which is a part of the upper half circle.

Examples E in Circles

Examples E in Circles

Constructing a curve in math, we can put it in a graph. Putting it in a graph though, we need the equation of it. So constructing it, we want to find the equation of it if we are not given the equation, of course. And finding the equation, it is said that we get it, too.

Suppose A and B are two points in the x-y plane, and G is an arbitrary point in a curve C in the plane. Suppose also, that the distance between G and A is twice the distance between G and B. Then:

0.   Find the equation of the curve C.

1.   Find the maximum magnitude of the slope of a line passing through a point in the curve C and the point A.

Suggestions or Solutions
To the Problem in the Example 0

Suppose A and B are two points in the x-y plane, and G is an arbitrary point in a curve C in the plane. Suppose also, that the distance between G and A is twice the distance between G and B. Then, find the equation of the curve C.

Suppose a is a nonzero constant, A is at $(-2a, 0)$, B is at $(a, 0)$, and $G(x, y)$ is the arbitrary point in the curve C.

Then, since we have: $\frac{\overline{AG}}{\overline{BG}} = 2$, by means of the distance formula between points, we get:

$$\sqrt{\{x - (-2a)\}^2 + (y - 0)^2} = 2\sqrt{(x - a)^2 + (y - 0)^2}.$$

So we get: $\sqrt{(x + 2a)^2 + y^2} = 2\sqrt{(x - a)^2 + y^2} \Rightarrow (x + 2a)^2 + y^2 = 4\{(x - a)^2 + y^2\}$

$$\Rightarrow 4\{(x - a)^2 + y^2\} - (x + 2a)^2 - y^2 = 4(x^2 - 2ax + a^2 + y^2) - x^2 - 4ax - 4a^2 - y^2$$

$$= 3x^2 - 12ax + 3y^2 = 0 \Rightarrow x^2 - 4ax + y^2 = 0$$

$$\Rightarrow x^2 - 4ax + 4a^2 - 4a^2 + y^2 = (x - 2a)^2 + y^2 - 4a^2 = 0 \Rightarrow (x - 2a)^2 + y^2 = 4a^2.$$

Therefore, the equation of C is: $(x - 2a)^2 + y^2 = 4a^2$, where a is a nonzero constant.

If not quite sure of the idea behind the processes above, follow the steps below:

To begin with, we know A and B are just two points in the x-y plane, and G is an arbitrary point in a curve C in the plane.

So assuming a, b, c, and d are constant, we can say: A is (a, b), B is (c, d), and G is (x, y).

So next, coming up with a connective equation between x and y, we get the equation of the curve C. How then, can we get the connective equation?

Let's first, put in a graph all the three points and some curve.

Fig. 0

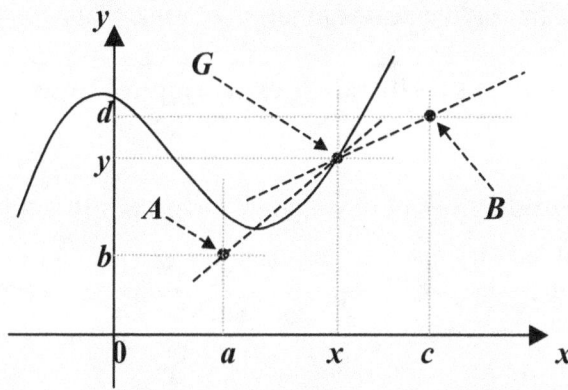

We have the distance ratio, $\frac{\overline{AG}}{\overline{BG}} = 2$. In other words, $\overline{AG} : \overline{BG} = 2 : 1$.

So using the distance formula, we can set up a relation among the three points in terms of the distances between them. However, putting the points in such a way as above, we get to end up with an expression $\sqrt{(x-a)^2 + (y-b)^2} = 2\sqrt{(x-c)^2 + (y-d)^2}$, which is not only highly unlikely to give us much help but quite complicated, too, due to too many constants involved.

It is important to note that the two points A and B are *just any two* points in a plane.

Thus, we are free to choose the positions of the two points A and B as long as A and B keep their positions (i.e., their relative positions are fixed). So let's try another graph.

Putting A and B on the *x*-axis, we could get a graph as below:

Fig. 1

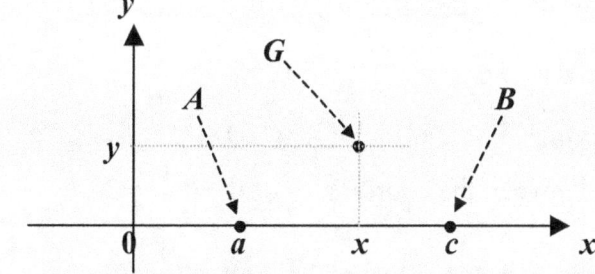

Now, the point A is $(a, 0)$, the point B is $(c, 0)$, and the arbitrary point G is (x, y).

Then, setting up the relation among the points in terms of the distances, we can get:

$$\sqrt{(x-a)^2+(y-0)^2}=2\sqrt{(x-c)^2+(y-0)^2} \Rightarrow \sqrt{(x-a)^2+y^2}=2\sqrt{(x-c)^2+y^2}.$$

It looks much simpler than the previous one, which however, is still not quite likely to give us much help either. So let's try another graph.

Fig. 2

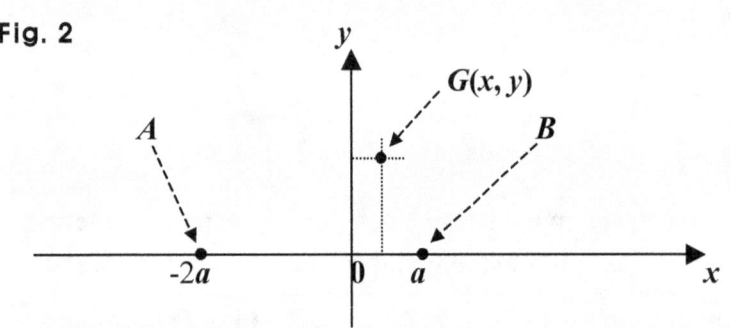

This time, the point A is $(-2a, 0)$, the point B is $(a, 0)$, and the arbitrary point G is (x, y).

The ratio is: $\frac{\overline{AG}}{\overline{BG}}=2$, which can give us: $\sqrt{\{x-(-2a)\}^2+(y-0)^2}=2\sqrt{(x-a)^2+(y-0)^2}$.

Thus, we get:

$$\sqrt{(x+2a)^2+y^2}=2\sqrt{(x-a)^2+y^2} \Rightarrow (x+2a)^2+y^2=4\{(x-a)^2+y^2\}$$

$$\Rightarrow 4\{(x-a)^2+y^2\}-(x+2a)^2-y^2=4(x^2-2ax+a^2+y^2)-x^2-4ax-4a^2-y^2$$

$$=3x^2-12ax+3y^2=0 \Rightarrow x^2-4ax+y^2=0$$

$$\Rightarrow x^2-4ax+4a^2-4a^2+y^2=(x-2a)^2+y^2-4a^2=0 \Rightarrow (x-2a)^2+y^2=4a^2,$$

which is the connective expression between the coordinates of the point G.

The connective expression indicates a circle of radius of $2a$ centered at $(2a, 0)$.

Therefore, the curve C is: $(x-2a)^2+y^2=4a^2$, where a is a nonzero constant.

Fig. 3

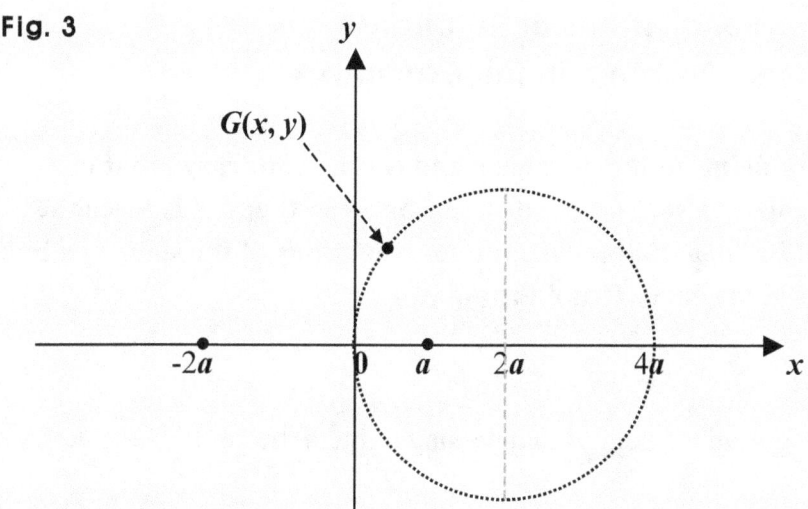

Such a circle as above is called Apollonius Circle.

Note:

If a problem says simply a point in a plane, we can put the point anywhere in the plane as long as we maintain its relative location to the other objects in the problem, and the location of the point should be a location where we can make the problem simpler. So we may want to begin such a problem with trying some several locations of such a point.

In short:

Suppose a is a nonzero constant, A is at $(-2a, 0)$, B is at $(a, 0)$, and $G(x, y)$ is the arbitrary point in the curve C.

Then, since we have: $\frac{\overline{AG}}{\overline{BG}} = 2$, by means of the distance formula between points, we get:

$$\sqrt{\{x-(-2a)\}^2 + (y-0)^2} = 2\sqrt{(x-a)^2 + (y-0)^2}.$$

So we get: $\sqrt{(x+2a)^2 + y^2} = 2\sqrt{(x-a)^2 + y^2} \Rightarrow (x+2a)^2 + y^2 = 4\{(x-a)^2 + y^2\}$

$\Rightarrow 4\{(x-a)^2 + y^2\} - (x+2a)^2 - y^2 = 4(x^2 - 2ax + a^2 + y^2) - x^2 - 4ax - 4a^2 - y^2$

$= 3x^2 - 12ax + 3y^2 = 0 \Rightarrow x^2 - 4ax + y^2 = 0$

$\Rightarrow x^2 - 4ax + 4a^2 - 4a^2 + y^2 = (x-2a)^2 + y^2 - 4a^2 = 0 \Rightarrow (x-2a)^2 + y^2 = 4a^2.$

Therefore, the equation of C is: $(x-2a)^2 + y^2 = 4a^2$, where a is a nonzero constant.

Suggestions or Solutions
To the Problem in the Example 1

Suppose A and B are two points in the x-y plane, and G is an arbitrary point in a curve C in the plane. Suppose also, that the distance between G and A is twice the distance between G and B. Then, find the maximum magnitude of the slope of a line passing through a point in the curve C and the point A.

Putting some rays emitting from the point A and passing through the circle C, we get:

Fig. 0

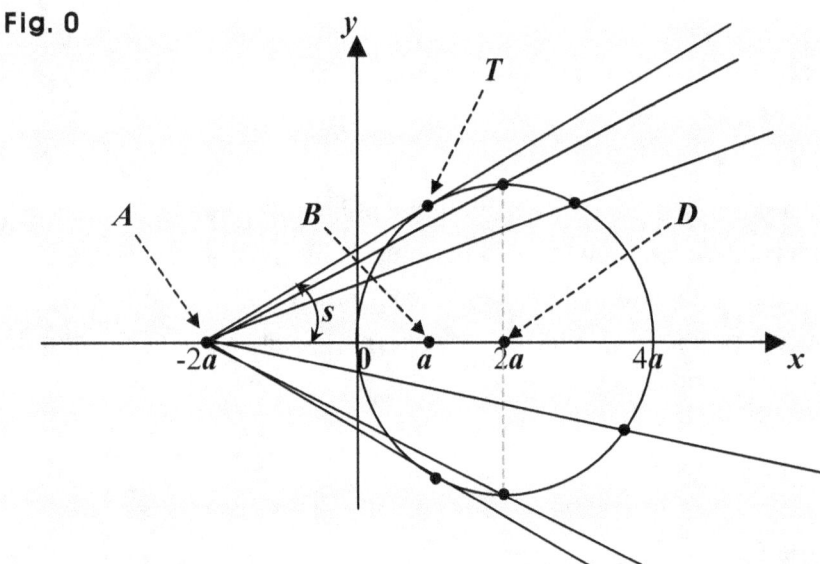

Then, when a ray emitting from the point A is tangent to the circle, the ray makes the largest angle s against the x-axis, and thus, has the maximum slope.

Suppose the center of C is D, and a ray from A is tangent to the circle C at a point T. Then, D is $(2a, 0)$, and $\overline{TD}$ is perpendicular to $\overline{TA}$, for $\overline{TA}$ is a part of the ray tangent to C, and $\overline{TD}$ is a part of the line passing through the center and the tangent point T.

We have: $\overline{TD} = |2a|$, and $\overline{AD} = |2a - (-2a)| = |4a|$.

So from the distance formula, we get:

$(\overline{AD})^2 = (\overline{TA})^2 + (\overline{TD})^2 \Rightarrow 16a^2 = (\overline{TA})^2 + 4a^2 \Rightarrow (\overline{TA})^2 = 12a^2 \Rightarrow \overline{TA} = |2a\sqrt{3}|$.

Therefore, the maximum magnitude of the slope is: $\frac{\overline{TD}}{\overline{TA}} = \frac{|2a|}{|2a\sqrt{3}|} = \frac{1}{\sqrt{3}} = \frac{\sqrt{3}}{3}$.

If not quite sure of the idea behind the processes above, follow the steps below:

In the previous problem, we've found that the curve C is a circle. So to begin with, let's put in a graph the circle C and the two points A and B. Then, putting some rays emitting from the point A and passing through the circle C, we can get a graph as below:

Fig. 1

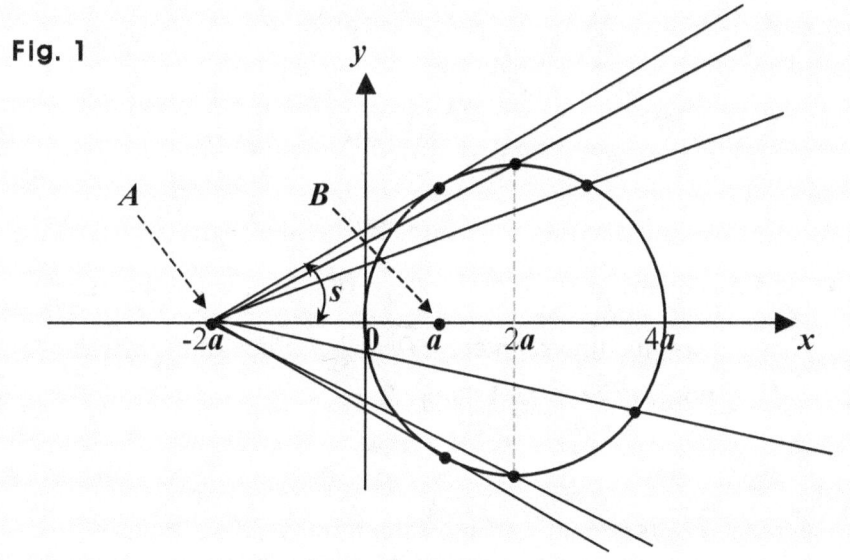

Then, we can see that when a ray emitting from the point A is tangent to the circle C, the ray makes the largest angle against the x-axis, and therefore, has the maximum slope.

Also, we can see a right triangle in the graph above because of the fact as follows:

• Suppose L is a line tangent to a circle at a point P. Then, the line L is perpendicular to the line passing through the center of the circle and the tangent point P.

Suppose now, that the center of the circle C is D, and that a ray emitting from the point A is tangent to the circle C at a point T as shown in the figure below.

Then, the center D is $(2a, 0)$, and a line segment $\overline{TD}$ is perpendicular to a line segment $\overline{TA}$, because the segment $\overline{TA}$ is a part of the ray tangent to the circle C, and the segment $\overline{TD}$ is a part of the line passing through the center and the tangent point T.

Fig. 2

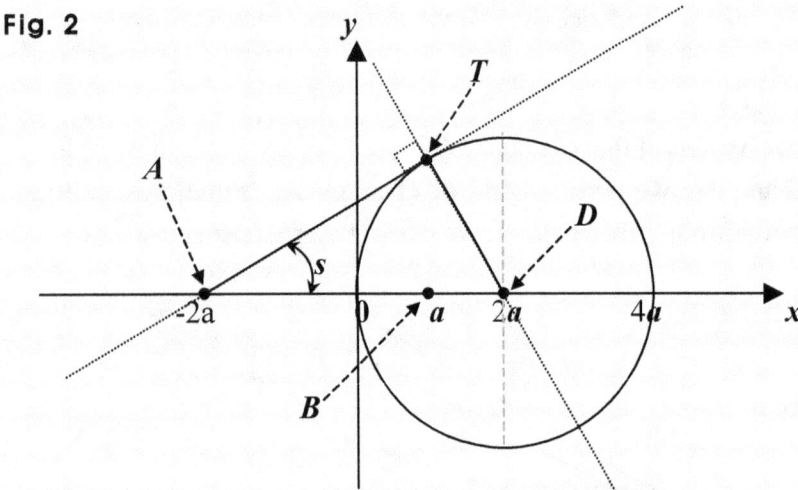

Therefore, we can see that the maximum slope is the slope of the line segment $\overline{TA}$, so we want to find the ratio $\frac{\overline{TD}}{\overline{TA}}$, which is the magnitude of the slope of a line, which includes the line segment $\overline{TA}$. To begin with, the line segment $\overline{TD}$ is the radius of the circle. So we have: $\overline{TD} = |2a|$. Why not just $\overline{TD} = 2a$, though?

That's because the line segment $\overline{TD}$ is a length, which cannot be negative, but a is just nonzero, and thus, can be negative.

Now, in the example 0, we found the Apollonius Circle as shown below:

Fig. 3

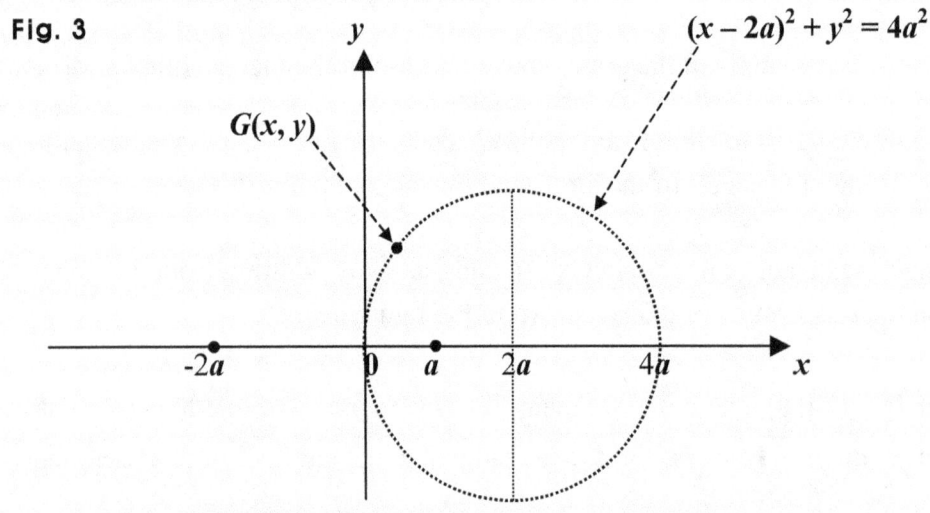

$(x - 2a)^2 + y^2 = 4a^2$

For instance, we can set *a* = -2. Then, we get another Apollonius Circle as shown below:

Fig. 4

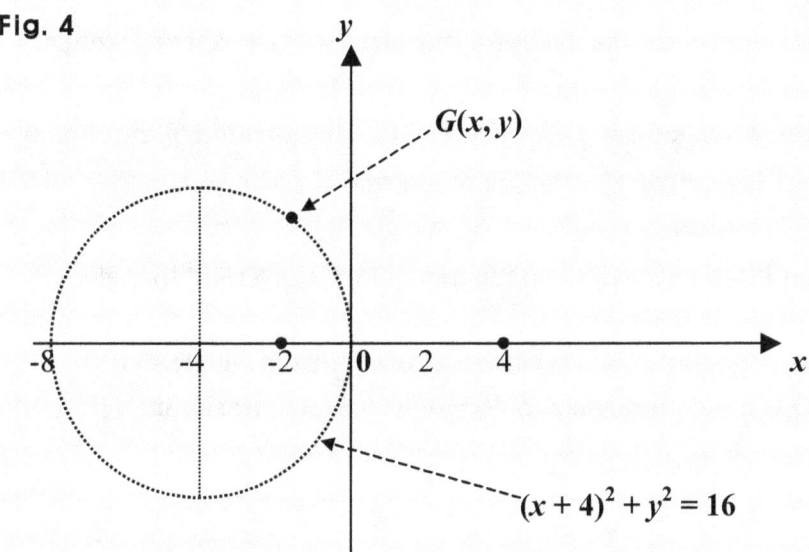

Now, getting back to the point where we left off, we get:

Fig. 5

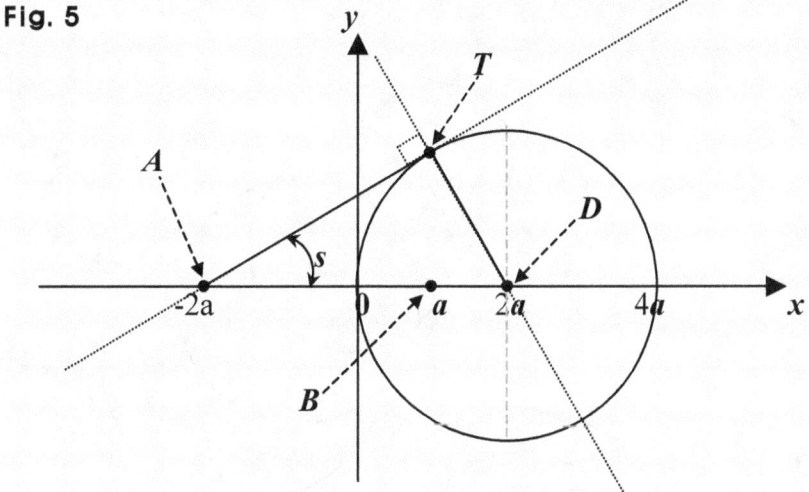

We can see that from the graph above, the line segment $\overline{AD}$ = $|2a - (-2a)|$ = $|4a|$.

Next, let's find the length of the segment $\overline{TA}$. Then, the distance formula (often called Pythagorean Theorem) comes in to play its part. Thus, we get:

$$(\overline{AD})^2 = (\overline{TA})^2 + (\overline{TD})^2 \Rightarrow 16a^2 = (\overline{TA})^2 + 4a^2 \Rightarrow (\overline{TA})^2 = 12a^2 \Rightarrow \overline{TA} = |2a\sqrt{3}|.$$

So we get: $\frac{\overline{TD}}{\overline{TA}} = \frac{|2a|}{|2a\sqrt{3}|} = \frac{1}{\sqrt{3}} = \frac{\sqrt{3}}{3}$, which is the maximum magnitude of the slope.

It seems that the point T is right above the point $(a, 0)$ when the slope is the maximum.

In fact, it is the case, because we get: $y = a\sqrt{3}$ when $x = a$ in the equation of the circle above, and the equation is: $(x - 2a)^2 + y^2 = 4a^2$. Thus, we get:

$(\overline{AT})^2 = 12a^2 = (\overline{AB})^2 + (\overline{TB})^2 = 9a^2 + 3a^2$, so the triangle ABT is a right triangle.

Of course, if familiar with trigonometry, you can see at once, that the angle s is $30°$ because the triangle ATD is a right triangle and $\frac{\overline{TD}}{\overline{AD}} = \frac{|2a|}{|4a|} = \frac{1}{2}$, which is: $\sin 30°$.

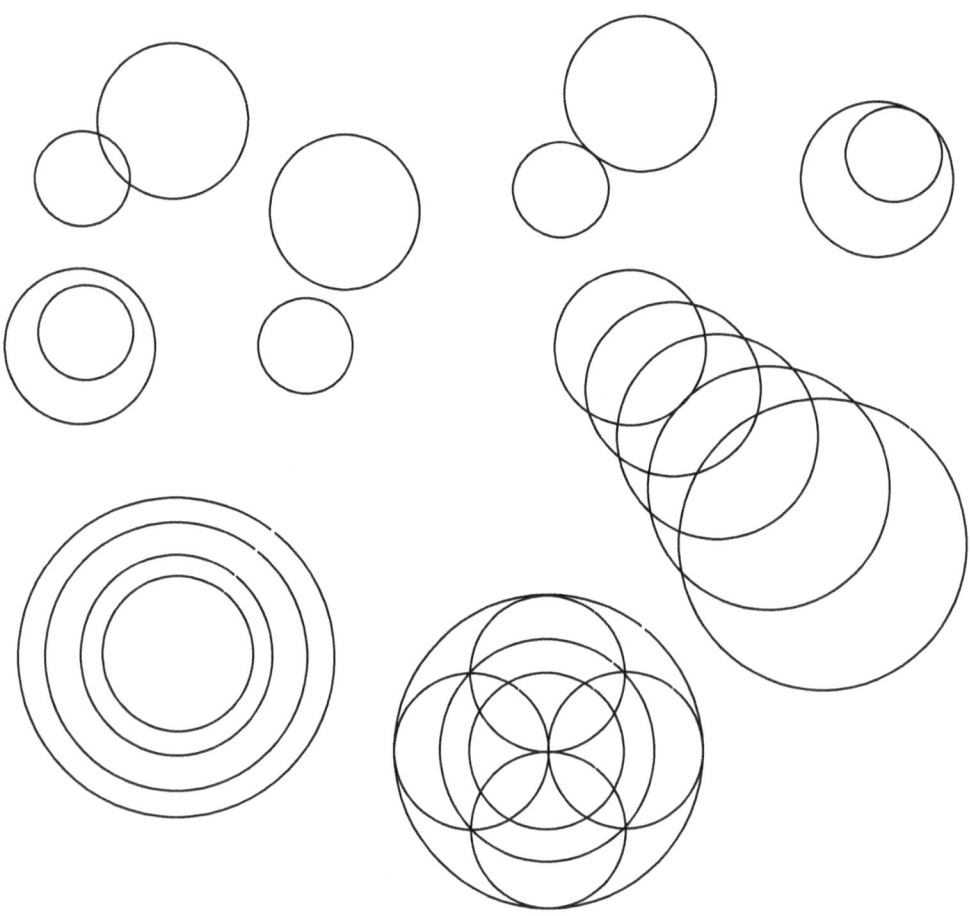